Lecture Notes in Computer Science 14070

Founding Editors

Gerhard Goos

Juris Hartmanis

Editorial Board Members

The series Lecture Notes in Computer Science (LNCS), including its subseries Lecture Notes in Artificial Intelligence (LNAI) and Lecture Notes in Bioinformatics (LNBI), has established itself as a medium for the publication of new developments in computer science and information technology research, teaching, and education.

LNCS enjoys close cooperation with the computer science R & D community, the series counts many renowned academics among its volume editors and paper authors, and collaborates with prestigious societies. Its mission is to serve this international community by providing an invaluable service, mainly focused on the publication of conference and workshop proceedings and postproceedings. LNCS commenced publication in 1973.

Domenico Cantone · Alfredo Pulvirenti
Editors

From Computational Logic
to Computational Biology

Essays Dedicated to Alfredo Ferro to Celebrate
His Scientific Career

 Springer

Editors
Domenico Cantone 🆔
University of Catania
Catania, Italy

Alfredo Pulvirenti 🆔
University of Catania
Catania, Italy

ISSN 0302-9743   ISSN 1611-3349 (electronic)
Lecture Notes in Computer Science
ISBN 978-3-031-55247-2   ISBN 978-3-031-55248-9 (eBook)
https://doi.org/10.1007/978-3-031-55248-9

This Springer imprint is published by the registered company Springer Nature Switzerland AG
The registered company address is: Gewerbestrasse 11, 6330 Cham, Switzerland

Paper in this product is recyclable.

Alfredo Ferro

Preface

This volume is published in honour of Alfredo Ferro, to celebrate his distinguished career as a computer scientist.

Born in Ramacca (Catania, Italy) on May 10, 1951, Alfredo Ferro commenced his mathematical studies at the University of Catania under the guidance of Carmelo Mammana, where he attained his bachelor's degree in 1973. While working as a research assistant in Higher Mathematics at the University of Catania, Alfredo had the privilege of meeting Prof. Martin Davis from New York University (NYU) during a summer school in Perugia in 1975. This encounter proved pivotal, as it fueled Alfredo's anticipation of the increasing potential of *Computer Science*. Driven by his innate vision and scientific curiosity, he made a crucial decision to shift from mathematics, dedicating the rest of his career to this burgeoning discipline. Venturing to the United States, he pursued a Ph.D. in Computer Science at NYU, working under the mentorship of the renowned Jacob Theodor (Jack) Schwartz. In 1981, Alfredo successfully received his Ph.D. in Computer Science at NYU, focusing on decision procedures for certain classes of unquantified set-theoretic formulae. His outstanding work earned him the Jay Krakauer Award for the best Ph.D. Dissertation in the School of Sciences of NYU during the academic year 1980–1981. After completing his Ph.D. studies, Alfredo returned to Italy and served as an associate professor in Computer Science from 1981 to 1987, subsequently attaining the position of full professor. His decision to return to Catania was primarily motivated by the desire to apply the scientific knowledge he had acquired, recognizing the potential of Computer Science to create new opportunities for young and brilliant minds in a region like Sicily. In fact, in 1990, he successfully established the Computer Science Undergraduate Program at the University of Catania!

Compelled by the age limit, in 2021 Alfredo concluded his formal career at the University of Catania, and in 2022 he was nominated *emeritus professor* in acknowledgment of his lifelong contributions to academia.

Alfredo's academic career as a computer scientist is characterized by two distinct research phases: *Computational Logic*, until approximately 1995, followed by a notable focus on *Data Mining* and *Bioinformatics*.

In the domain of Computational Logic, Alfredo Ferro made important contributions to the pioneering area of *Computable Set Theory*, as envisioned by Jack Schwartz, whose objective was the mechanical formalization of mathematics, utilizing a proof verifier based on the set-theoretic formalism. The preliminary fragment of set theory explored for decidability was MLS, an acronym for Multi-Level Syllogistic. MLS encompasses quantifier-free formulae of set theory, embodying the Boolean set operators ($\cup$, $\cap$, $\setminus$), as well as the relation symbols $=$ and $\in$, along with (existentially quantified) set variables. The satisfiability problem for MLS—essentially the problem of determining whether any given MLS formula is satisfied by some set assignment—has been established as decidable, as detailed in the seminal paper:

A. Ferro, E.G. Omodeo, and J.T. Schwartz. Decision procedures for elementary sublanguages of set theory. I: Multilevel syllogistic and some extensions. *Comm. Pure Appl. Math.*, 33:599–608, 1980.

This volume includes, as a historical document unpublished to date, the very first proof of the decidability of MLS, written by Alfredo Ferro and Eugenio Omodeo in 1979.

Subsequently, Alfredo significantly contributed to extending the core language MLS with set operators such as power set, general union set, a choice function, various constructs related to rank and maps, as well as predicates expressing finiteness and rank comparison. This effort culminated in Alfredo co-authoring the first monograph on Computable Set Theory. Additionally, Alfredo and his group made foundational contributions to one of the initial implementations of the prover envisioned by Jack Schwartz, named ETNA (Extensible Theorem Prover in NAtural Deduction). ETNA enabled users to create their own deduction environments for conducting formal proofs of statements expressed in a formalism that extended classical first-order set theory. We would like to note that Computable Set Theory still remains an active area of research, as highlighted in the contributions by **Cantone–Omodeo** and **Cristiá–Rossi** in the present volume.

From 1995 onward, computer science expanded into various new application areas such as distributed computing, database systems, and data mining. Alfredo swiftly transitioned to contribute to these emerging fields. He primarily employed graph matching to address problems in computer science and fostered collaborations with esteemed research scientists, including Roded Sharan (Tel Aviv University), Gary Bader (University of Toronto), and Dennis Shasha (NYU). Their focus was on subgraph matching algorithms, network querying, and motif discovery. In the early 2000s, Alfredo's collaboration with Dennis Shasha and Michele Purrello (University of Catania) led him into the realm of Computational Biology, establishing collaborations with researchers worldwide. His exploration of Bioinformatics also involved teaching a course for the Computer Science bachelor's degree. Notable accomplishments include the synthesis of artificial microRNA molecules to restore gene expression in cancer, achieved through collaborative efforts with researchers from NYU and Ohio State University.

Alfredo's research extended to further domains, including biological databases, pathway analysis, simulation models, and drug repurposing. His group focused on computational RNAi and network biology applications within the field of cancer biology. Recent initiatives involve developing computational methodologies for sequencing data analysis (NGS) in precision medicine, conducted in collaboration with researchers from Ohio State University and Queen Mary University of London. Additionally, Alfredo has been involved in the inference of knowledge graphs from biomedical texts, working alongside Paolo Ferragina (University of Pisa).

Beyond being active in research, Alfredo took the initiative to establish summer schools for Ph.D. students back in 1989. This initiative led to the creation of the so-called *Lipari School*, which has since evolved into the *J.T. Schwartz International School for Scientific Research*. Alfredo continues to serve as the director of this prestigious series, which includes schools focused on Computer Science, Complex Systems, and Computational Biology. The schools feature distinguished scientists as lecturers, with the Computational Biology series co-directed by Nobel Laureate Michael Levitt, Carlo Croce (Ohio State University), Alberto Mantovani (Humanitas University), and Søren Brunak (University of Copenhagen). Meanwhile, the Complex Systems series is co-directed by Paolo Ferragina, Dirk Helbing (ETH Zurich), and Carlo Ratti (MIT).

Throughout his career, Alfredo has held several notable roles in academia. Specifically, he served as the Coordinator of the Ph.D. Program in Computer Science at the University of Catania from 1996 to 2002, and later he was a co-founder of the Ph.D. program in Biology, Human Genetics, and Bioinformatics at the University of Catania. Furthermore, Alfredo took on leadership responsibilities as the coordinator of the Doctorates in Molecular and Computational Clinical Pathology and in Molecular Biomedicine.

Alfredo Ferro's impact on information technology has traversed diverse domains, encompassing computational logic, data mining, bioinformatics, and complex systems. His substantial contributions in each field are reflected in publications within esteemed international journals. Acknowledged not only as an outstanding scientist but also as a visionary, Alfredo was the first to recognize the transformative potential of Computer Science for a region like Sicily, thereby creating opportunities for talented young minds. Throughout his career, Alfredo seamlessly integrated theoretical findings into practical applications, extracting challenging questions from real-world scenarios to advance scientific knowledge.

With this Festschrift, we extend our heartfelt gratitude to Alfredo, a true master who has been a profound source of inspiration for us all.

January 2024   Domenico Cantone
                                                        Alfredo Pulvirenti

Grazie Alfredo!

Giovanni Gallo

Dipartimento di Matematica e Informatica, Università di Catania, Italy

Someone has said that in Science there are two kinds of 'animal': eagles and frogs. The 'eagles' are those researchers who are always straining to get higher to see beyond the horizon: they are the visionaries, always pushing boundaries and opening new chapters to research. The 'frogs', on the other side, live in ponds in love with beautiful water lilies and are not afraid to get dirty with the mud of tiny details to 'know' all about these flowers: they are the specialists, involved in investigating deeper and deeper the beauty of their elected topic. I believe that a third category should be added to this taxonomy: 'hunting dogs'. These are the scientists who 'hunt' for young talents to attract to their research fields, providing support, advice and challenging problems to work on.

As accurate as this scheme may be it fails if you wish to 'label' into a single one of these categories Alfredo Ferro: he is at the same time a brave 'eagle', a busy 'frog' and a restless 'hunting dog'. Here is not the place to demonstrate this claim with a complete analytic survey of the many relevant scientific and professional Alfredo's contributions to Science and to our community: I wish only to take this opportunity to shortly report about occasions where his 'three-faced nature' of scientist has shone and helped me.

I met Alfredo in 1986 and it was one of the most fortunate encounters of my life. Two years earlier I graduated in Mathematics, I was struggling to start my research career in abstract Mathematics. Sadly this was a sort of dead end for me: indeed I always loved computation and visual thinking and cold abstractness is almost the opposite of my inclinations. I was getting more and more frustrated and I was almost ready to give up the dream of getting actively involved in Science. Alfredo appeared at this critical juncture of my professional life. Alfredo's eagle view at the time was focusing on the need for our local academic community to open up from the traditional mathematical studies to create a high level Computer Science school in Catania. With his scouting, hunting dog attitude Alfredo asked me to join him in this new adventure. It was a great opportunity not to be missed and soon I found myself as a graduate student of Computer Science at New York University. It was an exciting and scary time of my life: professional and personal issues converged but eventually, in 1992, I was ready to come back to Catania and to start my career as a professor. Through all that I never missed Alfredo's encouragement, support, advice and patience. True: our personalities crashed at times but we were both blessed by the mediating presence of professor Pina Carrà, Alfredo's beloved wife. I enjoyed doing research in Computational Algebra with Pina and she treated me always as a trusted coworker and a dear friend. How to forget when during some animated discussion between me and Alfredo, Pina was affectionately saying: 'Giovanni is right, but ...' and so opening space for constructive and open dialogue?

Alfredo's eagle view helped me once more after the completion of my PhD in Computational Algebra. He encouraged me to go beyond purely theoretical studies and to do research in pragmatic areas of Computer Science, and so I found myself involved in 'Image Processing' soon to become 'Computer Graphics' and 'Computer Vision'. Following Alfredo's example I have worked to attract young talents to this new 'pond' filled with so many unexplored 'water lilies' and so many years later 'Catania Image Processing Lab' is an internationally recognized research center.

Around year 2000 Alfredo, after his many contributions in the pond of Computational Logic, lifted his eagle view to Data Processing and Analysis. There again his enthusiasm and ideas have been instrumental to open up for us, younger researchers, new horizons. This happened just in time to be ready for the great 'Data Revolution' of today and it was another of Alfredo's merits that a strong team of young researchers in this field has been able to successfully take up the new challenge. Within the new land of Data Processing, Bioinformatics has become the new 'lily pond' for Alfredo and his team of young talents, all of them coming from his relentless hunting search.

The story does not end here, of course, and I am sure that in the coming years Alfredo will point us towards new horizons, will help us to discover new species of beautiful lilies and will continue to attract to Science many young talents.

In closing this personal note I wish to publicly express to Alfredo my deepest gratitude, my great admiration for the scientist and, last but not least, my sincere love to the faithful friend that he has always been to me.

Reviewers of This Volume

Salvatore Alaimo
Vincenzo Bonnici
Agostino Dovier
Tiziana Di Matteo
Alessandro Laganà
Emanuele Martorana
Giovanni Micale
Amir Nakib
Eugenio G. Omodeo
Alberto Policriti
Andrea Rapisarda
Francesco Russo

Acknowledgements

We extend our gratitude to all the authors and reviewers who contributed to this Festschrift. We are also grateful to Springer Nature for providing us with the opportunity to express our thanks to Alfredo through the pages of this special collection.

Contents

Computational Logic

The Early Development of SETL

Edmond Schonberg[(✉)]

Courant Institute of Mathematical Science, New York University, New York, USA
schonberg@adacore.com

Introduction

I was privileged to meet Dr. Ferro in the incredibly fertile environment of Jacob Schwartz's research group at the Courant Institute of Mathematical Sciences (CIMS), New York University.

The informal remarks that follow describe Jack Schwartz's (JS) broad research program in various areas of Computer Science, and in particular the activity around the programming language SETL, its implementation, early applications, and impact on programming, algorithm specification, and formal methods. Several contributors to this volume were part of the inebriating atmosphere in which this work developed, and have enriched their chosen fields of research with the ideas incubated at CIMS under JS's guidance.

The Genesis of SETL

The first full description of SETL appears in "On Programming" [9]. JS's interest in programming had led him to a collaboration with John Cocke, an IBM Fellow and Turing Award recipient, who had guided the development of the first global optimizer for FORTRAN, and developed, with Fran Allen, the first algorithms for global program analysis: value tracing, live-dead analysis, constant propagation, peephole optimization, etc. This area involves novel and complex algorithms over graphs, and there was at the time no accepted formalism for describing them: one found in the literature various informal notations under the label "pseudo-Algol" or some such, and typically an algorithm was described in a mixture of natural language and low-level programming notions. The need for a standard notation with well-defined semantics was evident. A first sketch of such a formalism can be found in "Programming Languages and their Compilers" [1] which provided a remarkably broad description of the semantics and

---

The Computer Science Museum, thanks to the work of Paul McJones, has a comprehensive collection of documents related to the early development of SETL, including a full collection of the SETL newsletters.
See http://www.softwarepreservation.org/projects/SETL
Another large source of references to the work around SETL can be found in "The houses that Jack built: Essays in honor of Jack Schwartz", M. Davis, A. Gottlieb, and E. Schonberg (Eds.), Courant Institute of Mathematical Sciences, New York University, May 1995.

© The Author(s), under exclusive license to Springer Nature Switzerland AG 2024
D. Cantone and A. Pulvirenti (Eds.): *From Computational Logic to Computational Biology*,
LNCS 14070, pp. 3–8, 2024.
https://doi.org/10.1007/978-3-031-55248-9_1

implementation of different languages in use, including unconventional (for the time) programming paradigms such as pattern-matching in SNOBOL.

Coming from a stellar trajectory in pure Mathematics, JS found it natural to consider Set Theory as the best possible basis for a programming language that could also serve as a specification language, i.e. as a means of communication among programmers and software developers. It is interesting to note that two other high-level languages (i.e. languages with dynamic typing, garbage collection, and a minimal declarative machinery) were already available, namely LISP (inspired by the Lambda Calculus) and APL, based on multidimensional arrays and a rich algebraic structure to manipulate them. No doubt JS was partial to Set Theory because it is The Lingua Franca of Mathematics ever since its formalization by Zermelo, Fraenkel, and later von Neumann. In addition to stylistic choices, there were also pragmatic reasons that led JS to leave aside these two options. We can cite the following:

A) A purely functional model of programming, as offered by LISP, is inimical to the sequential execution that comes naturally to all who program. Whether this preference for sequential description has some "psychological" advantage is still debated, but the bulk of software development has a sequential flavor.

B) APL revolves around the manipulation of a particular data structure, and this choice (or bias) makes it ill-suited as a specification language with which to describe irregular data structures.

Be that as it may, JS found that an imperative language that incorporates the basic elements of Set Theory was a better vehicle for algorithm specification, and "SETL-like" descriptions became common in the description of new algorithms.

Set Theory and Executability: Where to Compromise?

JS used to state that SETL was designed "under the possibly intolerable constraint that it be executable". This meant that only finite sets could be built and manipulated, and constructed functions were needed to examine unbounded objects. Otherwise there were few concessions to performance: basic constructs were sets and mappings over arbitrary domains, and common operations on these, namely membership and map retrieval, relied on hash tables. One-dimensional arrays ("tuples") were also built-in data structures.

Arithmetic was also built-in (instead of the elegant but completely impractical von Neumann formalization of integers). Arbitrary precision arithmetic was built-in, but floating-point operations were those of the target machine. These pragmatic choices made it possible to implement the language by means of an interpreter and a garbage-collected environment.

Another concession to run-time performance was the presence of bit-vectors to encode sets of integers. Set operations on these (union, intersection, membership) benefited from boolean operations available in the hardware. The power-set operation was also a primitive of the language, albeit one used extremely rarely, for obvious reasons. Another concession to performance was the inclusion of a restricted power-set operation, which given a set S only constructed the subsets of S of a specific cardinality.

Finally, mutable objects and assignment made SETL into an imperative language, therefore congenial to most programmers of the time. To avoid issues of aliasing, assignment is by value, implying full copy of objects of all types. Address and pointers are completely absent from the language, but can be simulated by using maps whose domain is a set of generated internal tags (called "atoms" in SETL).

First Implementations

The first implementation of SETL was written by David Shields using FORTRAN. The next implementation, dubbed SETLB, was a parse-tree generator that produced source code for an elegant LISP/ALGOL hybrid called BALM, developed at NYU by Malcolm Harrison. Portability of the translator was an immediate concern: the FORTRAN code was hard to port to other machines because of hardware dependences on the main machine available at NYU at the time: a Control Data 6600, one of the most powerful machines at the time, but with an impractical architecture (60 bit words, bit- but not byte-addressable). JS started a separate language design effort around a low-level language, dubbed LITTLE, The language C began to spread at that time, and may have been a better choice for the long term, but LITTLE provided a convenient mechanism for specifying size and layout of run-time structures, and thus isolating machine dependencies in a few global declarations.

The substantial run-time library required by the language was written in LITTLE, by Hank Warren and others. The garbage-collector of the first implementation was radically improved By Robert Dewar, who joined the NYU CS Department in 1977 after participating in the design Of ALGOL68, and the construction of SPITBOL, a remarkably efficient implementation of the Pattern-matching language SNOBOL.

Algorithm Specification and Software Engineering

An aspect of modern Software Engineering that is absent from SETL is a mechanism for information hiding. In a sense, sets and maps are abstract data types, because they can be implemented in multiple ways, depending on performance considerations that should not influence the logical structure of an algorithm. This suggested a further development of the language, that would allow the (optional) specification of the concrete data structures to be used for sets and maps in a given program. Two approaches suggest themselves:

A) Add to the language a data-specification sub-language, where the user can specify how a given set or map should be implemented, in terms of a defined collection of built-in data-structures: arrays, linked lists, hash tables, ordered trees, graphs, etc. This sub-language allows the programmer to lower to semantic level of the language but remains optional and does not affect the structure of the original unannotated SETL program.

B) Develop optimization techniques to *infer* the most efficient data-structures from their use in a SETL program, thus relying on global program analysis to determine the types of objects being manipulated and the frequency of operations applied to them.

Transformational Programming and Optimization

The intrinsic inefficiency of SETL made it into an ideal testbed for optimization techniques, and led to the development of novel global program analysis algorithms. A summary of the multi-pronged attack on this challenge can be found in [10].

The first information needed for any SETL program analysis is the type of every object appearing in it. Having a dynamically typed language means that all runtime data must be tagged to be self-descriptive, and this imposes an overhead on every single operation. To determine types at compile-time one needs to use the types of explicit constants in the program (numeric literals, set constructors, etc.) and rely on global data-flow to propagate information from point of definition to points of use. The thesis of Aaron Tenenbaum [12] describes the structure of such a type analyzer. Gerald Weiss extended this technique to structures amenable to a recursive description, such as trees [13].

Once type-information is available, it becomes possible to choose specific data-structures for sets and maps appearing in a program, provided quantitative information is made available about frequency of use of different objects. This frequency can be estimated by examining the nesting structure of loops and calls in the program. The resulting technique is described in [8].

A much more ambitious set of optimizing transformations is Formal Differentiation, which is a generalization of Strength Reduction. This technique, first developed for FORTRAN, identifies constructs within an iterated construct that can be evaluated incrementally rather than recomputed each time. Robert Paige developed Formal Differentiation into a general Program transformation technique that is also amenable to formal proof [6].

Software Prototyping and "High-Level" Languages

The passage from specification to implementation is the central problem of software engineering, and SETL provided an ideal testbed for the use of a high-level language to construct executable specifications of complex systems. In 1980 a group at N.Y.U. started designing a translator for the new programming language Ada (winner of an international Competition commissioned by the U.S. Department of defense). The translator was intended to provide an executable specification of the language itself, and at the same time serve as a design tool for a proper translator. The first objective was realized by writing an interpreter resembling a denotational definition of the language; for example, iterative control structures were executed by run-time rewritings of the abstract syntax tree generated by the front-end of the compiler. This made the description of concurrency (at the time a bold innovation in language design) both clear and rigorous. Other formal definitions of Ada were proposed, but the SETL version had the enormous advantage of being executable, and thus amenable to systematic testing using the Ada Compiler Validation Capability that was developed independently of the language to ensure the conformance of different implementations of the language. The NYU system (named Ada/Ed, to indicate its pedagogical intent) was of course extremely inefficient (it was said to be intended for the "real-time simulation of paper-and-pencil computations") but was sufficient to execute correctly all ACVC tests, and thus become the first formally certified Ada compiler [7].

At the same time, the SETL code could serve as a design document for a proper language compiler, and it was subsequently translated into C (and later on into Ada itself [5]). The executability of the original was an enormous asset in the construction of efficient compilers, and the algorithmic details of some innovative aspects of Ada, such as overload resolution, used the Ada/Ed specification in Ada as a guideline in the construction of other compilers.

Correctness Proofs and Computational Logic

If the gap between specification and efficient implementation can be bridged through optimization and correctness-preserving program transformation, there remains to show that an executable SETL program containing imperative constructs (assignments, loops, constructors) is a correct implementation of a purely mathematical specification of an algorithm, in which there is no notion of order of execution, side-effects and other characteristics of a runnable program.

Martin Davis and JS established the foundations to make a proof system extensible [2] and this was used by E. Deak in her doctoral thesis [3]. This approach was then expanded into the extremely fertile topic of Computable Set Theory, leading among others to the series of articles by A. Ferro and others on Decision Procedures for Elementary Sublanguages of set theory (see [4] and references therein, and [11]). The capstone of this effort is a proof, ab-initio, of the Cauchy Integral theorem. This achievement is conclusive proof that combining the study of computability with pragmatic language design issues proved to be a remarkably fertile area at the intersection of Mathematics and Computer Science.

Conclusions

As the previous remarks indicate, the creation of SETL had a remarkably diverse impact on various areas of Computer Science, even though the original language did not come to widespread usage. It is worth mentioning that a number of aspects of Python were taken from SETL, by way of the language ABC (developed at Amsterdam's Mathematisch Centrum as a language to teach introductory programming). The fundamental inefficiency of SETL was a stimulus to original research rather than a programming hindrance. JS always emphasized that computers were bound to become faster and faster, and disregarding this temporary inconvenience led to a multitude of theoretical and pragmatic advances in Computer Science.

References

1. Cocke, J., Schwartz, J.T.: Programming Languages and their Compilers. Courant Institute of Mathematical Sciences Lecture Notes, New York University (1969)
2. Davis, M., Schwartz, J.T.: Metamathematical extensibility for theorem verifiers and proof checkers. Comput. Math. Appl. 5(3), 217–230 (1979)
3. Deak, E.: A transformational approach to the development and verification of programs in a very high-level language. Technical Report 22, Courant Institute of Mathematical Sciences, New York University (1980)
4. Ferro, A., Omodeo, E.G., Schwartz, J.T.: Decision procedures for some fragments of set theory. In: Bibel, W., Kowalski, R. (eds.) 5th Conference on Automated Deduction Les Arcs, France, July 8–11, 1980. Lecture Notes in Computer Science, vol. 87, pp. 88–96. Springer, Berlin (1980). https://doi.org/10.1007/3-540-10009-1_8
5. Kruchten, P., Schonberg, E., Schwartz, J.T.: Prototyping using the SETL programming language. IEEE Softw. 1(5), 66–75 (1984)
6. Paige, R., Koenig, S.: Finite differencing of computable expressions. ACM Trans. Program. Lang. Syst. 4(3), 402–454 (1982)
7. NYUADA group. An executable semantic model of Ada. Technical Report 84, Courant Institute of Mathematical Sciences, New York University (1983)
8. Schonberg, E., Schwartz, J.T., Sharir, M.: An automatic technique for selection of data structures in SETL programs. ACM TOPLAS 3(2), 126–143 (1981)
9. Schwartz, J.T.: On Programming: an interim report on the SETL project. New York University (1974)
10. Schwartz, J.T.: Optimization of very-high level languages. J. Comput. Lang. 1(2,3), 161–219 (1975)
11. Schwartz, J.T., Cantone, D., Omodeo, E.G.: Computational Logic and Set Theory – Applying Formalized Logic to Analysis. Springer, Cham (2011). Foreword by Martin D. Davis
12. Tenenbaum, A.: Type determination for very-high level languages. Technical Report 3, Courant Institute of Mathematical Sciences, New York University (1974)
13. Weiss, G.: Recursive data types in SETL: Automatic determination, data language description, and efficient implementation. Technical Report 102, Courant Institute of Mathematical Sciences, New York University (1986)

Onset and Today's Perspectives of Multilevel Syllogistic

Domenico Cantone[1]($\boxtimes$)(iD) and Eugenio G. Omodeo[2](iD)

[1] University of Catania, Catania, Italy
domenico.cantone@unict.it
[2] University of Trieste, Trieste, Italy
eomodeo@units.it

Abstract. We report on the initial phases of a systematic study (undertaken over forty years ago) on decidable fragments of Set Theory, to which Alfredo Ferro contributed and which later branched out in many directions. The impact that research has had so far and will continue to have, mainly in the areas of proof-checking, program-correctness verification, declarative programming—and, more recently, reasoning within description logics—is also highlighted.

Keywords: Decidability · Computable set theory · Proof verification

1 An Inference Mechanism for Membership

1.1 The Discovery of Multi-level Syllogisic, Dubbed **MLS**

March 1979 was a hectic time, at the Courant Institute of NYU, for Alfredo Ferro and for an author of this paper (Eugenio, provisionally dubbed 'I'): the former, who was in his second year of doctoral studies, was burnishing his comprehensive oral exam preparation; the latter had just become father for the first time. Professor Jacob T. Schwartz (universally dubbed 'Jack') came to us with the listing of a computer program which processed unquantified Boolean formulae and issued, for each of them, either a validity response or a counterexample. Jack intended to show that program to L. Wolfgang Bibel, a distinguished German computer scientist who would visit the Courant, and urged us to enhance it with the treatment of membership.

In the Boolean framework, a term τ is built up from variables by means of the dyadic operators $\cup, \cap$, and $\setminus$ designating union, intersection, and difference; inclusion and equality relators $\subseteq$ and $=$ are interposed between terms to form atomic formulae; and then the usual propositional connectives $\&$, $\vee$, $\neg$ are used to assemble compound formulae out of the atomic ones. A counterexample M to a formula φ must assign families of individuals to the Boolean variables so that φ evaluates to false under the replacement of each variable x by the corresponding family Mx within φ; validity means that no such falsifying M exists. Individuals (sometimes called 'urelements'), unlike families, can have no

D. Cantone and A. Pulvirenti (Eds.): *From Computational Logic to Computational Biology*,
LNCS 14070, pp. 9–55, 2024.
https://doi.org/10.1007/978-3-031-55248-9_2

members; a void family, $\emptyset$, is supposed to exist, but it is not regarded as an individual. The decision algorithm implemented by Jack originated from the syllogistic method discussed in [131].

In the set-theoretic framework which Alfredo and I would tackle upon Jack's request, the only syntactic novelty was that atomic formulae could also be of the form $\tau_1 \in \tau_2$; this seemingly small change entails a radical revision of the semantics, because one intends to deal with genuine nested sets, not just with flat families of individuals. Consider, for example, the negations $\neg\, (x_0 \in x_1 \,\&\, x_1 \in x_2 \,\&\, \cdots \,\&\, x_N \in x_{N+1})$ and $\neg\, (x_0 \in x_1 \,\&\, x_1 \in x_2 \,\&\, \cdots \,\&\, x_N \in x_0)$ of two membership chains, where the x_i's stand for distinct variables and N for an arbitrary natural number. The former of these admits a counterexample, because we can take as M an assignment such that $x_0 \overset{M}{\mapsto} \emptyset$, $x_1 \overset{M}{\mapsto} \{\emptyset\}$, $x_2 \overset{M}{\mapsto} \{\{\emptyset\}\}$, etc.; the latter, instead, should be regarded as valid if we are to comply with mainstream Set Theory, where membership is postulated to be well-founded and hence to be devoid of cycles. Once the decision algorithm entrusted to us by Jack came into existence, Alfredo and I named it 'multilevel syllogistic'—MLS for brevity—since it could model acyclic membership chains of any finite length.

It should be noted that Jack's prototype, as well as our first implementation of MLS, were developed in SETL [74], a language for executable algorithm-specifications based around the fundamental concepts of set, map, and sequence. As will emerge from Sects. 3.1–3.3 and 3.6, the fact that the logical language and the programming language revolved around the same mathematical notions was by no means accidental.

Another remarkable fact is that a working implementation of the decision algorithm for MLS pre-existed the mathematical proofs that the processed language admits, in fact, a validity/satisfiability decision algorithm and that the implemented testing method was correct.

This paper describes how the decidability of MLS has influenced and shaped much subsequent research on solvable and unsolvable classes of set-theoretic formulae. For the historically inclined reader, the seminal report on MLS, in its original form of June 1979, is published in this book for the first time.

A quick, modernized account of the decidability of that fragment of set theory is provided below, in Sect. 1.3.

1.2 '79/'80: Proof-Technology Seminar at NYU

In the morning of October 3, 1979, I (Eugenio) was in charge of the inaugural meeting of a seminar on computational logic which would gather some faculty of the Courant Institute with researchers of IBM's Thomas J. Watson Research Center. My task was the one of presenting the architecture of a proof-checker, provisionally named Yulog, which Jack had begun to develop at Yorktown during the summer. An electric adapter for the overhead projector was missing, and I could not find anybody to help me; surprisingly, even Alfredo was not around. When he at last arrived, he reported that he had remained trapped in a huge crowd cheering Paulus Johannes II, who was visiting New York City. Initially unaware of what was going on, Alfredo had pushed so strongly as to break all lines and to arrive very close to the pope.[1]

After this kickoff meeting, the series of events went on for several Wednesdays; sometimes the conference was followed by a working lunch, which Gregory J. Chaitin enriched with brilliant mathematical ideas. Regular attendees also included Martin D. Davis, Malcolm C. Harrison, and Vincent J. Digricoli; Michael Breban, who was already teaching at Yeshiva University and whose family was rapidly growing, could only seldom join the group. Jack once discussed a decision algorithm which had appeared in a seminal paper by Heinrich Behmann in 1922 [4,101,135], and posed the question: Is validity decidable in an unquantified first-order language encompassing Boolean constructs, membership, and the cardinality operator (as referred to finite as well as to infinite sets)? If such an enhancement of MLS was doable at all, it would have rimmed undecidability, because the satisfiability problem for this language seemed to get very close to Hilbert's 10th problem. A few days later, Jack announced that the sought decision algorithm had come into existence, and Alfredo was commissioned to present the result.

This was the milieu in which various decidable extensions of multi-level syllogistic came to light. Jack, Alfredo, and I started preparing the results for publication. The first journal which would host an account of MLS was *Communications on Pure and Applied Mathematics* and the first conference would be CADE 5 (cf. [88,89]). Then, during the winter holidays, by reflecting on a simple extension of MLS which Jack had devised, I got the idea of expanding the language with a construct somehow related to the restricted quantifier $\forall x \in y$. Jack found of no interest my proposal, which I had drafted in rather convoluted terms; on the opposite, when Alfredo returned from vacationing in Italy and gave a look at my draft, he got very excited about it: he glimpsed, in fact, the possibility of capturing with my method the truth of various results on ordinal numbers, which he had found in a paper by the French mathematician Dominique Pastre, [129]. So my idea was examined more closely by Jack. On the technical level, Alfredo had dropped a clanger; however, he rightly saw the potential inherent in a smart treatment of restricted quantifiers. Among others, the Boolean operators can easily be eliminated through them: e.g., $x = y \cup z$ can be rewritten as $\forall v \in x \, (v \in y \vee v \in z) \, \& \, \forall v \in y \, (v \in x) \, \& \, \forall v \in z \, (v \in x)$, and $x = y \setminus z$ as $\forall v \in x \, (v \in y \, \& \, v \notin z) \, \& \, \forall v \in y \, (v \in z \vee v \in x)$. One can also express the singleton constraint $x = \{y\}$ as $y \in x \, \& \, \forall v \in x \, (v = y)$. As for the property $\mathsf{On}(x)$, meaning 'x is an ordinal', it can be formulated as $\forall v \in x \, \forall w \in x \, (v = w \vee v \in w \vee w \in v) \, \& \, \forall v \in x \, \forall w \in v \, (w \in x)$ [132]; but unfortunately, the 'knot' visible in the quantification pattern $\forall v \in x \, \forall w \in v$ where the bound variable v also occurs on the right of '$\in$', called for a sophisticated treatment which my original idea did not contemplate. "One only learns from one's mistakes", commented Jack; and in fact, after a while, without us being

---

[1] This happening left a very profound impression on Alfredo, who afterward developed a strong religious involvement.

able to treat knotted quantifiers in full generality, some useful cases—capturing, in particular, the notion of ordinal—could be adjoined to MLS (cf. [12]).

1.3   The Decidability of MLS in a Nutshell

By using disjunctive normal form, the satisfiability problem (*s.p.*, for brevity) for MLS can be reduced to the s.p. for conjunctions of literals of the forms

$$t_1 = t_2, \quad t_1 \neq t_2, \quad t_1 \in t_2, \quad t_1 \notin t_2,$$

where t_1 and t_2 stand for possibly nested set terms built up from set variables and the Boolean operators $\cup$, $\cap$, and $\setminus$.

Let us call 'flat' all literals of the following types

$$x = y \cup z, \quad x = y \cap z, \quad x = y \setminus z, \quad x = y, \quad x \neq y, \quad x \in y, \quad x \notin y,$$

where x, y, z are set-variables.

Any term t of the form $y \cup z$, $y \cap z$, or $y \setminus z$ (with y and z variables) occurring in some non-flat literal of a given conjunction C of MLS can be replaced by a fresh variable, say ξ_t, while adding the literal $\xi_t = t$ to C as a new conjunct. Such replacements are clearly satisfiability preserving and so, by applying them repeatedly to C, one can end up with an equisatisfiable conjunction, dubbed *normalized conjunction*, of 'flat' literals.

Hence, to solve the s.p. for generic MLS formulae it is enough to exhibit a decision procedure for the s.p. for normalized conjunctions of MLS.

So, let C be a normalized conjunction of MLS and let V be the collection of the set variables occurring in C. Let us also assume that C be satisfiable, and let M be a set assignment over V satisfying C. Let $\sim_M$ be the equivalence relation over $D_M := \bigcup_{x \in V} Mx$ such that

$$a \sim_M b \leftrightarrow_{\text{Def}} (\forall x \in V)(a \in Mx \Longleftrightarrow b \in Mx),$$

for all $a, b \in D_M$.

The equivalence classes of $\sim_M$ are the *Euler-Venn blocks* of the partition induced by the collection $\{Mx : x \in V\}$, namely the non-empty sets A_S of the form

$$A_S := \bigcap_{x \in S} Mx \setminus \bigcup_{y \in V \setminus S} My,$$

where $\emptyset \neq S \subseteq V$.

Let $\mathcal{V}_M$ be the set of the Euler-Venn blocks induced by the Mx's. For each $B \in \mathcal{V}_M$, let us pick any $\emptyset \neq S_B \subseteq V$ such that $B = A_{S_B}$, and form their collection Π_M. Also, let $V_M := \{x \in V \mid Mx \in D_M\}$. Then, for each $x \in V_M$ there exists a unique $S_x \in \Pi_M$ such that $Mx \in A_{S_x}$. In addition, let

$$A_\emptyset := \{Mx : x \in V \setminus V_M\}$$

and, for each $x \in V \setminus V_M$, define $S_x := A_\emptyset$.

Thus, the statement

$$\underset{S \in \Pi_M}{\&} \left(A_S = \bigcap_{x \in S} Mx \setminus \bigcup_{y \in V \setminus S} My \ \& \ A_S \neq \emptyset \right)$$

$$\& \ \underset{x \in V}{\&} \ Mx = \bigcup_{x \in S \in \Pi_M} A_S \ \& \ \underset{x \in V}{\&} \ Mx \in A_{S_x}$$

is plainly true and therefore the following MLS-conjunction is satisfiable

$$C \ \& \ \underset{S \in \Pi_M}{\&} \left(\alpha_S = \bigcap_{x \in S} x \setminus \bigcup_{y \in V \setminus S} y \ \& \ \alpha_S \neq \emptyset \right)$$

$$\& \ \underset{x \in V}{\&} \ x = \bigcup_{x \in S \in \Pi_M} \alpha_S \ \& \ \underset{x \in V}{\&} \ x \in \alpha_{\mathrm{at}_M(x)}, \quad (\dagger)$$

where the α_S's, for $S \in \Pi_M \cup \{\emptyset\}$, are pairwise distinct set variables associated with the members of Π_M, which are new to the conjunction C, and $\mathrm{at}_M \colon V \to \Pi_M \cup \{\emptyset\}$ is the map such that $\mathrm{at}_M(x) = S_x$, for each $x \in V$. Indeed, $(\dagger)$ is satisfied by the extension M' of M to the newly introduced variables α_S, for $S \in \Pi_M$, where

$$M'z := \mathbf{if} \ z = \alpha_S \ \text{for some} \ S \in \Pi_M \cup \{\emptyset\} \ \mathbf{then} \ A_S \ \mathbf{else} \ Mz.$$

The conjunction $(\dagger)$ belongs to a finite collection, of size a priori known, of MLS-conjunctions $C_{\Pi,\mathrm{at}}$, where

$$C_{\Pi,\mathrm{at}} \leftrightarrow_{\mathrm{Def}} C \ \& \ \underset{S \in \Pi}{\&} \left(\alpha_S = \bigcap_{x \in S} x \setminus \bigcup_{y \in V \setminus S} y \ \& \ \alpha_S \neq \emptyset \right)$$

$$\& \ \underset{x \in V}{\&} \ x = \bigcup_{x \in S \in \Pi} \alpha_S \ \& \ \underset{x \in V}{\&} \ x \in \alpha_{\mathrm{at}(x)},$$

each of which is fully specified once one fixes a set $\Pi \subseteq \mathcal{P}(V) \setminus \{\emptyset\}$ and a map $\mathrm{at} \colon V \to \Pi \cup \{\emptyset\}$. Hence, the above argument proves that if the conjunction C is satisfiable, then so is some conjunction $C_{\Pi,\mathrm{at}}$, for a suitable choice of $\Pi \subseteq \mathcal{P}(V) \setminus \{\emptyset\}$ and $\mathrm{at} \colon V \to \Pi \cup \{\emptyset\}$.

The reverse implication is trivially true, namely it holds that C is equisatisfiable to the disjunction

$$\bigvee_{\substack{\Pi \subseteq \mathcal{P}(V) \setminus \{\emptyset\} \\ \mathrm{at} \in (\Pi \cup \{\emptyset\})^V}} C_{\Pi,\mathrm{at}} \,,$$

where $(\Pi \cup \{\emptyset\})^V$ is the set of all maps from V into $\Pi \cup \{\emptyset\}$. We call this formula the *disjunctive decomposition* of C.

What have we gained from such an elaborate decomposition? As we will hint at next, given a conjunction $C_{\Pi,\mathrm{at}}$, for given set $\Pi \subseteq \mathcal{P}(V) \setminus \{\emptyset\}$ and map $\mathrm{at} \colon V \to \Pi \cup \{\emptyset\}$, we can test in linear time whether it is satisfiable or not. Hence, the decidability of the s.p. for MLS follows, since, given a normalized conjunction C of MLS to be tested for satisfiability, we can effectively construct its disjunctive decomposition and then test each of its disjuncts for satisfiability (see below), until a satisfiable one is found, if any, in which case C will be declared 'satisfiable'. Otherwise, C is declared 'unsatisfiable'. In addition, by pruning the disjunctive decomposition of C from its unsatisfiable disjuncts, one obtains an exhaustive description of all the models of C, if any (see [34]).

As mentioned above, the satisfiability of any component $C_{\Pi,\mathrm{at}}$ of the disjunctive decomposition of C can be tested in linear time. In fact, it can be proved that the following three conditions are necessary and sufficient for the conjunction $C_{\Pi,\mathrm{at}}$ to be satisfiable.

Condition C1: The assignment $I_\Pi : V \to \mathcal{P}(\Pi)$ over V defined by

$$I_\Pi\, x := \{S \in \Pi \mid x \in S\}, \quad \text{for all } x \in V,$$

satisfies all the Boolean literals of C, namely the literals in C of any of the types

$$x = y \cup z, \quad x = y \cap z, \quad x = y \setminus z, \quad x = y, \quad x \neq y.$$

Condition C2: For all $x, y \in V$,

– if $x \in S \iff y \in S$ holds for every $S \in \Pi$, then $\mathrm{at}(x) = \mathrm{at}(y)$;
– if $x \in y$ occurs in C, then $y \in \mathrm{at}(x)$;
– if $x \notin y$ occurs in C, then $y \notin \mathrm{at}(x)$.

Condition C3: There exists a total ordering $\prec$ of V such that

$$y \in \mathrm{at}(x) \implies x \prec y, \quad \text{for all } x, y \in V.$$

In particular, when the conditions C1–C3 are true, letting

– $S \mapsto \ddot{\imath}_S$ be any injective map from Π into the universe of all sets such that $|\ddot{\imath}_S| > |V| + |\Pi|$ for each $S \in \Pi$, and
– $\prec$ be any total ordering of V satisfying the condition C3,

it can be shown that the algorithm in Fig. 1 constructs a model M for $C_{\Pi,\mathrm{at}}$ in a finite number of steps.

Initialization Phase

1. **for each** $S \in \Pi$ **do**
2. $M\alpha_S := \{\ddot{\imath}_S\}$;

Stabilization Phase

3. **for each** $x \in V$, following the ordering $\prec$, **do**
4. $Mx := \displaystyle\bigcup_{x \in S \in \Pi} M\alpha_S$;
5. $M\alpha_{\mathrm{at}(x)} := M\alpha_{\mathrm{at}(x)} \cup \{Mx\}$;
6. **return** the assignment M over $V \cup \{\alpha_S : S \in \Pi\}$.

Fig. 1. Model construction algorithm for $C_{\Pi,\mathrm{at}}$.

The approach just illustrated to the s.p. for MLS can be traced back to the original treatment in [88], subsequently refined in [19, 30] (among others). Specifically, the sets $S \in \Pi$ (and the variables α_S associated with them and constrained by the clauses $\alpha_S = \bigcap_{x \in S} x \setminus \bigcup_{y \in V \setminus S} y$ and $\alpha_S \neq \emptyset$) correspond to the *singleton models* of [88], later on called *places* (see, for instance, [11, 16, 19]). So, Π can be regarded as a collection of the *old* places. In addition, the map at: $V \rightarrow \Pi \cup \{\emptyset\}$ sends each variable $x \in V$ to the correponding *place at the variable x*, according to the definition given in [11, 16, 19], provided that $\text{at}(x) \neq \emptyset$.[2]

The algorithm in Fig. 1 is the prototype upon which the construction algorithms for several extensions of MLS are based: there is an initialization phase followed by a stabilization phase. In the initialization phase, the unrestrained places are preliminarily set to suitable non-empty, pairwise disjoint sets (in the MLS case, these are all the places α_S in Π), whereas the remaining places are interpreted as the empty set. During the stabilization phase, the values of the variables α_S increase monotonically, while remaining at all times pairwise disjoint, eventually becoming all non-empty. Once the value of a variable x in the input conjunction is set to $\bigcup_{x \in S \in \Pi} \alpha_S$ during the stabilization phase, the value of the places α_S that form x cannot change anymore, so that the value of x will not change either. In addition, it must be assured that such a value of x, which will be put in the place $\alpha_{\text{at}(x)}$ (at x) as an element, be new to the places α_S, to enforce their mutual disjointness. In the case of MLS, the above requirements are guaranteed by conditions C2 and C3, and by the fact that the cardinality of the sets $\ddot{\imath}_S$'s used in the assignment at line 2 of Fig. 1 is larger than the cardinality that any set x can ever reach, and so the membership and non-membership literals of the input conjunction are correctly modeled by the returned set assignment M. In addition, in view of the mutual disjointness and the non-emptyness of the places α_S at the end of the stabilization phase, the condition C1 will maintain that also the remaining literals of the input conjunction will all be satisfied by the set assignment M.

Notice that we have proved more than the mere decidability of the s.p. for MLS. Indeed, we have also shown that for any satisfiable MLS-formula we can effectively construct a satisfying assignment.[3]

It is plain that in presence of additional constructs to the initial endowment of MLS (much) more involved satisfiability conditions than the above C1–C3 may be needed. Also, the initialization and stabilization phases of the related construction algorithms (such as the one in Fig. 1) are expected to be (much) more intricate.

As a simple illustration, let us consider the case of the extension MLSS of MLS with singleton literals of the form $y = \{x\}$. Let, therefore, C be a normalized MLSS-conjunction, namely a conjunction of literals of the following types:

---

[2] The case $\text{at}(x) = \emptyset$ has been introduced here just to make the map at total on V.

[3] The problem of producing a concrete model for any satisfiable formula (being also able to flag when none exists) is sometimes called *satisfaction problem*: hence, for MLS, this problem is algorithmically solvable.

$$x = y \cup z, \quad x = y \cap z, \quad x = y \setminus z, \quad x = y, \quad x \neq y, \quad x \in y, \quad x \notin y, \quad y = \{x\}$$

(see, among many, [62, 141]). It turns out that the satisfiability conditions for MLSS can be obtained from the ones for MLS just by adding to condition C2 the following subcondition

– if $y = \{x\}$ occurs in C, then
  • $y \in S \iff S = \text{at}(x)$ holds for every $S \in \Pi$, and
  • it $\text{at}(x) = \text{at}(x')$, for some x' in C, then $x \in S \iff x' \in S$ holds for every $S \in \Pi$

to the three subconditions already in C2.

Concerning the model construction algorithm in Fig. 1, it is enough to replace line 2 in the initialization phase by the following one:

2′. $M\alpha_S := \textbf{if } S = \text{at}(x) \text{ for some } x \in V_{\text{sing}} \textbf{ then } \emptyset \textbf{ else } \{\ddot{\imath}_S\};$

where

$$V_{\text{sing}} := \{x \in V \mid y = \{x\} \text{ is in } C \text{ for some } y\}.$$

In Sect. 2.3 we will further exemplify the above point, when treating the extension of MLS with the union-set operator and then with the power-set operator.

Complexity Issues. The decision procedure for MLS outlined above requires one to test for satisfiability each component $C_{\Pi,\text{at}}$ of the disjunctive decomposition of the input conjunction C, until a satisfiable one is found. Checking whether the conditions C1–C3 are fulfilled for a given $C_{\Pi,\text{at}}$ takes linear time in the size of $C_{\Pi,\text{at}}$, which may be up to exponential in the size of C. Hence, the overall computation takes non-deterministic exponential time in the size of C. However, it can be shown that it is enough to test for satisfiability only those components $C_{\Pi,\text{at}}$ such that $|\Pi| \leqslant 2 \cdot |V| - 2$, where as before V denotes the set of the distinct variables of C. In addition, since the s.p. for propositional formulae in conjunctive normal form, whose clauses contain at most three disjuncts (3SAT), is NP-complete and is readily polynomially reducible to the s.p. for MLS-formulae, it follows that the s.p. for MLS is NP-complete too (see [53]).

2 Early Offsprings of Multilevel Syllogistic

2.1 "Any New Results?"

When, in the corridors of the Courant Institute, Jack met me (Eugenio) or Alfredo, he invariably asked: "Any new results?". This mantra went on all over the year 1980 and the first trimester of 1981. Extending MLS with the singleton construct[4] had been a relatively easy task—this is the decidable extension

---

[4] Set-terms of the form $\{s_0, \ldots, s_N\}$, which can be seen as abbreviating $\{s_0\} \cup \cdots \cup \{s_N\}$, enter into play for free with singleton.

MLSS briefly discussed above—, but Jack expected that sooner or later we would succeed in strengthening MLS with the treatment of much more challenging constructs, such as the union-set operator $\bigcup$ or the power-set operator $\mathcal{P}$ (the terms $\bigcup x$ and $\mathcal{P}(x)$ designate the set of all members of members of x and the set of all subsets of x, respectively). Jack also envisaged an enhancement of MLS with a 'global choice' operator arb such that $\mathsf{arb}(x)$ selects arbitrarily from each set x an element of x, if any (for definiteness, when $x = \varnothing$ holds, one takes $\mathsf{arb}(x) = \varnothing$).

After many efforts, it was then shown that decidability was not disrupted by allowing either two terms of the form $\mathcal{P}(x)$, or one of the form $\bigcup x$, to appear in the formula to be tested for satisfiability. The former of these results was achieved by Alfredo, the latter by Michael Breban: they appear in their PhD theses [10, 83]. The extension of MLS with $\mathsf{arb}(x)$, whose decidability was found somewhat later, to a large extent owing to Alfredo, would be treated in my PhD thesis [104].[5] The very recent result [60] makes it plausible that MLS extended with the Cartesian product operator $\times$, a question raised in 1985 by Domenico, is decidable (as usual, $x \times y$ designates the set of all ordered pairs $\langle u, v \rangle$ with $u \in x$ and $v \in y$).

2.2 Enhancements of Multilevel Syllogistic in Catania

Just back from graduating in Computer Science at New York University, from late 1981 to early 1982 Alfredo addressed a series of memorable seminars on multilevel syllogistic and some of its extensions in front of a small group made by Giuseppe (dubbed 'Peppino') Sorace, the late Biagio (dubbed 'Gino') Micale, and myself (Domenico). All meetings took place in a small office of the new site of the "Seminario Matematico" of the University of Catania, which had been inaugurated by the president of Italy Sandro Pertini just two years before, on November 10, 1979. I was on my fourth (and last) year of my undergraduate studies in Mathematics, and that was just the right time for Alfredo to come back to Italy![6]

In his very enjoyable seminars, starting just from scratch, Alfredo provided a very detailed presentation of the state-of-the-art decidability results for multilevel syllogistic and some of its extensions, among which, most notably, the two extensions with one occurrence of the power-set operator $\mathcal{P}$ and with one

---

[5] See also [85, 87] and cf. footnote 14 below. Early decidability-related results concerning arb relied on the assumption that a strict well-ordering $\lhd$ of the universe of all sets, complying with various conditions, is available and that $\mathsf{arb}(s)$ chooses the $\lhd$-least element of s from each non-empty set s. One of the conditions imposed on $\lhd$ entails that $x \lhd y$ holds when $x \in y$; in addition, $\lhd$ must be *anti-lexicographic* over finite sets, which means: $\{x_{m+p}, \ldots, x_m, \ldots, x_1\} \lhd (\{y_{q+1}, \ldots, y_1\} \cup \{x_m, \ldots, x_1\})$ follows from $x_{m+p} \lhd \cdots \lhd x_m \lhd \cdots \lhd x_1$ & $y_{q+1} \lhd \cdots \lhd y_1 \lhd x_m$ & $(p = 0 \lor (p > 0$ & $x_{m+1} \lhd y_1))$.

[6] I had been introduced to Alfredo by the late Filippo Chiarenza, one of his very close friends and fellow students, at a beach in the Catania shoreline, a couple of years before in the summer of 1979, when Alfredo was temporarily back to Catania to get married to his beloved wife Pina Carrà.

occurrence of the general union-set operator $\bigcup$, respectively. I still remember the pleasure we had during those meetings and the great contagious enthusiasm Alfredo put in them! My bachelor dissertation in Mathematics, titled "*Questioni di decidibilità in teoria degli insiemi*",[7] to be delivered in the summer of 1982, was essentially based in what I had been taught during those seminars: I learnt a lot with a lot of fun!

Our group, led by Alfredo, continued to meet on a quite regular basis for the rest of 1982 and the first half of 1983, with the aim of spotting out decision procedures for fragments of set theory extending multilevel syllogistic in various directions.

Our efforts culminated in the solution of the satisfiability problem for two notable extensions of multilevel syllogistic.

The first one, MLSRC, was the extension of multilevel syllogistic with the dyadic operator $\mathsf{RC}(x, y)$, whose intended meaning is $\mathrm{rank}(x) \leqslant \mathrm{rank}(y)$. For the sake of self-containedness, we recall that the rank of a set s in the well-known *von Neumann cumulative hierarchy*

$$\mathcal{V} := \bigcup_{\alpha \in \mathsf{On}} \mathcal{V}_\alpha,$$

cf. [102], is the least (ordinal) $\rho \in \mathsf{On}$ such that $s \subseteq \mathcal{V}_\rho$, where one defines $\mathcal{V}_\alpha := \bigcup_{\beta < \alpha} \mathcal{P}(\mathcal{V}_\beta)$ by transfinite recursion on $\alpha \in \mathsf{On}$. As it turns out, $\mathrm{rank}(x) = x$ holds if and only if x is an ordinal number; hence the extension of multilevel syllogistic with ordinal numbers is encompassed by this MLSRC.

The second decidability result concerned the extension of multilevel syllogistic with one occurrence of the unary '*singletor*' Σ, defined by

$$\Sigma(s) := \left\{ \{t\} : t \in s \right\},$$

for every set s. As already noted in the paper [28] that later collected such results, co-authored by Alfredo, Gino, Peppino, and myself, the Σ operator is somehow related both to the Cartesian product and to the power-set operator. The importance of those operators in 'computable set theory' was soon recognized (as recalled above, in Sect. 2.1), and in fact their inclusion in decidable extensions of multilevel syllogistic has subsequently characterized an important line of research, still active to these days.

Semi-decidability of Finite Satisfiability. A rather straightforward exploitation of Kőnig's infinity lemma shows that if one shifts focus from mere satisfiability of a given formula φ to its *finite satisfiability* problem, then the problem turns out to be *semi-decidable* even when the demanding constructs $\mathcal{P}(\cdot)$, $\mathsf{arb}(\cdot)$, and $\cdot \times \cdot$ enter into play along with $\cdot \cap \cdot$, $\cdot \cup \cdot$, $\cdot \setminus \cdot$, $\{\cdot\}$, with the relators $=, \in$, and with the propositional connectives $\&$, $\vee$, $\neg$. 'Finite satisfiability' means that one insists that each one of the sets substituted for the variables of φ in order to make φ true must have finitely many members. 'Semi-decidability'

---

[7] "Decidability issues in set theory".

means that the systematic search for the satisfying substitution must be ensured to terminate when φ is finitely satisfiable, but may go on forever in the contrary case. Membership is assumed to be a well-founded relation over the universe of all sets, which—as usual—also consists of infinite sets along with the finite ones (the latter being the only sets entitled to be chosen as values of the variables).

It was first proved in [104, Chapter 4] that the finite satisfiability problem for the so-called 'safe' formulae—the adjective in quotes refers to certain syntactic restraints that must be placed on the occurrences of arb—is semi-decidable; then, in [29], it was also proved that this problem and the *hereditarily-finite* satisfiability problem for safe formulae amount to one another. In the latter problem, the only sets entitled to be chosen as values for the variables are the ones belonging to the sub-universe

$$\mathcal{H} =_{\text{Def}} \overbrace{\underbrace{\mathcal{P}(\emptyset)}_{\mathcal{H}_1} \cup \underbrace{\mathcal{P}(\mathcal{P}(\emptyset))}_{\mathcal{H}_2}}^{\mathcal{H}_3} \cup \mathcal{P}(\mathcal{P}(\mathcal{P}(\emptyset))) \cup \cdots$$

of $\mathcal{V}$ whose levels, $\mathcal{H}_0 = \emptyset$ and $\mathcal{H}_{n+1} = \mathcal{P}(\mathcal{H}_n)$, are indexed by natural numbers.

Hence we have a semi-decision method for the finite satisfiability and the hereditarily finite satisfiability problem for a wide collection of unquantified set-theoretic formulae. In some favorable cases, the search space can be so pruned that it becomes finite, without loss of correctness: one then obtains a decision algorithm. A couple of examples of this kind are treated in [30, Chapter 5].

To frame in historical context what precedes, let me add that I (Eugenio) discovered the semi-decision method for finite satisfiability of safe formulae during a visit that Martin Davis made, upon invitation of Alfredo Ferro, to the University of Catania in the winter of 1982/83. I conjectured (cf. [104]) that my method worked for hereditarily finite satisfiability as well, and presented my conjecture in a talk at the University of Catania when Alfredo invited me during the summer of 1985. Domenico, who then was in the audience, promptly solved my conjecture, as Alfredo reported to me a few days later: this led to our joint publication [29]. The decidability by-products mentioned at the end of the preceding paragraph arrived months later; they were attained by Alberto Policriti and Franco Parlamento.

2.3   Multilevel Syllogistic in New York Again

Full Treatments of Power-Set, Union-Set, and More. After the decidability results found in the early 80's by Alfredo (MLS plus two power-set terms) and Michael Breban (MLSS plus one union-set term) in their Ph.D. theses (see [83, 10, 11]), it took some years before the satisfiability problems for the extensions of MLS with *any* number of either power-set terms or union-set terms were settled.

Thus, in the late Spring term of 1984 at Courant Institute (NYU), while I (Domenico) was completing my first Ph.D. year, there was a great excitement at the announcement of a special Computer Science Colloquium in which Jack would have reported his breakthrough in the satisfiability problem for the extension MLSU of MLS with any number of union-set terms.

Some weeks later, Jack handed me a research report he wrote on that, asking me to read it very carefully. That was the right time for me, since I had just completed my Spring term examinations.

The novelty of Jack's idea was that the decidability conditions were expressed in terms of certain connectivity properties of a graph G_C, named *Ugraph*, suitably associated with the conjunction C whose satisfiability must be demonstrated. Provided such a graph exist, it would steer the construction of a model for C, as resulting from the stepwise filling of the disjoint places of the variables of C.[8] After the initialization of the places of C with certain *secondary elements*, the connectivity properties of G_C would allow a (possibly infinite) stabilization phase to take place, by using an infinite endowment of *auxiliary elements* initially associated with each place of C. As in the case of MLS, the fact that places remain mutually disjoint throughout stabilization, would ensure the correct modeling of all literals of the forms $x = y \cup z$, $x = y \cap z$, and $x = y \setminus z$ in C. In addition, thanks to the connectivity properties of the Ugraph G_C and to the properties enjoyed by the auxiliary and secondary elements associated with the places of C, it turned out that the final assignment constructed by the stabilization phase modeled correctly also all the union-set clauses $u = \bigcup y$ and the membership clauses $x \in y$ and $x \notin y$ present in C.

Unfortunately, as I discovered, in his report Jack overlooked the case in which a *null node*, whose corresponding place is constrained to contain as element just the empty set $\emptyset$, was present in the Ugraph. Similar constraints would be inherited also by the so-called *trapped nodes*, namely those nodes of the Ugraph whose sufficiently long paths forward from them eventually reach the null node. By letting the height of a trapped node be the length of the longest path from it to the null node, the place associated with any trapped node ν turned out to be constrained to contain as elements only sets of rank at most the height of ν, in particular preventing it from being infinite and thus calling for more thoughtful conditions on the Ugraph and a more sophisticated model construction.

Jack invited me as a co-author, with the task of addressing the combinatorial difficulties due to the presence of trapped nodes, and I was lucky enough to succeed in handing him a complete solution just before flying back to Italy for the summer.

The summer of 1984 was by no means a vacation time for me. On the one hand, I had to prepare for the quite challenging written comprehensive examination that I was supposed to take in September, when back to NYU; on the other hand, I had very frequent meetings with Alfredo in an all-out effort to adapt Jack's Ugraph technique to the case of the extension MLSP of MLS with any number of power-set terms. At that time, we were in fact working under a

---

[8] Places have been defined in Sect. 1.3.

certain pressure as the initial plan of Jack and Alfredo was to collect the two decidability results, for MLSU and MLSP, in the same paper.

It soon turned out that the adaptation to MLSP of the technique developed for MLSU was not at all an easy task.[9]

The *P-graph* for a conjunction C of MLSP with set of places Π was defined as a graph whose set of vertices is $\Pi \cup \mathcal{P}(\Pi)$ and whose edges are of two types:

- *membership edges* $\alpha_i \to \{\alpha_1, \ldots, \alpha_\ell\}$, where $1 \leqslant i \leqslant \ell$ and $\alpha_1, \ldots, \alpha_\ell \in \Pi$; and
- P-edges $\{\alpha_1, \ldots, \alpha_\ell\} \to \beta$, where $\alpha_1, \ldots, \alpha_\ell, \beta \in \Pi$.

In particular, P-edges were intended to represent the semantic relationship

$$\mathcal{P}^*(\{\sigma_{\alpha_1}, \ldots, \sigma_{\alpha_\ell}\}) \cap \sigma_\beta \neq \emptyset,$$

where $\sigma_{\alpha_1}, \ldots, \sigma_{\alpha_\ell}, \sigma_\beta$ are the mutually disjoint sets associated with the places $\alpha_1, \ldots, \alpha_\ell, \beta$ and $\mathcal{P}^*(\cdot)$ is the so-called *intersecting power-set operator* defined by

$$\mathcal{P}^*(\{s_1, \ldots, s_m\}) =_{\mathrm{Def}} \left\{ t \subseteq \bigcup_{1 \leqslant i \leqslant m} s_i \; \middle| \; t \cap s_i \neq \emptyset, \text{ for } i = 1, \ldots, m \right\}.$$

Provided that certain conditions are satisfied, P-graphs behave like a sort of flow graph in the realm of set theory that, through simple rules, allows one to construct a model for C in a number of steps, within an initialization and a stabilization phase, that turns out to be finite.

The initialization phase was intended to produce mutually disjoint non-empty sets $\overline{\alpha}$ to be associated with the places α of (the MLS part of) the conjunction C in such a way that the temporary assignment

$$\overline{x} := \bigcup_{x \in \alpha} \overline{\alpha}, \quad \text{for } x \text{ in } C,$$

modeled correctly all the inclusions $x \subseteq \mathcal{P}(y)$, for every power-set clause $x = \mathcal{P}(y)$ in C. Instead, the subsequent stabilization phase was intended to force the correct modeling of the power-set clauses $x = \mathcal{P}(y)$ and the membership clauses $x \in y$ and $x \notin y$ in C.[10]

While the stabilization phase was somewhat inspired by the one for MLSU, the initialization phase was problematic, since the production of new elements to be added to the sets $\overline{\alpha}$, starting from the bare empty set $\emptyset$, was greatly constrained and could easily terminate with failure if wrong choices were made.

---

[9] Actually, power-set clauses impose stronger constraints than union clauses do, insofar as the implication

$$x = \mathcal{P}(y) \implies y = \bigcup x$$

is true for all sets x and y, while the converse implication does not hold in general.

[10] Correct modeling of also the literals of the forms $x = y \cup z$, $x = y \cap z$, and $x = y \setminus z$ was ensured since the sets $\overline{\alpha}$, with $\alpha \in \Pi$, were maintained mutually disjoint during the stabilization phase too.

Eventually a solution emerged for handling the combinatorial difficulties alluded to above. Specifically, so long as our conjunction C was satisfiable, it turned out that the initialization phase can non-deterministically be accomplished in at most $(2^m - 1)2^{2^m-1}\rho$ steps, where m is the number of distinct variables occurring in C and ρ is any positive integer such that $2^{\rho-m} \geqslant \rho$. Up to secondary technicalities, this was enough to settle positively the decision problem for MLSP, and so ... I could fully concentrate on the examination I would have taken in a few weeks at NYU!

Because of their lengths, the two decidability results for MLSU and MLSP were finally split into distinct papers: [33] (which, due to publication delays, appeared only in 1987) and [32], respectively.

The threshold $(2^m - 1)2^{2^m-1}\rho$ played a major role also in the extensions of MLSP with singleton terms $\{y\}$ (my Ph.D. thesis; see [15,17,61]) and later also with the predicate Finite(y) (stating that the set y has finitely many elements) and its negation (see [58]).

I soon conjectured that the double-exponential threshold $(2^m - 1)2^{2^m-1}\rho$ could be lowered to a linear expression, but so far this remains an open question.

Another notable decidability result discovered at the end of the 80's, while I was a Visiting Assistant Professor at NYU, concerned the extension of MLS with various map constructs, such as domain, range, image and inverse image of sets, inverse map, map restriction, and predicates about single-valuedness and injectivity (see [57], preliminarily published in 1988 as Technical Report n. 374, New York University, Department of Computer Science).

As early as 1980, a similar extension of MLS had been proved decidable in [89]. However, it involved two types of variables, one for sets and one for functions, and therefore combinations of function variables such as $f_1 = f_2 \cup f_3$, $f_1 = f_2 \cap f_3$, and $f_1 = f_2 \setminus f_3$ were not admitted there. On the other hand, the extension of MLS addressed in [57] involved just one type of variables, to be interpreted as generic sets.

Once a specific set encoding of ordered pairs $\langle \cdot, \cdot \rangle$ was fixed, the above-mentioned map-related constructs were supposed to apply only to the subset of the ordered pairs belonging to each of their arguments. For instance,

- Domain$(s) = u \leftrightarrow_{\mathrm{Def}} (\forall a)\big(a \in u \Longleftrightarrow (\exists b)\langle a, b\rangle \in s\big)$,
- Singlevalued$(s) \leftrightarrow_{\mathrm{Def}}$
$$(\forall a, b, c, d)\big((\langle a, b\rangle \in s \,\&\, \langle c, d\rangle \in s \,\&\, a = c) \Longrightarrow b = d\big),$$
- Injective$(s) \leftrightarrow_{\mathrm{Def}} (\forall a, b, c, d)\big((\langle a, b\rangle \in s \,\&\, \langle c, d\rangle \in s \,\&\, b = d) \Longrightarrow a = c\big)$,
- Restriction$(s, t, u) \leftrightarrow_{\mathrm{Def}} (\forall a, b)\big(\langle a, b\rangle \in s \Longleftrightarrow (\langle a, b\rangle \in t \,\&\, a \in u)\big)$,
- Inverse$(s, t) \leftrightarrow_{\mathrm{Def}} (\forall a, b)\big(\langle a, b\rangle \in s \Longleftrightarrow \langle b, a\rangle \in t\big)$,

where s, t, and u stand for generic sets *not* restricted to contain as elements just ordered pairs. Membership clauses of the form $\langle x, y\rangle \in z$ and their negations were admitted too. However, a major limitation that so far has not been overcome is that terms of type Pairs($\cdot$) were not expressible, where, for any set s, Pairs(s) is the set of all ordered pairs in s. If terms of such form were admitted, one could also express the singleton operator $\{\cdot\}$. Indeed, by letting for instance

$$\langle x, y \rangle =_{\text{Def}} \big\{ \{0, \{x\}\}, \{2, \{y\}\} \big\}$$

as was done in [57], where 0 and 2 are encoded as von Neumann integers, namely

$$0 =_{\text{Def}} \varnothing \qquad \text{and} \qquad 2 =_{\text{Def}} \{\varnothing, \{\varnothing\}\},$$

then

$$z = \{x\} \iff (\exists u, v, w)(v \in \mathsf{Pairs}(u) \ \& \ w \in v \ \& \ \varnothing \in w \ \& \ z \in w \ \& \ x \in z).$$

Hence, the clause $z = \{x\}$ would be equisatisfiable to the conjunction

$$v \in \mathsf{Pairs}(u) \ \& \ w \in v \ \& \ \varnothing \in w \ \& \ z \in w \ \& \ x \in z,$$

where u, v, w are new variables just introduced to express the clause $z = \{x\}$.

We finally mention that in the second half of the 80's, the three of us, namely Alfredo, Eugenio, and myself, were also very busy with the first monograph on computable set theory, [30], which appeared in print in 1989. As Jack kindly wrote in his forward,

> "The [···] book reports on an exceptionally successful continuing series of investigations of [···] elementary set theory. The techniques that it describes combine model-theoretic with syntactic considerations in interesting ways, which, as will be seen, grow increasingly challenging and intricate as successively wider sub-languages of set theory are considered."

Table 1. A yardstick weak set theory.

Extensionality: $\forall x \, \forall y \, \big(\, \forall v \, (v \in x \iff v \in y) \implies x = y \, \big)$
Empty set: $\exists z \, \forall v \, \big(\, \neg v \in z \, \big)$
Adjunction: $\forall x \, \forall y \, \exists w \, \forall v \, \big(\, v \in w \iff (v \in x \lor v = y) \, \big)$
Single removal: $\forall y \, \forall x \, \exists \ell \, \forall v \, \big(\, v \in \ell \iff (v \in x \ \& \ \neg v = y) \, \big)$
Foundation (FA): $\forall x \, \exists v \, \forall y \, \big(\, y \in x \implies (v \in x \ \& \ \neg y \in v) \, \big)$

Limiting Results. An issue had been addressed readily after the implementation of the satisfiability tester for MLS: how far can we go without disrupting decidability? After arriving at NYU in 1986, Alberto Policriti—who had met Alfredo in Sicily while performing his civilian service as a conscientious objector—gave impulse to the research on limiting results regarding Set Theory. At the University of Turin (Italy), after graduating he had started with Franco Parlamento an investigation on undecidability in Set Theory; in New York, this line of research about limiting results culminated in significant discoveries:

1. An axiomatic system as simple as the one shown in Table 1, or even weaker, suffices to ensure essential undecidability (cf. [124]).[11]
2. The existence of infinite sets amounts to the existence of two sets satisfying a certain formula, statable in terms of propositional connectives and restricted universal quantifiers—see Table 2. This fact, first shown in [125–127], explains why knotted quantifiers pose a serious challenge to the invention of satisfiability testers; on a mathematical plan, the challenge becomes unsurmountable [128], unless narrow syntactic restraints are imposed on the formulae to be analyzed by the tester.
3. Hilbert's 10th problem [103] can be reduced to the satisfiability problem for small fragments of set theory, not much richer that the language of MLS; those fragments hence have an unsolvable decision problem [23].

Table 2. Two ways of stating, by means of restricted universal quantifiers, the existence of disjoint infinite sets ω_0, ω_1. The shorter formulation presupposes the foundation axiom, the longer one works under no axiomatic commitment concerning the well-foundedness of $\in$. Concretely, one can take $\omega_0 := \{\omega_{1,i} : i \in \mathbb{N}\}$ and $\omega_1 := \{\omega_{0,i} : i \in \mathbb{N}\}$, where, recursively, $\omega_{1,i} := \{\omega_{0,j} : 0 \leqslant j \leqslant i\}$ and $\omega_{0,i} := \{\omega_{1,j} : 0 \leqslant j < i\}$ for all $i \in \mathbb{N} = \{0, 1, 2, 3 \ldots\}$.

$$\omega_0 \neq \omega_1 \quad \& \quad \forall x \in \omega_0 \, \forall y \in \omega_1 \, (x \in y \lor y \in x) \quad \&$$
$$\mathop{\&}_{i \in \{0,1\}} \left(\omega_i \notin \omega_{1-i} \ \& \ \forall x \in \omega_i \, \forall y \in x \ y \in \omega_{1-i} \right)$$

$$\omega_0 \neq \omega_1 \quad \& \quad \forall x \in \omega_0 \, \forall y \in \omega_1 \, (x \in y \lor y \in x) \quad \&$$
$$\forall x_1, x_2 \in \omega_0 \, \forall y_1, y_2 \in \omega_1 (x_2 \in y_2 \in x_1 \in y_1 \implies x_2 \in y_1) \ \&$$
$$\mathop{\&}_{i \in \{0,1\}} \left(\omega_i \notin \omega_{1-i} \ \& \ \forall x \in \omega_i \, \forall y \in x \ y \in \omega_{1-i} \right)$$

An interesting issue was addressed at the time: How crucial is the role played by the *foundation axiom* FA—which (in combination with other axioms) ensures that no infinite descending membership chain $s_0 \ni s_1 \ni s_2 \ni \ldots$ exists—in the decidability of fragments of set theory? Simply staying neutral on the issue of the well-foundedness of membership does not pay off. A disadvantage arising from such an attitude is that certain formulae of MLS turn out to be valid in von Neumann's above-cited privileged universe of sets (namely the cumulative hierarchy), but false in many non-standard universes which comply with all postulates about abstract sets save one: the foundation axiom. Just a couple of years later, Peter Aczel would propose in [2] an elegant theory of sets encompassing

---

[11] Compared with the very weak aggregate theory Z_1 treated by Robert L. Vaught, which is generated by the empty set axiom and the adjunction axiom alone (cf. [143, p. 21]), the systems studied by Parlamento and Policriti ensure a more transparent syntax arithmetization and a tighter connection with computability theory.

a postulate, AFA, antithetic to foundation:[12] as we were to discover a few years later, the adoption of AFA yields, much as the adoption of FA does, the completeness of the axiomatic system with respect to MLSS and to other fragments of Set Theory; however, such a shift calls for a redesign of the decision methods (cf. [109] and [115, Sec. 3.5]), and one hence has the duty of reassessing their algorithmic complexities.

2.4   A Downsizing and Broadenings of MLS

An important downsizing of MLS, coeval to it, was named *two-level syllogistic*, 2LS for brevity. In 2LS there are terms of two disjoint sorts: *individuals* and *sets*. Each membership atom $x \in \tau$ relates an individual-variable, x, with a set-term τ whose construction involves: set-variables; a constant, $\mathbb{1}$, designating the collection of all individuals; and the usual Boolean constructs $\cup$, $\cap$, and $\setminus$. These constructs enable one to specify the complement of the set designated by any set-term θ, as $\mathbb{1} \setminus \theta$. Atoms of the forms $x = y$, $\tau = \theta$, and $\tau \subseteq \theta$, where x and y stand for individual-variables and τ and θ stand for set-terms, are also available. Much as in MLS, the propositional connectives $\&$, $\vee$, $\neg$ are used to assemble compound formulae out of the atomic ones. Decidability can be proved either by direct means—as done in [86]—or by embedding 2LS into MLS—as done in [30, Sec. 6.7].

The implementation, named Hilberticus, of a decision algorithm for a language very similar to 2LS was carried out by Jörge Lücke and described in [100], and then further developed into an 'abacus for sets', as documented at https:// zbmath.org/software/1264.

The satisfiability problems for several common extensions of MLS, including MLSS, can be solved by way of straightforward reductions to the satisfiability problem for MLS. As a way of exemplification, we consider here the case of the extension of MLSS with the predicates (and their negations):

- Finite(x) (which states that x is finite),
- Countable(x) (which states that x is countable, i.e., either finite or denumerable), and
- Continuum(x) (stating that x has the cardinality of the continuum).

Given a conjunction C in this extension of MLSS, we introduce three new variables, Fi, Co, and $\mathfrak{c}$, and obtain the MLSS-conjunction C' by:

- first adding to C the following statements

  - $Co \supseteq Fi$ $\&$ $\mathfrak{c} \cap Co = \emptyset$,
  - $Fi \supseteq x$, for each variable x for which either a statement $x = \{y\}$ or a statement Finite(x) is present in C,
  - $Co \supseteq x$, for each x for which Countable(x) is present in C,

---

[12] The term *hyperset* was first proposed in [3] to mean—in a universe modeling Aczel's AFA—a 'set' x_0 whose transitive closure is potentially ill-formed, in the sense that membership may form an infinite descending chain $x_0 \ni x_1 \ni x_2 \ni \cdots$.

- $\mathfrak{c} \cap x \neq \emptyset$, for each x for which $\mathsf{Continuum}(x)$ is present in C,
- $\neg\,(Fi \supseteq x)$, for each x for which $\neg\mathsf{Finite}(x)$ is present in C,
- $\neg\,(Co \supseteq x)$, for each x for which $\neg\,\mathsf{Countable}(x)$ is present in C,
- $(Co \cup \mathfrak{c} \supseteq x) \implies (Co \supseteq x)$, for each x for which $\neg\,\mathsf{Continuum}(x)$ is present in C,

- and then dropping all statements of the forms $\mathsf{Finite}(x)$, $\mathsf{Countable}(x)$, $\mathsf{Continuum}(x)$, $\neg\,\mathsf{Finite}(x)$, $\neg\,\mathsf{Countable}(x)$, and $\neg\,\mathsf{Continuum}(x)$.

If C is satisfied by a model M, then the extension of M to the variables Fi, Co, and $\mathfrak{c}$ of C', where

- Fi is interpreted as the union of all *finite* Euler regions of the collection $\{M\,x : x \text{ in } C\}$,
- Co is interpreted as the union of all *countable* Euler regions of $\{M\,x : x \text{ in } C\}$,
- $\mathfrak{c}$ is interpreted as the union of all Euler regions of $\{M\,x : x \text{ in } C\}$ having the cardinality of the continuum,

plainly satisfies C'. To get the converse, namely that if C' is satisfiable so is C, one needs a slightly more involved technique (of the kind presented in [19, Sec. 4.13]) for transforming any given model M' of C' into a model M of C.

For more applications of the reduction technique just exemplified, the reader is referred to [19, Sec. 4].

3 Applications of Multilevel Syllogistic

3.1 Algorithm Correctness Verification

The following tiny example, drawn from [30, pp. 24–25], shows in watermark potential uses of automatic inference mechanisms in the proofs of correctness of imperative SETL-like programs (concerning SETL, see [74, 137]).

Consider the problem, given a set S_0 and a dyadic relation F, of finding the smallest superset S of S_0 that is closed with respect to F in the following sense: Every y such that $\langle x, y \rangle \in F$ holds for some $x \in S$ belongs in its turn to S. After making the definition $F[X] =_{\mathrm{Def}} \{y : x \in X \ \&\ \langle x, y \rangle \in F\}$, we propose the following specification of how to solve the problem in question:

> **hypothesis:** $S_1 \supseteq S_0 \ \&\ F[S_1] \subseteq S_1$;
> $S := S_0$;
> **while** $F[S] \not\subseteq S$
>   **invariant:** $S_1 \supseteq S \ \&\ S \supseteq S_0 \ \&\ F[S_1] \subseteq S_1$;
> $S := S \cup F[S]$;
> **end while**;
> **claim:** $S_1 \supseteq S \ \&\ S \supseteq S_0 \ \&\ F[S] \subseteq S$.

This annotated program gets translated straightforwardly into the collection of the following four statements, which are all valid if and only if the specified procedure is partially correct (viz., it produces the right answer when its execution terminates):

1) $S_1 \supseteq S_0 \ \& \ F[S_1] \subseteq S_1 \ \& \ S = S_0 \ \& \ F[S] \subseteq S \implies$
$$S_1 \supseteq S \ \& \ S \supseteq S_0 \ \& \ F[S] \subseteq S ;$$

2) $S_1 \supseteq S_0 \ \& \ F[S_1] \subseteq S_1 \ \& \ S = S_0 \ \& \ F[S] \not\subseteq S \implies$
$$S_1 \supseteq S \ \& \ S \supseteq S_0 \ \& \ F[S_1] \subseteq S_1 ;$$

3) $S_1 \supseteq S \ \& \ S \supseteq S_0 \ \& \ F[S_1] \subseteq S_1 \ \& \ \bar{S} = S \cup F[S] \ \& \ F[\bar{S}] \not\subseteq \bar{S} \implies$
$$S_1 \supseteq \bar{S} \ \& \ \bar{S} \supseteq S_0 \ \& \ F[S_1] \subseteq S_1 ;$$

4) $S_1 \supseteq S \ \& \ S \supseteq S_0 \ \& \ F[S_1] \subseteq S_1 \ \& \ \bar{S} = S \cup F[S] \ \& \ F[\bar{S}] \subseteq \bar{S} \implies$
$$S_1 \supseteq \bar{S} \ \& \ \bar{S} \supseteq S_0 \ \& \ F[\bar{S}] \subseteq \bar{S} .$$

These statements can be submitted directly to one of the syllogistic deciders presented in [89], which will demonstrate the correctness of the procedure by proving them valid.

Nowadays, set-based proof-technology has made enough progress that an automated proof verifier embodying MLSS was exploited to establish the correctness of the Davis-Putnam-Logemann-Loveland satisfiability test for CNF-formulae of propositional logic (cf. [118]).

3.2   Program Transformation

"When formal differentiation is applied to algorithms expressed as high-level, lucid, but inefficient problem statements, the transformed algorithms materialize as more complex but efficient program versions."   *(Robert Paige and Shaye Koenig, 1982)*

"We believe that the utility of transformational systems to program development will ultimately rest on a practical program correctness technology."
(Robert Paige and Jean-Pierre Keller, 1995)

Program-transformation techniques can obtain efficient code out of very high-level but poorly performing executable code; this was very convincingly shown by applying such techniques to SETL algorithm specifications (see, in particular, [122] and [75, Chapter V]). It is critically important that each program transformation is semantics preserving, so that it does not jeopardize the compliance of an algorithm with the specification of a problem that the algorithm is supposed to solve: this crucial correctness issue calls for a reliable proof technology.

In [96], a program development methodology based on proof-checked program transformations was illustrated through derivations of a bisimulation algorithm and a minimum-state DFA algorithm. In order to assess the viability of the proposed approach, the authors built a prototype program-correctness checker apt to the needs, but they found that the labor involved in their experiment was almost overwhelming. Their effort could today benefit of the advances made. in proof-correctness verification based on set theory. We will come back to this point in Sect. 3.6, where we will also discuss why a clever implementation of a broadened MLS plays a crucial role in the design of a proof-checker based on set theory.

3.3   Declarative Set-Based Programming

ENI, the Italian national hydrocarbon group, promoted and funded in 1985 a five-year project (whose beginnings are documented by [30,105] and by [138]) coordinated by Francesco Zambon and myself (Eugenio), aiming at the development of a set-based declarative programming language. Ron Sigal, Domenico Cantone, Alberto Policriti, Marco Pellegrini and Raimondo Sepe [130], and Gianfranco Rossi, were among the participants in that project, within whose framework two students of the university of Udine, Agostino Dovier and Enrico Pontelli (who today are full professors, respectively at the University of Udine and at the New Mexico State University), began to develop around 1990 an extension of Prolog named {log} (read 'setlog'), cf. [77,78]. Set-unifiers and, more generally, set-constraint solvers entering in the design of {log} and shaping a long stream of research stimulated by that programming language (see, e.g., [7,79,80]) were devised independently of MLS and extensions thereof, nonetheless they pertain to the same circle of ideas.

3.4   Hybrid Syllogistics

Some decidable fragments of set theory comprise constructs belonging to a specific field of mathematics. A very basic example has just been mentioned: the set-unification algorithm involved in the parameter-passing mechanism of {log} refers to a blended universe generated from the set-theoretic constant $\varnothing$ and an indefinite number of uninterpreted constants by means of the adjunction operator with such that x with y designates the set $x \cup \{y\}$ (see Table 1), together an indefinite number of free Skolem function symbols.[13]

In some other cases, the enrichment is made to the sublanguage 2LS of MLS described in Sect. 2.4, as for instance in the cases of the extensions of Tarski's and Presburger's arithmetics with sets, of topological syllogistics, and of a fragment of graph theory with acyclicity and reachability predicates, which we briefly review.

Tarski's arithmetic with sets, $\mathcal{TS}$, is the propositional closure of atomic formulae of the form

$$x = 0, \qquad x = 1, \qquad x = y + z, \qquad x = y \cdot z, \qquad x \leqslant y, \qquad x < y,$$
$$X = Y \cup Z, \qquad X = Y \setminus Z, \qquad x \in X,$$
$$x = \sup X, \qquad x = \inf X, \qquad interval(X),$$

where lower-case variables range over real numbers and upper-case variables range over sets of real numbers. The intended meaning of the above literals should be clear. In particular, $interval(X)$ states that X is a bounded interval. Other constructs that are readily expressible in terms of the above ones are:

---

[13] One often dubs as 'free' or 'uninterpreted' those symbols that are not subdued to any axiomatic constraint; this is not the case of with, which must comply with the laws x with y with $z = x$ with z with y and x with y with $y = x$ with y.

$\cap$, $\subseteq$, $\{x\}$, $x \star X$ and $X \star Y$ (with $\star \in \{<, \leqslant, \geqslant, >\}$), $open\_interval(X)$, and $closed\_interval(X)$. The decision procedure for $\mathcal{TS}$ described in [24] is based on the places technique and uses as a subroutine the celebrated Tarski's algorithm for real geometry [142], in the form developed by Collins with the cylindrical decomposition technique [68]; therefore it runs in non-deterministic double-exponential time. However, when multiplication is absent, by applying the ellipsoid method for linear programming in place of Tarski's algorithm, the decision problem for the resulting theory $\mathcal{TS}^-$ turns out to be NP-complete.

The similar theory of Presburger's arithmetic with sets, $\mathcal{PS}$, has also been proved to be decidable in [24]. The theory $\mathcal{PS}$ involves the same literals as $\mathcal{TS}$, save the ones of type $x = y \cdot z$ and $interval(X)$. However, the numeric and the set variables in $\mathcal{PS}$ range over the set $\mathbb{Z}$ of the integers and the collection of the subsets of $\mathbb{Z}$, respectively. The decidability of $\mathcal{PS}$ follows from the combination of the 'place' technique with integer linear programming (ILP). The NP-completeness of 2LS and ILP (cf. [123]) is inherited by the theory $\mathcal{PS}$.

Notably, multiplication is expressible in the extension $\mathcal{PS}_{\forall^1}$ of $\mathcal{PS}$ with universal quantification over numeric variables, even if the quantifier prefixes are restricted to have length 1. Hence, Hilbert's 10th problem [103] can be reduced to $\mathcal{PS}_{\forall^1}$, yielding its undecidability.

Analogously, the alike extension $\mathcal{TS}^-_{\forall^1}$ of $\mathcal{TS}^-$ is undecidable. Indeed, using the Archimedean property of the reals, one can characterize the set $\mathbb{N}$ of the natural numbers in $\mathcal{TS}^-_{\forall^1}$, and so, as before, Hilbert's 10th problem can be reduced to the decision problem of $\mathcal{TS}^-_{\forall^1}$.

Topological syllogistics have been investigated in [22, 49, 50]. To accommodate for multiple topological spaces, these are represented as a multi-sorted extension of 2LS, designated by $\mathcal{L}^{2,\infty}$, with continuous and closed maps between distinct spaces with their own Kuratowski closure operators and complementation operators. Besides terms denoting direct images of closed maps and inverse images of continuous maps, $\mathcal{L}^{2,\infty}$ admits predicated stating the injectivity and the surjectivity of maps, cardinality comparison, and connectedness.

Several useful topological terms and predicates are expressible in $\mathcal{L}^{2,\infty}$ with the above constructs and the 2LS apparatus. These include interior and boundary of a set, derived set (namely the set of all the accumulation points of its argument), and the 'open' and 'closed' predicates. However, $\mathcal{L}^{2,\infty}$-formulae must obey a technical restriction that prevents formation of function cycles.

The s.p. for $\mathcal{L}^{2,\infty}$ is solved by reducing it to the s.p. for 2LS. Specifically, given an $\mathcal{L}^{2,\infty}$-formula F, an equisatisfiable 2LS-formula F^* is effectively constructed. As it often happens, showing that F^* admits a set model whenever F has a topological model is quite an easy matter. On the other hand, the inverse implication is more difficult to prove, since one has to construct suitable topologies (or at least attest their existence) fulfilling a number of constraints.

As argued in [22], most of the theorems in the first chapters of introductory textbooks in topology can be expressed by $\mathcal{L}^{2,\infty}$-formulae, and therefore they can be automatically proved by the decision procedures for $\mathcal{L}^{2,\infty}$. Some of them are:

- if a set is connected, so is its topological closure;
- every dense subset of a closed and connected set is connected;
- every point of an open set is an accumulation point of its interior;
- the union of two connected sets A and B such that $\overline{A} \cap B \neq \emptyset$ holds is connected;
- given two topological spaces U and U^* such that $U = f^{-1}[U^*]$, for some continuous and surjective map $f : U \to U^*$, if U is connected, so is U^*.

We finally mention here that the decision problem for syllogistics involving product topologies has been addressed in [21].

The decision problem for a fragment of lattice theory with monotone functions, called POSMF, has been studied in [31], among other decidability results about fragments of real analysis. The normalized conjunctions of POSMF involve literals of the following form:

$$x \leqslant y, \quad x \not\leqslant y, \quad x = y, \quad x \neq y, \quad x = f(y), \quad \mathsf{up}(f), \quad \mathsf{down}(f),$$

where the variables x, y are supposed to range over a lattice S and the map symbol f ranges over the functions from S into S. In particular, in presence of a clause $\mathsf{up}(f)$ (resp. $\mathsf{down}(f)$), the map symbol f is supposed to range over the non-decreasing (resp. non-increasing) functions from S into S.

Given a normalized conjunct C of the theory POSMF, initially we make a copy C_{temp} of C and complete it with all possible consequences of its map clauses

$$x = f(y), \quad \mathsf{up}(f), \quad \mathsf{down}(f)$$

in combination with the clauses in C of type

$$x \leqslant y, \quad x = y.$$

For instance, for every pair of clauses in C of the form $x = f(y)$ and $x' = f(y')$, the implication $y = y' \implies x = x'$ is added to C_{temp}. Subsequently, all map clauses are dropped from C_{temp}, obtaining a formula C^*, which is satisfiable whenever so is C.

The formula C^* is then transformed into an equisatisfiable 2LS formula C^{**} by replacing every occurrence in C^* of the relation symbols $\leqslant$ and $\not\leqslant$ by $\subseteq$ and $\not\subseteq$, respectively.

Finally, we observe that any set model M for C^{**} can be extended to a model M^+ that interprets also the function symbols in C in such a way that all the map clauses in C are correctly modeled by M^+, thus showing that M^+ is indeed a model for C. Thus, C and C^{**} are equisatisfiable, proving that the decision problem for POSMF can be reduced to the one for 2LS.

The result outlined above has later been further extended in [64].

Decidable extensions of 2LS with various graph constructs have been addressed in [20] and in [63]. Specifically, in [20] the decidability is proved of a quite expressive fragment of a theory of graphs, both directed and undirected, in which properties such transitivity, completeness, bipartiteness, being a clique or an independent set, can be stated.

The theory DGRA (Directed Graphs with Reachability and Acyclicity), proved decidable in [63], involves three sorts of variables (for nodes, sets of nodes, and graphs). In addition to the singleton operator and the set operators of 2LS, which can be used to construct complex terms of types node and graph, DGRA includes a predicate Reachable(a, b, G), stating that there is a directed path in the graph G from a to b, and a predicate Acyclic(G), stating that the graph G contains no cycles. A small model property is proved for the theory DGRA that allows one to show the NP-completeness of its satisfiability problem. In particular, a decision procedure for DGRA is provided in the form of a decidable tableau calculus.

These decidable extensions of 2LS, along with many decidable fragments of mathematical analysis (see, among several, [13]) and of specific algebraic theories (e.g., ordered Abelian groups [56, Sec. 2]) indicate how inference mechanisms embodying specific knowledge pertaining to various branches of mathematics can become part of broad-gauge proof-assistants, as Woody Bledsoe advocated in [8] many years ago; moreover, such specialized reasoners can benefit from techniques developed in connection with set-based syllogistics.

3.5 Reasoners for Description Logics

Description logics (DLs) are a family of logic-based languages intended to represent knowledge about a domain of interest in a structured way, specifically in terms of *individuals*, designating specific domain elements, of *concepts*, denoting sets of elements, and of *roles*, representing dyadic relations between elements. Starting with concept names, role names, and individual names, complex terms can be formed by means of the constructors present in the DL. The domain structure of a DL *knowledge base* is suitably constrained by finite sets of *axioms* and *assertions*.

Most DLs are just decidable fragments of first-order logic. Several results and decision procedures designed in the context of DLs have found notable applications also in the area of *Semantic Web*.

In the last decades, many efforts have been devoted to find a convenient trade-off between DLs expressivity and the computational complexity of their fundamental reasoning tasks, such as knowledge base consistency, instance checking, and concept satisfiability, subsumption and equivalence. We recall that the basic description logic $\mathcal{ALC}$ is ExpTime-complete.

The first application of Computable Set Theory to DLs can be traced back in 2010 with the papers [37,40], where a two-sorted extension $\mathsf{MLSS}^{\times}_{2,m}$ of MLSS, with set and map variables, and its DL counterpart $\mathcal{DL}\langle\mathsf{MLSS}^{\times}_{2,m}\rangle$ have been proposed, respectively.

$\mathsf{MLSS}^{\times}_{2,m}$ involves two types of variables, namely set and map variables. Set terms are the usual ones, as in MLSS. Map terms have the form $X \times Y$ (where

X and Y are set terms and '$\times$' is the Cartesian product), $F_{X|.}$, $F_{.|Y}$, $F_{X|Y}$ (left, right, and left-right restrictions of map terms), F^{-1} (map inversion), and Boolean combinations of map terms $F \cup G$, $F \cap G$, $F \setminus G$ (where F, G stand for map terms). Atomic formulae are of the following types:

$$X \in Y, \quad X \subseteq Y, \quad X = Y, \quad [X, Y] \in F, \quad F \subseteq G, \quad F = G,$$
$$\text{Injective}(F), \quad \text{Singlevalued}(F), \quad \text{Bijective}(F),$$

where, as above, X, Y stand for set terms and F, G for map terms.

It turns out that the decision problem for $\mathsf{MLSS}_{2,m}^{\times}$ is NP-complete, and so is the complexity of reasoning in $\mathcal{DL}\langle \mathsf{MLSS}_{2,m}^{\times} \rangle$. Notice that, for efficiency reasons, $\mathsf{MLSS}_{2,m}^{\times}$ involves neither a map image nor a domain operator since, as proved in [37], the ExpTime-satisfiability problem for $\mathcal{ALC}$ is reducible to that of any extension of MLS with a map image operator. However, as argued empirically, the lack of an image operator in $\mathcal{DL}\langle \mathsf{MLSS}_{2,m}^{\times} \rangle$ is not crucial for a large portion of real-world knowledge bases.

A significantly more expressive fragment of set theory, named $\forall_0^{\pi}$ and involving restricted universal quantification over one-sorted variables and pairs, has been later on proved decidable in [38]. Specifically, quantifiers in $\forall_0^{\pi}$-formulae have the form $(\forall u \in \overline{\pi}(w))$ or $(\forall [x, y] \in z)$, where $\overline{\pi}(w)$ denotes the collection of the members of y that are not pairs and $[x, y]$ denotes the ordered pair of x, y. In addition, in any quantifier prefix

$$(\forall u_1 \in \overline{\pi}(w_1)) \cdots (\forall u_n \in \overline{\pi}(w_n))(\forall [x_1, y_1] \in z_1) \cdots (\forall [x_m, y_m] \in z_m),$$

the sets

$$\{u_1, \ldots, u_n, x_1, \ldots, x_m, y_1, \ldots, y_m\} \quad \text{and} \quad \{w_1, \ldots, w_n, z_1, \ldots, z_m\}$$

of variables are required to be disjoint (*knot-freeness* condition).

The description logic corresponding to the theory $\forall_0^{\pi}$, denoted $\mathcal{DL}\langle \forall_0^{\pi} \rangle$, extends $\mathcal{DL}\langle \mathsf{MLSS}_{2,m}^{\times} \rangle$ with several new constructs, such as role transitivity, self-restriction, and role identity. It also admits *finite* existential restrictions of the form $\exists R.\{a_1, \ldots, a_n\}$, which can be used with no limitations. Despite its great expressivity, the consistency problem for $\mathcal{DL}\langle \forall_0^{\pi} \rangle$-knowledge bases is still NP-complete, since it can be reduced to the satisfiability problem for $(\forall_0^{\pi})^{\leqslant 2}$-formulae, namely $\forall_0^{\pi}$-formulae whose quantifier prefixes have length at most 2. The description logic $\mathcal{DL}\langle \forall_0^{\pi} \rangle$ can also be extended with SWRL rules (where SWRL stands for Semantic Web Rule Language, see [144]) without disrupting the decidability of its knowledge-base consistency problem. These are Horn-style rules of the form

$$B_1 \,\&\, \cdots \,\&\, B_n \implies H,$$

where $B_1, \ldots, B_n, H$ are atoms of the forms $A(x)$, $P(x, y)$, $x = y$, $x \neq y$, with A a concept name, P a role name, and x, y either SWRL-variables or individual names.

The theories $\mathsf{MLSS}_{2,m}^{\times}$ and $\forall_0^{\pi}$ have been subsequently combined and extended in the two-sorted theory $\forall_{0,2}^{\pi}$, studied in [35], which involves restricted quantifiers of the forms $(\forall x \in y)$, $(\forall [x, y] \in f)$ (still subject to the knot-freeness condition) and literals of the forms

$$x \in y, \quad [x, y] \in f, \quad x = y, \quad f = g,$$

where x, y are set variables and f, g are map variables.

Though $\forall_{0,2}^{\pi}$ cannot express inclusions such as

$$x \subseteq \mathsf{Domain}(f), \quad x \subseteq \mathsf{Range}(f), \quad x \subseteq f[y], \quad h \subseteq f \circ g, \tag{‡}$$

but only those in which the operators domain, range, (multi-)image, and map composition occur in the left-hand side of inclusions, the theory $\forall_{0,2}^{\pi}$ allows one to express considerably many set-theoretic constructs. The satisfiability problem for $\forall_{0,2}^{\pi}$ is in general in NExpTime. However, under suitable conditions it is NP-complete.

It turns out that the theory $\forall_{0,2}^{\pi}$ is very close to the undecidability boundary. Indeed, by extending the family of $\forall_{0,2}^{\pi}$-conjunctions with two positive literals both of which are of any of the types (‡) decidability is lost.

The description logic $\mathcal{DL}\langle \forall_{0,2}^{\pi} \rangle$ corresponding to the fragment $\forall_{0,2}^{\pi}$, which extends $\mathcal{DL}\langle \forall_0^{\pi} \rangle$ with some metamodeling features, is presented in [36]. Specifically, the distinction between concepts and individuals is relaxed in $\mathcal{DL}\langle \forall_{0,2}^{\pi} \rangle$, so that concepts can either participate to relations or be instances of other concepts, as individuals do. In spite of this, the consistency problem for $\mathcal{DL}\langle \forall_{0,2}^{\pi} \rangle$-knowledge bases remains NP-complete.

Table 3. Axioms of the minimal set theory Ω for reasoning in non-classical logics.

$$
\begin{aligned}
&\forall x \, \forall y \, \forall v \, (\, v \in x \cup y \iff (v \in x \vee v \in y) &&) \\
&\forall x \, \forall y \, \forall v \, (\, v \in x \setminus y \iff (v \in x \,\&\, v \notin y) &&) \\
&\forall x \, \forall y \, (\quad x \subseteq y \iff \forall v (v \in x \implies v \in y) \,) \\
&\forall x \, \forall y \, (\, x \in \mathcal{P}(x) \iff x \subseteq y &&)
\end{aligned}
$$

More far-reaching metamodeling capabilities are present in the extension $\mathcal{ALC}^{\Omega}$ of $\mathcal{ALC}$ proposed in [91], where concepts are interpreted as sets of the very weak set theory Ω, consisting of just the four axioms shown in Table 3, intended to characterize set inclusion and the operators of dyadic union, set difference, and power-set. Power-set concepts in $\mathcal{ALC}^{\Omega}$ make interactions between concepts and metaconcepts possible. In addition, the lack of a foundation axiom in Ω permits circularity of membership in $\mathcal{ALC}^{\Omega}$, even between concepts. The description logic $\mathcal{ALC}^{\Omega}$ is proved decidable via a polynomial encoding into the ExpTime-decidable description logic $\mathcal{ALCOI}$.

Since 2015, the research moved towards regarding computable set theory as a formal knowledge representation language for Semantic Web ontologies. The fragment 4LQS^{R}, a multi-sorted stratified syllogistic admitting variables of four sorts and a restricted form of quantification over variables of the first three sorts (see [42]), is leveraged in [39] to represent the decidable description logic $\mathcal{DL}\langle 4LQS^R\rangle(\mathbf{D})$ ($\mathcal{DL}_\mathbf{D}^4$ for short). We recall that $\mathcal{DL}_\mathbf{D}^4$ provides several features such as min/max cardinality constructs on the left-/right-hand sides of inclusion axioms, role chain axioms, and datatypes—a simple form of concrete domains that are relevant in the ontology realm. $\mathcal{DL}_\mathbf{D}^4$ turns out to be quite expressive in comparison with $\mathcal{SROIQ}(\mathbf{D})$ (see [94]). The description logic $\mathcal{SROIQ}(\mathbf{D})$ underpins the Web Ontology Language OWL (OWL; see [145]), namely the standard language for representing ontologies. The consistency problem for knowledge bases of $\mathcal{DL}_\mathbf{D}^4$ is proved decidable via a suitable translation process from 4LQS^R to $\mathcal{DL}_\mathbf{D}^4$ formulae: under some not very restrictive constraints, the consistency problem for knowledge bases of $\mathcal{DL}_\mathbf{D}^4$ is proved to be **NP**-complete. Most notably, in [39], the set-theoretic fragment 4LQS^R is leveraged to extend $\mathcal{DL}_\mathbf{D}^4$ with a form of Horn-like rules stemming from the Semantic Web Rule Language (see [144]). This shows the suitability of 4LQS^R as a theoretic foundation for representing ontologies, as it combines in the same formalism the capabilities of both DLs and rule languages. For instance, rules with negated atoms such as

$$Person(?p) \;\&\; \neg\, hasPartner(?p, ?c) \implies SinglePerson(?p).$$

can be represented in 4LQS^R but not in SWRL.

An extension of $\mathcal{DL}_\mathbf{D}^4$ called $\mathcal{DL}\langle 4LQS^{R,\times}\rangle(\mathbf{D})$ ($\mathcal{DL}_\mathbf{D}^{4,\times}$ for short) is presented in [43]. The description logic $\mathcal{DL}_\mathbf{D}^{4,\times}$ supports, among other features, Boolean operations on concepts and roles, role constructs such as the product of concepts and role chains on the left-hand side of inclusion axioms, and role properties such as transitivity, symmetry, reflexivity, and irreflexivity, and valued, datatyped, and self-concept existential quantification. Full existential quantification and minimum (qualified) cardinality restriction are admitted only on the left-hand side of inclusion axioms, whereas full universal quantification and maximum (qualified) cardinality restriction are admitted on the right-hand side. Hence, even though $\mathcal{DL}_\mathbf{D}^{4,\times}$ is less powerful than logics such as $\mathcal{SROIQ}(\mathbf{D})$, as it cannot generate new individuals, it is more liberal in the definition of role inclusion axioms, since the roles involved are not required to be subject to any ordering relationship, and the notion of simple role is not needed. In [43], the *Conjunctive Query Answering* (CQA) problem [14] for $\mathcal{DL}_\mathbf{D}^{4,\times}$ is presented. CQA is a powerful way to query ABoxes particularly relevant in the context of DLs and for real world applications based on Semantic Web, as it provides a mechanism that allows users and applications to interact with ontologies and data. The CQA problem for $\mathcal{DL}_\mathbf{D}^{4,\times}$ is addressed by rewriting both knowledge bases and conjunctive queries of $\mathcal{DL}_\mathbf{D}^{4,\times}$ to suitable 4LQS^R-formulae. Then, the CQA problem for $\mathcal{DL}_\mathbf{D}^{4,\times}$ is solved by the decision procedure for 4LQS^R, implemented in the form of a decidable KE-tableau. The complexity of the KE-tableau-based procedure for $\mathcal{DL}_\mathbf{D}^{4,\times}$ turns out to be in ExpTime, which is comparable with the one for $\mathcal{SROIQ}(\mathbf{D})$, proved to be 2-ExpTime-complete. The KE-tableau is a refutation system inspired to Smullyan's semantic tableaux [139] that additionally includes an analytic cut rule (PB-rule) minimizing the inefficiencies of semantic tableaux.

The decidability results in [43] are generalized in [44] to achieve a solution for the most common ABox reasoning services for the description logic $\mathcal{DL}\langle\mathsf{4LQS}^{\mathsf{R},\times}\rangle(\mathbf{D})$. These include: a) *instance checking*, the problem of deciding whether or not an individual a is an instance of a concept C; b) *instance retrieval*, the problem of retrieving all the individuals that are instances of a given concept; c) *role filler retrieval*: the problem of retrieving all the fillers x such that the pair (a, x) is an instance of a role R; d) *concept retrieval*, the problem of retrieving all concepts which an individual is an instance of; e) *role instance retrieval*, the problem of retrieving all roles which a pair of individuals (a, b) is an instance of. Hence, the CQA problem for $\mathcal{DL}_{\mathbf{D}}^{4,\times}$ is generalized so as to involve conjunctive queries admitting variables of three sorts, namely individual and data type variables, concept variables, and role variables. The resulting problem, called *Higher Order Conjunctive Query Answering* (HOCQA) problem for $\mathcal{DL}_{\mathbf{D}}^{4,\times}$ is proved to be decidable by reducing it to the analogous problem for $\mathsf{4LQS}^{\mathsf{R}}$. To this purpose, the KE-tableau calculus presented in [43] is extended in such a way as to calculate the answer set to higher order $\mathcal{DL}_{\mathbf{D}}^{4,\times}$-conjunctive queries on $\mathcal{DL}_{\mathbf{D}}^{4,\times}$-knowledge bases, without affecting computational complexity.

The first implementation of the calculus introduced in [44] is presented in [45]. The procedure, implemented in C++, has been included within a reasoner that checks the consistency of $\mathcal{DL}_{\mathbf{D}}^{4,\times}$-knowledge bases represented in terms of the set-theoretical fragment $\mathsf{4LQS}^{\mathsf{R}}$ and deals with OWL ontologies serialized in the OWL/XML format.

The KE-tableau system presented in [44] is extended and improved in [46, 47] so as to deal with a *skeptical* approach [134], in which queries are matched over all the models of the given knowledge base—in contrast to credulous approaches where answer sets are derived only from some models. The resulting calculus is a variant of the KE-tableau system for sets of universally quantified clauses, in which the standard KE-elimination rule is replaced by a novel elimination rule, called E^{γ}-rule, that incorporates the standard γ-rule to handle universally quantified formulae. The correctness of the new KE-tableau calculus, which is shown to be in 2-ExpTime, is proved in [47].

A second more efficient release of the reasoner, which also solves the HOCQA problem for $\mathcal{DL}_{\mathbf{D}}^{4,\times}$ and in which the E^{γ}-rule does not need to store the instances of the universally quantified formulae on the KE-tableau, has been presented in [48, 69].

3.6   Proof-Verification: Yulog, AXL, ..., ÆtnaNova

"The correctness of any program ultimately rests upon various mathematical facts implicit in it; and, depending upon the sophistication of the program, these mathematical facts may be arbitrarily deep. Thus a powerful, general proof checker will be a necessary part of any mature program verification system. Such a proof checker is an interactive programmed system into which one can enter sequences of logical/mathematical formulae, which it will accept as long as it can perform some computation which guarantees that each new formula is a logical consequence of preceding formulae. Such a verifier ensures rigorously against logical error, [···]"   *(Jacob T. Schwartz, 1978)*

The studies on decidable fragments of set theory surveyed in this paper were intended, since their inception in 1978 [135], as contributions to the then emerging field of proof technology; as such, they were repeatedly accompanied by prototype implementations and reports on the architecture of a proof-checker (see, e.g., [72,73,117]). Bits and pieces of a proof assistant provisionally named Yulog were implemented in SETL during the summer of 1979, as recalled above; they could not be transferred directly from IBM to NYU, hence a new implementation named AXL ('AXiomatic Language') was undertaken in 1980 at the Courant Institute. The proof-checker ETNA (outlined in [27, Sec. 6]), and its companion algorithm-verifier LAVA inherited from that stream of prototype systems in the mid-1990's.

A more sustained effort to develop a broad-gauge proof-checker was jointly undertaken around the year 2000 by Jack, Eugenio, and Domenico, and led to a system, once more implemented in SETL, initially named Ref[eree] but later also called ÆtnaNova in reminiscence of ETNA. As told in [106], Jack intended to demonstrate the usability of ÆtnaNova through the buildout of a large-scale,

---

$$\text{Partition}(P) \leftrightarrow_{\text{Def}} (\forall b \in P \mid \{k \in P \mid k \cap b \neq \emptyset\} = \{b\})$$
$$\text{ChSet}(C, T) \leftrightarrow_{\text{Def}} \{\{x\} : x \in C\} = \{C \cap b : b \in T\}$$

Theorem ch_0 : [Every partition has a choice set] $\text{Partition}(P) \to (\exists c \mid \text{ChSet}(c, P))$.
Proof : Suppose_not$(p_0) \Rightarrow$ Stat0 : $(\neg \exists c \mid \text{ChSet}(c, p_0))$ & $\text{Partition}(p_0)$

For, suppose that p_0 makes a counterexample. In particular, the inequality $\{\{\text{arb}(b)\} : b \in p_0\} \neq \{\{\text{arb}(b) : b \in p_0\} \cap b : b \in p_0\}$ must hold, in view of the definition of $\text{ChSet}(c, p_0)$.

$\{\text{arb}(b) : b \in p_0\} \hookrightarrow$ Stat0 $\Rightarrow \neg \text{ChSet}(\{\text{arb}(b) : b \in p_0\}, p_0)$
Use_def(ChSet) $\Rightarrow \{\{x\} : x \in \{\text{arb}(b) : b \in p_0\}\} \neq \{\{\text{arb}(b) : b \in p_0\} \cap b : b \in p_0\}$
SIMPLF $\Rightarrow$ Stat1 : $\{\{\text{arb}(b)\} : b \in p_0\} \neq \{\{\text{arb}(b) : b \in p_0\} \cap b : b \in p_0\}$
Therefore, some block b_0 of the partition p_0 exists which witnesses the said inequality. Since blocks are non-null, $\text{arb}(b_0) \in b_0$.

Use_def(Partition) $\Rightarrow$ Stat2 : $(\forall b \in p_0 \mid \{k \in p_0 \mid k \cap b \neq \emptyset\} = \{b\})$
$b_0 \hookrightarrow$ Stat1 $\Rightarrow$ Stat3 : $\{\text{arb}(b_0)\} \neq (\{\text{arb}(b) : b \in p_0\} \cap b_0)$ & $b_0 \in p_0$
$b_0 \hookrightarrow$ Stat2(Stat2*) $\Rightarrow$ Stat4 : $\{k \in p_0 \mid k \cap b_0 \neq \emptyset\} = \{b_0\}$
Consequently, $\text{arb}(b_0) \in \{\text{arb}(b) : b \in p_0\} \cap b_0$ holds. This enables simplification of the inequality $\{\text{arb}(b_0)\} \neq \{\text{arb}(b) : b \in p_0\} \cap b_0$ into $\{\text{arb}(b) : b \in p_0\} \cap b_0 \not\subseteq \text{arb}(b_0)$; therefore, an a_0 other than $\text{arb}(b_0)$ belongs to both of b_0 and $\{\text{arb}(b) : b \in p_0\}$.

Suppose $\Rightarrow \text{arb}(b_0) \notin (\{\text{arb}(b) : b \in p_0\} \cap b_0)$
$k_0 \hookrightarrow$ Stat4(Stat4*) $\Rightarrow$ Stat5 : $\text{arb}(b_0) \notin \{\text{arb}(b) : b \in p_0\}$
$b_0 \hookrightarrow$ Stat5 $\Rightarrow$ AUTO
(Stat3*)Discharge $\Rightarrow$ AUTO
$a_0 \hookrightarrow$ Stat3(Stat3*) $\Rightarrow$ Stat6 : $a_0 \in \{\text{arb}(b) : b \in p_0\}$ & $a_0 \in b_0$ & $a_0 \neq \text{arb}(b_0)$
Such an a_0 can be rewritten as $\text{arb}(b_1)$ for some b_1 other than b_0 in p_0, but this contradicts the fact that any two blocks in p_0 are disjoint.

$b_1 \hookrightarrow$ Stat6(Stat6, Stat4) $\Rightarrow$ Stat7 : $b_1 \notin \{k \in p_0 \mid k \cap b_0 \neq \emptyset\}$ & $b_1 \in p_0$ &
$$a_0 = \text{arb}(b_1)$$
$b_1 \hookrightarrow$ Stat2(Stat7*) $\Rightarrow$ Stat8 : $b_1 \in \{k \in p_0 \mid k \cap b_1 \neq \emptyset\}$
$() \hookrightarrow$ Stat8(Stat7) $\Rightarrow a_0 \in b_1$
$b_1 \hookrightarrow$ Stat7 $\Rightarrow$ AUTO
(Stat6*)Discharge $\Rightarrow$ QED

Fig. 2. ÆtnaNova's proof of the controversial Zermelo's principle [65].

automatically validated proof script, culminating in the derivation of the celebrated Cauchy integral theorem of mathematical analysis from the bare rudiments of set theory. This goal was not reached, mainly because of Jack's death in 2009, but the experience with ÆtnaNova was all in all a success, ending in a reliable feasibility study. It culminated in the monograph [136] and in a series of automatically checked, detailed proof scenarios, discussed in [55, 65–67, 119–121, etc.].

A variant of the Zermelo-Fraenkel set theory, encompassing axioms of foundation and global choice, reflects into the semantics of many of the 15 or so mechanisms constituting the inferential armory of ÆtnaNova: those comprise an inference mechanism, named ELEM, which implements an enhanced variant of the MLS decision algorithm. By means of that algorithm, the ÆtnaNova verifier can identify many cases in which a conjunction constructed by negating a statement ϑ of a proof and conjoining a selection Θ of earlier statements of the same proof is unsatisfiable, so that the preceding Θ yields ϑ. Should some of the constructs appearing in the context $(\neg\,\vartheta)\,\&\,(\&\,\Theta)$ be unamenable to ÆtnaNova's built-in syllogistic, a preprocessing phase replaces all the parts of this context whose lead operators cannot be treated by the decision algorithm by new variables designating either sets (when they occur as terms) or propositions (when they occur as sub-formulae), replacing syntactically identical (or recognizably equal) parts by the same variable.

ELEM is the most central of all inference mechanisms, because its use is often tacitly combined with other forms of inference. In fact, in ÆtnaNova proofs are lists of two-portion lines (see, e.g., Fig. 2, which comes from [65]): the second portion of each line is the *assertion* being derived while the first part, named the *hint*, references the basic inference mechanism enabling that derivation. The case when the assertion of a proof line is sharply determined by the hint and by the lines that precede it in the proof is rather rare, though: such indeterminacy in the outcome of an inference step stems, by and large, from the pervasiveness of ELEM as a 'behind-the-scenes' inference mechanism. A hint often designates a specific assertion which, however, is shrouded in a 'penumbra' of obvious consequences of its conjunction with the context preceding it in the proof. From among those consequences, derivable via ELEM, the user who develops the proof is left free to pick one or another.

The syllogistic embodied in the ELEM mechanism extends MLSS, and the way Jack implemented it inside ÆtnaNova was inspired by [62] (whose method is analyzed in depth in [140]). Constructs which ELEM normally handles, but whose treatment the ÆtnaNova user can suspend desultorily to improve verification performance, are: a pairing operator $\langle\,\cdot\,,\,\cdot\,\rangle$ along with its associated projections $\cdot^{[1]}$ and $\cdot^{[2]}$, and a global choice operator $\mathsf{arb}(\,\cdot\,)$. Those constructs are subject to the laws

$$\langle x,\,y\rangle = p \iff p^{[1]} = x \;\&\; p^{[2]} = y\,,$$
$$\mathsf{arb}(x) \in x \cup \{x\} \quad \& \quad \mathsf{arb}(x) \cap x = \varnothing\,.$$

It is worth stressing that nothing more is assumed than what these laws entail; hence ELEM can neither prove that $\langle x, y \rangle = \{\{x,y\}, \{x\}\}$ nor that[14] $\text{arb}(\{\varnothing, \{\{\varnothing\}\}\}) = \varnothing$ (whereas ELEM can prove that $\text{arb}(\{\varnothing, \{\varnothing\}\}) = \varnothing$).

The extensions just mentioned of MLSS by the operators $\langle \cdot, \cdot \rangle$, $\cdot^{[1]}$, $\cdot^{[2]}$, and $\text{arb}(\cdot)$ rely upon reduction techniques of the type illustrated in Sect. 2.4.

The reader is referred to [65] for a 'panoramic tour' of ÆtnaNova, to [120, Sec. 3] for a crash course on ÆtnaNova, and to [136] for a much broader introduction to this proof-verifier and its underlying logic. A quick comparison of this system with other set-oriented proof-assistants can be found at [120, Sec. 6]; moreover, [117, Sec. 6] carries out a comparison of ÆtnaNova's THEORYs—an important construct designed to support reusability of proofware components—with various related modularization constructs available, in particular, in the OBJ family of languages (see [93]) and in the Interactive Mathematical Proof System (IMPS) described in [82].

A SETL implementation of ÆtnaNova has been exploitable online from 2008 till recently; its source files are linked at the page
        https://www.softwarepreservation.org/projects/SETL
of the Computer History Museum's Software Preservation Group.

4   Evolving Trends

The way in which the MLS-related issues are addressed has changed a lot, today, compared to the onset of MLS in 1979. Among other things, the decision methods discovered in the period 1985/1988 have been revisited, and their applicability realm has widened, achieving a gain in uniformity of approach thanks to a new notion, the so-called *formative processes* (cf. [59,61]).

When a branch of theoretical computer science reaches maturity, its issues are addressed with a closer eye on algorithmic complexity. This happened, in particular, with studies on solvable and unsolvable classes of quantificational formulae, which after the historical achievements reported in [1,81,98] were partly redirected towards complexity issues [9,99]. In a similar attitude, as we will report in Sect. 4.4, recent studies on decidable and undecidable fragments of set theory have bestowed more attention to complexity. We report on these and other new tendencies in this area in what follows.

4.1   Boundaries of the Undecidable and of Infinitude

"Extensions of the decision procedure to handle additional predicates and operations are currently under study, in an attempt to discover how far one can push an unquantified theory of sets without crossing the boundary of the undecidable."

                              (A. Ferro, E. G. Omodeo, and J. T. Schwartz, 1980)

---

[14] In this respect, the treatment of the global choice operator is simpler-minded—but also considerably more efficient—than the one proposed in [85,87,104].

A selection of decidable extensions of MLS is shown in Table 4, each of whose rows represents an unquantified set-theoretic language, namely the fragment of set theory involving only the operators and relators selected by the occurrences of '$\star$' on that row (along with a full endowment of propositional connectives).

Table 4. Some decidable extensions of MLS in well-founded set theory. The weak Cartesian product $x \otimes y$ designates the set $\{\{u,v\} : u \in x \,\&\, v \in y\}$ consisting of singletons and doubletons. The intersection and union operations $x = y \cap z$ and $x = y \cup z$ can be seen as abbreviating, respectively, $x = y \setminus (y \setminus z)$ and $x \setminus y = z \setminus y \,\&\, x \setminus z = y \setminus z \,\&\, x \setminus (z \setminus y) = y$.

$\cdot\setminus\cdot$	$\{\cdot\}$	$\cup\cdot$	$\mathcal{P}(\cdot)$	arb$(\cdot)$	$\cdot\otimes\cdot$	$\cdot=\cdot$	$\cdot\in\cdot$	Finite$(\cdot)$	RC$(\cdot,\cdot)$	Bib. sources
$\star$	$\star$	$\star$				$\star$	$\star$	$\star$		[58]
$\star$	$\star$		$\star$			$\star$	$\star$			[85] vs [19, Sec. 4.5]
$\star$		$\star$				$\star$	$\star$			[33]
$\star$						$\star$	$\star$		$\star$	[28]
$\star$					$\star$	$\star$				[60]

It should be recalled that the historical classifications of solvable and unsolvable cases of the decision problem for predicate logic (see [1,81,98]) mostly refer to languages whose signatures only involve relators (no operators or function symbols). In those accounts, each (solvable or unsolvable) class collects all sentences whose quantificational prefix—once they are brought to prenex normal form—follows a certain pattern.

Sometimes, a similar approach has been adopted to deal with the decision problem for fragments of set theory. From this perspective, the decision problem for MLSS gets embedded in the slightly richer problem: can a given sentence of the form $\exists x_1 \cdots \exists x_N \forall y\, \mu$, where μ is a propositional combination of atomic formulae of the forms $u \in v$ and $u = v$ (u, v variables) be derived under the axioms seen in Table 1 (or in a weaker or variant axiomatic system)? The answer in this case is affirmative (cf. [92,107–109]); but what happens if we tackle the more general form $\exists x_1 \cdots \exists x_N \forall y_1 \cdots \forall y_\kappa\, \mu$, or an even more general form permitting multiple quantifier alternations? Concerning the form $\exists x_1 \cdots \exists x_N \forall y_1 \forall y_2\, \mu$, a positive answer to the decision problem is supplied in [5]; concerning the form $\exists x_1 \cdots \exists x_N \forall y_1 \cdots \forall y_\kappa\, \mu$, where κ ranges over all positive numbers, a positive answer to the decision problem was supplied through [110,111]. When κ exceeds 1, decidability presupposes a strengthening of the axioms shown in Table 1, typically including the infinity axiom: things are so because the existence of infinite sets can be stated by a sentence in the class at hand, as exemplified by either one of the formulae in Table 2 (we will say more on this in Sect. 4.1).

Remark 1 (Reducing propositional 3CNF to set-theoretic 2CNF). In [108], the NP-*completeness of the set-theoretic* $\exists^*\forall$-*satisfiability problem was proved.*[15] *As regards the* NP-*hardness of this problem, that paper showed how to translate 3CNF-formulae of propositional logic—whose satisfiability problem is per se an* NP-*complete problem, cf. [90]—into sentences of the form* $\exists x_1 \cdots \exists x_N \, \forall y \, \mu$, *where* μ *stands for a conjunction* $\&_{i=1}^m C_i$, *each clause* C_i *being either a single literal or the disjunction of two literals of the forms* $v \in w$, $v \notin w$, $v = w$, *and* $v \neq w$. *(See [54, pp. 167–168]).* ⊣

Essentially Undecidable, Weak Set Theories. Some important results on unsolvable cases of the set-theoretic decision problem regard the first two levels, Δ_0 and Σ_1, of Azriel Lévy's hierarchy of formulae in Set Theory [97]: Δ_0 consists of all formulae in which quantifiers occur only in the restricted forms $\forall x \in y$ and $\exists x \in y$; unsolvability of the decision problem affects the Σ_1-sentences, i.e., the existential closures of Δ_0-formulae. A Δ_0-formula is said to be a $\forall \exists \cdots Q_n$-formula ($Q_n$ being either $\forall$ or $\exists$ according to whether n is odd or even) if it can be transformed via a Tarski-Kuratowski computation (cf. [133, Sec.14.3]) into a logically equivalent prenex formula consisting of n alternating batches of unrestricted alike quantifiers.

Table 5. According to von Neumann's conception of ordinal numbers: the first ordinal, zero, is the void set; the successor of each ordinal o equals the set $o \cup \{o\}$; the first limit ordinal, $\mathbb{N}$, is the set of all natural numbers; and each limit ordinal λ equals $\bigcup_{o \in \lambda} o$. The properties of being a transitive set, an ordinal, or a natural number are captured by the definitions shown here. Within a very weak axiomatic set theory, a more elaborate definition of $\mathsf{Nat}(\cdot)$ may be needed in order that the sets m such that $\mathsf{Nat}(m)$ holds behave as intended; but even in a theory where the definition as proposed here works fine, it may be necessary to resort to a less direct one in order to achieve the least possible number of quantifier alternations in this crucial definition.

$$
\begin{aligned}
&\mathsf{Trans}(X) \;\leftrightarrow_{\mathrm{Def}} \forall v \in X \; \forall u \in v \;\; u \in X, \\
&\quad \mathsf{On}(X) \;\leftrightarrow_{\mathrm{Def}} \mathsf{Trans}(X) \;\wedge\; \forall u, v \in X \left(u \in v \vee v \in u \vee v = u \right); \\
&\mathsf{Nat}(M) \;\leftrightarrow_{\mathrm{Def}} \forall y \in M \left[\exists z \in M \big(\right. \\
&\qquad\qquad \forall w \in z \;\; w \in M \;\; \& \;\; \forall v \in M \, (v = z \vee v \in z) \big) \;\; \& \\
&\qquad\qquad \forall w \in y \;\; \exists z \in M \, (z \in y \;\; \& \\
&\qquad\qquad \forall w \in z \;\; w \in y \;\; \& \;\; \forall v \in y \, (v = z \vee v \in z)) \big) \left. \right].
\end{aligned}
$$

The result of [124] already cited in Sect. 2.3 showed how to apply arguments *à la* Gödel to a very weak axiomatic set theory and to the class of Σ_1-sentences

---

[15] We mean here the problem of determining whether or not a sentence of the said form $\exists x_1 \cdots \exists x_N \, \forall y \, \mu$ (with $N \geqslant 0$) is satisfiable in the standard universe of sets. We will be more explicit on the notation $\exists^*\forall$ in footnote (see footnote 16).

whose Δ_0-matrix is a $\forall\exists\forall$-formula; i.e., the class consisting of existential closures of formulae that involve only restricted quantifiers but are writable in prenex form with at most two alternations of unrestricted quantifiers (the leftmost symbol being a '$\forall$'). In the recent revisitation [52] of the cited work by Parlamento and Policriti, Mattia Panettiere produced slightly less weak theories each of whose consistent, recursively axiomatizable extensions suffers from incompleteness relative to the existential closures of Δ_0-matrices that can be written as $\forall\exists$-formulae, hence with one fewer quantifier alternation. Table 5 and its caption aim at conveying a feel of how ingenious a balance must be sought between the strength of the axioms and the syntactic complexity of the definitions entering in the arithmetization of syntax.

Undecidable Unquantified Languages. The above-recalled results of [23] (see Sect. 2.3) have evolved into the recent paper [51], which shows that the satisfiability problems for various fragments of ZF—namely, the Zermelo-Fraenkel first-order set theory with the foundation axiom, as treated, e.g., in [95]—are algorithmically unsolvable.

For each fragment Φ taken into account, the undecidability result stems from a uniform translation method which turns every instance $D = 0$ (concerning a multi-variate polynomial $D(x_1, \ldots, x_m)$ with coefficients in $\mathbb{Z}$) of Hilbert's 10th problem into a formula φ of Φ so that the existential closure $\varphi^{\exists}$ of φ is provable in ZF if and only if the Diophantine equation $D = 0$ has solutions in $\mathbb{N}$. Through this translation, the algorithmic unsolvability of Hilbert's 10th problem carries over to the satisfiability problem for Φ.

By way of first approximation, let us disallow quantification. The undecidable Φ's then are slight extensions of a core language consisting of all conjunctions of literals of the forms $x = y \cup z$, $x = y \otimes z$, $x \cap y = \varnothing$, and $|x| = |y|$, where x, y, z stand for variables, $y \otimes z$ is the variant of Cartesian product specified in the caption of Table 4, and $|x| = |y|$ designates equinumerosity between x and y. Additional conjuncts entering into play can be, e.g.: one literal of the form $\mathsf{Finite}(x)$, taken along with one literal of the form $x \neq \varnothing$. Another option would be to extend the said syntactic core by allowing three *negated* equinumerosity literals of the form $|x| \neq |y|$ to appear in the conjunction. These two, and similar, slim undecidable set-theoretic languages lie extremely close to fragments of ZF which are either known [54, Sec. 11.1], or conjectured [58, p. 239], to be decidable.

Coarsely speaking, the cardinality operator $|\cdot|$ brings in the essential amount of number theory that makes it possible to mimic addition and multiplication of natural numbers; in fact:

- $|x| + |y| = |x' \cup y|$ and
- $|x| \cdot |y| = |x' \otimes y|$

when $|x| = |x'|$ and $x' \cap y = \varnothing$; and such an x' can always be found, for any given pair x, y of sets.

Let us now move on into a language where restricted quantification is admitted.

Undecidable Quantified Languages. One can recast the undecidable fragments of ZF which are devoid of quantification but encompass function symbols such as $\cdot \cup \cdot$, $\cdot \otimes \cdot$, $|\cdot|$ in a terse set-theoretic language which only involves the relators $\in$ and $=$, propositional connectives, and also the constructs $\forall x \in y \; \psi$ and $\exists x \in y \; \psi$ involving restricted quantifiers, subject to a very restrained usage.

Definition 1. *We dub* $(\forall\exists)_0$-FORMULA *any conjunction* Ψ *of the form*

$$\underset{j \in \{0,\dots,M\}}{\&} (\forall y_{j1} \in y'_{j1}) \cdots (\forall y_{jp_j} \in y'_{jp_j}) (\exists x_{j1} \in x'_{j1}) \cdots (\exists x_{jq_j} \in x'_{jq_j}) \psi_j$$

where, for each j, *the formula* ψ_j *is devoid of quantifiers and either* $p_j > 0$, $q_j \geqslant 0$ *or* $p_j = q_j = 0$ *holds.*

Definition 2. *We dub* $(\forall\exists)_0$ SPECIFICATION *of an* m-place relationship R over sets a $(\forall\exists)_0$ formula Ψ with free variables $a_1, \dots, a_m, x_1, \dots, x_\kappa$ (where $\kappa \geqslant 0$) such that, under the axioms of ZF, one can prove:

$$R(a_1, \dots, a_m) \iff (\exists x_1, \dots, x_\kappa) \, \Psi .$$

Examples can be drawn from some operator eliminations seen in Sect. 1.2: e.g., the right-hand sides of the bi-implications

$$a = b \setminus c \iff (\forall t \in a)(t \in b \, \& \, t \notin c) \, \& \, (\forall t \in b)(t \in c \vee t \in a),$$
$$\mathsf{Sngl}(a) \iff (\exists x)\big(x \in a \, \& \, (\forall y \in a)(y = x)\big)$$

are $(\forall\exists)_0$ specifications of the 3-place relationship $a = b \setminus c$ and, respectively, of the property "being a singleton set".

As shown in [51], rather sophisticated notions like "being an ordinal", "being the first limit ordinal $\mathbb{N}$", and "having a finite cardinality", admit $(\forall\exists)_0$ specifications, which gives evidence of the high expressive power of restricted quantification in the context of ZF.

In a full-fledged language for set theory, $(\forall\exists)_0$ specifications are only seldom used in order to define mathematical notions. Hence, to support the correctness of the proposed $(\forall\exists)_0$-specifications of $\mathbb{N}$, equinumerosity, and finitude, the cited paper proves that they are equivalent to more direct and practical characterizations of the same notions. This formal accomplishment was carried out with the aid of the proof-checker ÆtnaNova discussed above (see Sect. 3.6), embodying a computational version of ZF.

"Infinity, in Short". Far from being a mere curiosity, the statement of existence of two infinite sets shown at the top of Table 2 offered the clue (cf. [5,6]) for solving the truth problem for $\exists^*\forall\forall$-sentences[16] over $\mathcal{V}$, the cumulative hierarchy often indicated as the intended model for ZF. A variant of that statement,

---

[16] To stay aligned with the literature cited in this section, we slightly depart from the notation used above to classify prenex formulae according to their quantificational prefix. Here and below, $\exists^*$ and $\forall^*$ denote batches (possibly empty) of alike quantifiers, while $\exists\exists$ and $\forall\forall$ denote two consecutive quantifiers of existential, respectively universal, type.

involving $n + 2$ existential variables where n can be any natural number, plays a likewise crucial role in [110] and [111], which solve the truth problem for the entire class of $\exists^*\forall^*$-sentences (still over the standard universe $\mathcal{V}$).

Willing to establish whether the truth problem for $\exists^*\forall^*$-sentences is also solvable in a universe of hypersets that models ZF $-$FA $+$AFA (see Sect. 2.3), it seemed reasonable to start by pinpointing sentences logically equivalent to the infinity axiom. A first step in this direction was taken in [112,113], which produced $\exists\exists\forall^*$-sentences stating the existence of hypersets which, besides being infinite, also are ill-founded.

Then, in [114], two variants of the Parlamento–Policriti's infinity axioms seen on Table 2, were pinpointed. They are shown in Table 6: one of the two works under FA, von Neumann's foundation axiom, as well as under its opponent AFA, Aczel's anti-foundation axiom; the other one works even under no axiomatic commitment concerning the well-foundedness of $\in$.

As of today, the decidability problem regarding hypersets whose study led to the novel statements of infinity remains unsolved.

Table 6. Improved ways of stating, by means of restricted universal quantifiers, the existence of a doubly stranded infinite spiral ω_0, ω_1 (see caption of Table 2). The shorter formulation presupposes either FA or AFA; the longer one presupposes neither axiom.

$$\omega_0 \neq \omega_1 \quad \& \quad \forall x \in \omega_0 \, \forall y \in \omega_1 \, (x \in y \vee y \in x) \quad \&$$
$$\forall y \in \omega_0 \, y \notin \omega_1 \quad \&$$
$$\underset{i \in \{0,1\}}{\&} \left(\omega_i \notin \omega_{1-i} \; \& \; \forall x \in \omega_i \, \forall y \in x \, y \in \omega_{1-i} \right)$$

$$\omega_0 \neq \omega_1 \quad \& \quad \forall x \in \omega_0 \, \forall y \in \omega_1 \, (x \in y \vee y \in x) \quad \&$$
$$\forall y_1, y_2 \in \omega_1 (y_1 \in y_2 \implies \forall z \in y_1 \, z \in y_2) \quad \&$$
$$\underset{i \in \{0,1\}}{\&} \left(\omega_i \notin \omega_{1-i} \; \& \; \forall x \in \omega_i \, \forall y \in x \, y \in \omega_{1-i} \right)$$

The following moral can be drawn from comparing the technical difficulties associated with the decidability results about $\exists^*\forall$-sentences on the one hand and about $\exists^*\forall^*$-sentences on the other: as for the discovery of a decision algorithm, at times it makes a dramatic difference whether or not, within a fragment of set theory, any satisfiable formula is also hereditarily finitely satisfiable. As recalled above, the formulae of MLS and MLSS enjoy this property. The formulae of MLSP extended with the singleton operator $\{\cdot\}$, whose satisfiability problem calls for a considerably more challenging syllogistic (cf. [61]), also enjoy it; extending the latter fragment with literals of the forms Finite$(\cdot)$ and $\neg$ Finite$(\cdot)$ (as done in [58]) does not disrupt decidability, but makes proving it even more challenging.

There are satisfiable formulae which can only be modeled by infinite sets in MLSU (consider, e.g., the formula $y \in x \mathbin{\&} x = \bigcup x$), as well as in the extension of MLS with map-related constructs discussed in Sect. 2.3 (as, for instance, the formula $y \in x \mathbin{\&} x = \mathsf{Domain}(x)$).

The sets entering into play in the modeling of an $\exists^*\forall^*$ set-theoretic sentence, regardless of whether they are finite, can always be associated with a finite digraph which patterns their construction: it hence turns out that the infinitudes associated with $\exists^*\forall^*$ statements of existence of infinite sets provide a revealing insight on Frank P. Ramsey's celebrated theorem (ca. 1928), fundamental in finite combinatorics. The collection of all prenex sentences with quantificational prefix $\exists^*\forall^*$ of a first-order predicate language $\mathcal{L}$ whose signature consists of an arbitrary number of relators (in any number of arguments), one of its relators being equality, is known as the Bernays-Schönfinkel-Ramsey class after the names of three mathematicians who studied it in depth at the beginning of the 19th century. Ramsey's theorem shows that the so-called *spectrum* of any sentence α in that class—i.e., the set of all cardinalities of (the support domains of) finite models of α—is either a finite subset of $\mathbb{N}$ or the complement of a finite subset of $\mathbb{N}$; moreover, a threshold number $\mathfrak{r}(\alpha)$ can be computed so that if α has a finite model of cardinality exceeding $\mathfrak{r}(\alpha)$, then its spectrum is infinite. In [116] and in [115, Sec. 8.3] a quadratic-time translation algorithm is proposed that, given an $\exists^*\forall^*$-sentence α of $\mathcal{L}$, translates it into an $\exists^*\forall^*$-sentence α^σ of the set-theoretic language $\mathcal{L}_{=,\in}$ devoid of function symbols so that α is satisfiable by arbitrarily large models if and only if α^σ is provable in standard ZF (including FA).

4.2   Syllogizing About Individuals and Nested Sets

Most of the decision algorithms deal with sets (or, occasionally, with hypersets) whose construction is ultimately based on the empty set. In practical applications, however, it is common to assume that certain primitive objects (sometimes called individuals or urelements) can be members or sets but behave differently from them. An approach to modeling such atoms, that allows one to retain the original formulation of the extensionality axiom, was proposed by Quine: atoms are self-singletons, i.e., they solve the equation $x = \{x\}$. In [76] this approach was adopted in coping with the satisfiability problem: the decidability of this problem relativized to $\exists^*\forall$ sentences over well-founded sets was proved, and a redesign of the $\{\mathsf{log}\}$ set-unification algorithm consonant with the view that atoms are self-singletons was proposed.

4.3   The Long-Lasting Fortune of $\{\mathsf{log}\}$

Over the years, the $\{\mathsf{log}\}$ language has been improved in many respects and enjoyed a long-lasting fortune. This motivated the authors in putting together, along with Alberto Policriti, the monograph [54]. Some early examples of use of {log}—e.g., Agostino Dovier's generation of jazz chords —had a vaguely recreational flavor; however, in the last decade Maximiliano Cristiá and Gianfranco Rossi have coauthored a wealth of papers (to cite just a pioneering one, [70]) concerning the exploitation of formal methods based on finite sets and {log} to validate programs and to generate test-cases in a model-based testing setting.

4.4   The Long March Towards *Computational* Set Theory

Concerning the complexity of the decision problem for MLS and for solvable broadenings of it, the main facts known when the monograph [30] was published were that MLS and MLSS have NP-complete satisfiability problems (see [53]). This is why [30] was titled *"Computable set theory"* rather than "Computational set theory".

On the theoretical side the situation did not improve much until recently; albeit the two papers [18,62] regarding fast set-theoretic tableaux promoted, on the implementation side, a great speed up.

A systematic study on the said complexity issue was undertaken lately by a team at the University of Catania, who has already supplied the complexity taxonomies shown in Tables 7 and 8 (cf. [26,41]). Table 7 regards a downsizing of MLS, BST (acronym for 'Boolean set theory'), which is even narrower than 2LS in two respects: individual variables—and, hence, membership atoms—are disallowed; moreover, the only formulae available in BST are conjunctions of literals. The emptyness and non-emptyness monadic relators $\cdot = \varnothing$ and $\cdot \neq \varnothing$, as well as the dyadic relators $\mathsf{Disj}(\cdot,\cdot)$ and $\neg\,\mathsf{Disj}(\cdot,\cdot)$ designating disjointness and overlap, are expressible in merely Boolean terms and they are allowed in BST. Table 8 regards MST (acronym for 'membership set theory'), which is a narrower language than MLS: its formulae, in fact, do not involve equality—their only relator is $\in$—; moreover, they just are conjunctions of literals.

Each row of Table 7 represents a fragment of BST, namely the one involving only the operators and relators selected by the occurrences of '$\star$' on that row. Correspondingly, the last column indicates the asymptotic time-complexity of the best decision algorithm known for the satisfiability problem of that fragment, when the complexity is polynomial. In many cases, the last column flags that the decision problem of that fragment (and, consequently, of any broader fragment of BST) is NP-complete. Any sub-fragment of a fragment whose decision problem has polynomial complexity is, in turn, polynomial; thus, altogether, Table 7 informs us that, out of the $2040 = 2^3 \cdot (2^8 - 1)$ distinct fragments of BST obtainable by selecting a set of operators and at least one of the available relators, 1278 have an NP-complete problem while the remaining 762 have a problem decidable in polynomial time. Interpreting Table 8 is even more straightforward.

It is worth mentioning that a quadratic-time algorithm has been recently proposed (see [25]) for translating conjunctions of literals of the forms $x = y \setminus z$, $x \neq y \setminus z$, and $z = \{x\}$, where x, y, z stand for variables ranging over the von Neumann universe of sets, into unquantified Boolean formulae of a rather simple conjunctive normal form. Put differently, the satisfiability problem for MLSS is reducible in quadratic time to the corresponding problem for propositional combinations of purely Boolean literals. This reduction hence translates all of MST (and more, as the singleton former appears in the source language) into a language very close to BST.

Table 7. Taxonomy of minimal NP-complete and maximal polynomial fragments of BST.

∪	∩	\	=∅	≠∅	Disj	¬Disj	⊆	⊄	=	≠	Complexity
★	★	★	★		★		★		★		$\mathcal{O}(1)$
★	★			★		★	★		★		$\mathcal{O}(1)$
★			★	★	★	★		★		★	$\mathcal{O}(n^3)$
★			★	★		★	★	★	★	★	$\mathcal{O}(n^3)$
	★		★	★	★	★	★	★	★	★	$\mathcal{O}(n^3)$
	★									★	NP-complete
	★							★			NP-complete
	★					★					NP-complete
	★			★							NP-complete
★	★									★	NP-complete
★	★							★			NP-complete
★	★		★	★							NP-complete
★	★				★	★					NP-complete
★	★		★			★					NP-complete
★	★			★	★						NP-complete
★					★	★		★			NP-complete
★				★	★			★			NP-complete
★					★	★	★				NP-complete
★					★				★	★	NP-complete
★					★			★	★		NP-complete
★			★	★			★				NP-complete
★					★		★			★	NP-complete
★					★		★	★			NP-complete

Table 8. Full taxonomy of MST. Maximal polynomial and minimal NP-complete fragments are in black.

$\cup$	$\cap$	$\setminus$	$\in$	$\notin$	Complexity	$\cup$	$\cap$	$\setminus$	$\in$	$\notin$	Complexity
⋆			⋆	⋆	$\mathcal{O}(n)$				⋆	⋆	**NP-complete**
⋆			⋆		$\mathcal{O}(n)$	⋆			⋆	⋆	NP-complete
			⋆	⋆	$\mathcal{O}(n)$			⋆	⋆	⋆	NP-complete
			⋆		$\mathcal{O}(n)$	⋆		⋆	⋆	⋆	NP-complete
	⋆		⋆	⋆	$\mathcal{O}(n^2)$		⋆		⋆	⋆	NP-complete
	⋆		⋆		$\mathcal{O}(n^2)$	⋆	⋆		⋆	⋆	NP-complete
⋆	⋆	⋆		⋆	$\mathcal{O}(1)$		⋆	⋆	⋆	⋆	NP-complete
⋆		⋆		⋆	$\mathcal{O}(1)$	⋆	⋆	⋆	⋆	⋆	NP-complete
⋆	⋆			⋆	$\mathcal{O}(1)$	⋆	⋆	⋆			**NP-complete**
	⋆	⋆		⋆	$\mathcal{O}(1)$	⋆	⋆		⋆	⋆	NP-complete
⋆				⋆	$\mathcal{O}(1)$						
	⋆			⋆	$\mathcal{O}(1)$						
		⋆		⋆	$\mathcal{O}(1)$						
				⋆	$\mathcal{O}(1)$						

Acknowledgments. We are grateful to Daniele Santamaria for contributing to Sect. 3.5.

We gratefully acknowledge partial support from project "STORAGE—Università degli Studi di Catania, Piano della Ricerca 2020/2022, Linea di intervento 2", and from INdAM-GNCS 2019 and 2020 research funds.

References

1. Ackermann, W.: Solvable Cases of the Decision Problem. North-Holland Pub. Co., Amsterdam (1954)
2. Aczel, P.: Non-Well-Founded Sets, vol. 14. CSLI Lecture Notes, Stanford (1988)
3. Barwise, J., Moss, L.: Hypersets. Math. Intell. **13**(4), 31–41 (1991)
4. Behmann, H.: Beiträge zur Algebra der Logik insbesondere zum Entscheidungsproblem. Math. Ann. **86**, 163–220 (1922)
5. Bellè, D., Parlamento, F.: Truth in V for $\exists^*\forall\forall$-sentences is decidable. J. Symb. Logic **71**(4), 1200–1222 (2006)
6. Bellè, D., Parlamento, F.: Decidability of $\exists^*\forall\forall$-sentences in HF. Notre Dame J. Formal Logic **49**(1), 55–64 (2008)
7. Bergenti, F., Monica, S., Rossi, G.: Constraint logic programming with polynomial constraints over finite domains. Fundam. Informaticae **161**(1–2), 9–27 (2018)
8. Bledsoe, W.W.: Non-resolution theorem proving. Artif. Intell. **9**(1), 1–35 (1977)
9. Börger, E., Grädel, E., Gurevich, Yu.: The Classical Decision Problem. Perspectives in Mathematical Logic. Springer, Heidelberg (1997)
10. Breban, M.: Decision algorithms for a class of set-theoretic formulae involving one occurrence of union-set operator. Ph.D. thesis, NYU/GSAS (1982)

11. Breban, M., Ferro, A.: Decision procedures for elementary sublanguages of set theory. III. Restricted classes of formulas involving the power set operator and the general set union operator. Adv. in Appl. Math. **5**(2), 147–215 (1984)

12. Breban, M., Ferro, A., Omodeo, E.G., Schwartz, J.T.: Decision procedures for elementary sublanguages of set theory. II. Formulas involving restricted quantifiers, together with ordinal, integer, map, and domain notions. Comm. Pure Appl. Math. **34**, 177–195 (1981)

13. Buriola, G., Cantone, D., Cincotti, G., Omodeo, E.G., Spartà, G.T.: A decidable theory involving addition of differentiable real functions. Theor. Comput. Sci. **940**(Part A), 124–148 (2023)

14. Calvanese, D., Eiter, T., Ortiz, M.: Answering regular path queries in expressive description logics: an automata-theoretic approach. In: Proceedings of the Twenty-Second AAAI Conference on Artificial Intelligence, Vancouver, British Columbia, Canada, 22–26 July 2007, pp. 391–396. AAAI Press (2007)

15. Cantone, D.: A decision procedure for a class of unquantified formulae of set theory involving the powerset and singleton operators. Ph.D. thesis, NYU/GSAS (1987)

16. Cantone, D.: A survey of computable set theory. Le Matematiche XLII **I**(1), 125–194 (1988)

17. Cantone, D.: Decision procedures for elementary sublanguages of set theory. X. Multilevel syllogistic extended by the singleton and powerset operators. J. Autom. Reason. **7**(2), 193–230 (1991)

18. Cantone, D.: A fast saturation strategy for set-theoretic tableaux. In: Galmiche, D. (ed.) TABLEAUX 1997. LNCS, vol. 1227, pp. 122–137. Springer, Heidelberg (1997). https://doi.org/10.1007/BFb0027409

19. Cantone, D.: Decision procedures for elementary sublanguages of set theory. XVII. Commonly occurring extensions of multi-level syllogistic. In: Davis, M., Schonberg, E. (eds.) From Linear Operators to Computational Biology, pp. 47–85. Springer, London (2013). https://doi.org/10.1007/978-1-4471-4282-9_5

20. Cantone, D., Cutello, V.: A decidable fragment of the elementary theory of relations and some applications. In: Watanabe, S., Nagata, M. (eds.) Proceedings of the 1990 International Symposium on Symbolic and Algebraic Computation, ISSAC 1990, Tokyo, Japan, 20–24 August 1990, pp. 24–29. ACM Press, New York (1990)

21. Cantone, D., Cutello, V.: On the decidability problem for a topological syllogistic involving the notion of topological product. Technical Report TR-91-029, International Computer Science Institute, Berkeley, CA, May 1991

22. Cantone, D., Cutello, V.: Decision algorithms for elementary topology. I. Topological syllogistic with set and map constructs, connectedness and cardinality comparison. Comm. Pure Appl. Math. **47**(9), 1197–1217 (1994)

23. Cantone, D., Cutello, V., Policriti, A.: Set-theoretic reductions of Hilbert's tenth problem. In: Börger, E., Büning, H.K., Richter, M.M. (eds.) CSL 1989. LNCS, vol. 440, pp. 65–75. Springer, Heidelberg (1990). https://doi.org/10.1007/3-540-52753-2_32

24. Cantone, D., Cutello, V., Schwartz, J.T.: Decision problems for tarski and presburger arithmetics extended with sets. In: Börger, E., Kleine Büning, H., Richter, M.M., Schönfeld, W. (eds.) CSL 1990. LNCS, vol. 533, pp. 95–109. Springer, Heidelberg (1991). https://doi.org/10.1007/3-540-54487-9_54

25. Cantone, D., De Domenico, A., Maugeri, P., Omodeo, E.G.: Complexity assessments for decidable fragments of set theory. IV: A quadratic reduction of con-

straints over nested sets to Boolean formulae (2021). https://arxiv.org/abs/2112.04797

26. Cantone, D., De Domenico, A., Maugeri, P., Omodeo, E.G.: Complexity assessments for decidable fragments of set theory. I: a taxonomy for the Boolean case. Fundam. Informaticae **181**(1), 37–69 (2021)

27. Cantone, D., Ferro, A.: Techniques of computable set theory with applications to proof verification. Comm. Pure Appl. Math. **48**(9), 901–945 (1995)

28. Cantone, D., Ferro, A., Micale, B., Sorace, G.: Decision procedures for elementary sublanguages of set theory. IV. Formulae involving a rank operator or one occurrence of $\Sigma(x) = \{\{y\}|y \in x\}$. Comm. Pure Appl. Math. **40**(1), 37–77 (1987)

29. Cantone, D., Ferro, A., Omodeo, E.G.: Decision procedures for elementary sublanguages of set theory VIII: A semidecision procedure for finite satisfiability of unquantified set-theoretic formulae. Comm. Pure Appl. Math. **41**(1), 105–120 (1988)

30. Cantone, D., Ferro, A., Omodeo, E.G.: Computable set theory, volume no.6 Oxford Science Publications of International Series of Monographs on Computer Science. Clarendon Press (1989). Foreword by Jacob T. Schwartz

31. Cantone, D., Ferro, A., Omodeo, E.G., Schwartz, J.T.: Decision algorithms for some fragments of analysis and related areas. Comm. Pure Appl. Math. **40**(3), 281–300 (1987)

32. Cantone, D., Ferro, A., Schwartz, J.T.: Decision procedures for elementary sublanguages of set theory. VI. Multilevel syllogistic extended by the powerset operator. Comm. Pure Appl. Math. **38**(5), 549–571 (1985)

33. Cantone, D., Ferro, A., Schwartz, J.T.: Decision procedures for elementary sublanguages of set theory. V. Multilevel syllogistic extended by the general union operator. J. Comput. Syst. Sci. **34**(1), 1–18 (1987)

34. Cantone, D., Ghelfo, S., Omodeo, E.G.: The automation of syllogistic. I: Syllogistic normal forms. J. Symb. Comput. **6**(1), 83–98 (1988)

35. Cantone, D., Longo, C.: A decidable quantified fragment of set theory with ordered pairs and some undecidable extensions. In: Faella, M., Murano, A. (eds.) Proceedings Third International Symposium on Games, Automata, Logics and Formal Verification, Napoli, Italy, 6–8th June 2012, volume 96 of Electronic Proceedings in Theoretical Computer Science, pp. 224–237. Open Publishing Association (2012). http://rvg.web.cse.unsw.edu.au/eptcs/Published/GandALF2012/Papers/628/arXiv.pdf

36. Cantone, D., Longo, C.: A decidable two-sorted quantified fragment of set theory with ordered pairs and some undecidable extensions. Theoret. Comput. Sci. **560**, 307–325 (2014)

37. Cantone, D., Longo, C., Nicolosi-Asmundo, M.: A decision procedure for a two-sorted extension of multi-level syllogistic with the Cartesian product and some map constructs. In: Faber, W., Leone, N. (eds.) Proceedings of the 25th Italian Conference on Computational Logic (CILC 2010), Rende, Italy, 7–9 July 2010, vol. 598, pp. 1–18. CEUR Workshop Proc., June 2010. ISSN 1613-0073

38. Cantone, D., Longo, C., Nicolosi-Asmundo, M.: A decidable quantified fragment of set theory involving ordered pairs with applications to description logics. In: Bezem, M. (ed.) Computer Science Logic (CSL 2011) - 25th International Workshop/20th Annual Conference of the EACSL. Leibniz International Proceedings in Informatics (LIPIcs), Dagstuhl, Germany, vol. 12, pp. 129–143. Schloss Dagstuhl-Leibniz-Zentrum fuer Informatik (2011)

39. Cantone, D., Longo, C., Nicolosi-Asmundo, M., Santamaria, D.F.: Web ontology representation and reasoning via fragments of set theory. In: ten Cate, B., Mileo, A. (eds.) RR 2015. LNCS, vol. 9209, pp. 61–76. Springer, Cham (2015). https://doi.org/10.1007/978-3-319-22002-4_6

40. Cantone, D., Longo, C., Pisasale, A.: Comparing description logics with multi-level syllogistics: the description logic $\mathcal{DL}\langle\mathsf{MLSS}_{2,m}^\times\rangle$. In: Traverso, P. (ed.) 6th Workshop on Semantic Web Applications and Perspectives (Bressanone, Italy, 21–22 September 2010), pp. 1–13 (2010)

41. Cantone, D., Maugeri, P., Omodeo, E.G.: Complexity assessments for decidable fragments of set theory. II: a taxonomy for 'small' languages involving membership. Theor. Comput. Sci. **848**, 28–46 (2020)

42. Cantone, D., Nicolosi-Asmundo, M.: On the satisfiability problem for a 4-level quantified syllogistic and some applications to modal logic. Fundam. Informaticae **124**(4), 427–448 (2013)

43. Cantone, D., Nicolosi-Asmundo, M., Santamaria, D.F.: Conjunctive query answering via a fragment of set theory. In: Proceedings of ICTCS 2016, Lecce, 7–9 September. CEUR Workshop Proc., vol. 1720, pp. 23–35 (2016). ISSN 1613-0073

44. Cantone, D., Nicolosi-Asmundo, M., Santamaria, D.F.: A set-theoretic approach to ABox reasoning services. In: Costantini, S., Franconi, E., Van Woensel, W., Kontchakov, R., Sadri, F., Roman, D. (eds.) RuleML+RR 2017. LNCS, vol. 10364, pp. 87–102. Springer, Cham (2017). https://doi.org/10.1007/978-3-319-61252-2_7

45. Cantone, D., Nicolosi-Asmundo, M., Santamaria, D.F.: A C++ reasoner for the description logic $\mathcal{DL}_\mathbf{D}^{4,\times}$. In: Proceedings of CILC 2017, Naples, Italy, 26–29 September 2017. CEUR Workshop Proc., vol. 1949, pp. 276–280 (2017). ISSN 1613-0073

46. Cantone, D., Nicolosi-Asmundo, M., Santamaria, D.F.: An optimized KE-tableau-based system for reasoning in the description logic $\mathcal{DL}_\mathbf{D}^{4,\times}$. In: Benzmüller, C., Ricca, F., Parent, X., Roman, D. (eds.) RuleML+RR 2018. LNCS, vol. 11092, pp. 239–247. Springer, Cham (2018). https://doi.org/10.1007/978-3-319-99906-7_16

47. Cantone, D., Nicolosi-Asmundo, M., Santamaria, D.F.: A set-theoretic approach to reasoning services for the description logic $\mathcal{DL}_\mathbf{D}^{4,\times}$. Fundam. Informaticae **176**, 349–384 (2020)

48. Cantone, D., Nicolosi-Asmundo, M., Santamaria, D.F.: An improved set-based reasoner for the description logic $\mathcal{DL}_\mathbf{D}^{4,\times}$. Fundam. Informaticae **178**(4), 315–346 (2021)

49. Cantone, D., Omodeo, E.G.: On the decidability of formulae involving continuous and closed functions. In: Sridharan, N.S. (ed.) Proceedings of the 11th International Joint Conference on Artificial Intelligence, Detroit, MI, USA, August 1989, vol. 1, pp. 425–430. Morgan Kaufmann (1989)

50. Cantone, D., Omodeo, E.G.: Topological syllogistic with continuous and closed functions. Comm. Pure Appl. Math. **42**(8), 1175–1188 (1989)

51. Cantone, D., Omodeo, E.G., Panettiere, M.: From Hilbert's 10th problem to slim, undecidable fragments of set theory. In: Cordasco, G., Gargano, L., Rescigno, A.A. (eds.) Proceedings of the 21st Italian Conference on Theoretical Computer Science, Ischia, Italy, 14–16 September 2020, volume 2756 of CEUR Workshop Proc., pp. 47–60. CEUR-WS.org (2020)

52. Cantone, D., Omodeo, E.G., Panettiere, M.: Very weak, essentially undecidable set theories. In: Monica, S., Bergenti, F. (eds.) Proceedings of the 36th Italian

Conference on Computational Logic, Parma, Italy, 7–9 September 2021, volume 3002 of CEUR Workshop Proc., pp. 31–46. CEUR-WS.org (2021)

53. Cantone, D., Omodeo, E.G., Policriti, A.: The automation of syllogistic. II: optimization and complexity issues. J. Autom. Reason. **6**(2), 173–187 (1990)

54. Cantone, D., Omodeo, E.G., Policriti, A.: Set Theory for Computing. From Decision Procedures to Declarative Programming with Sets. Texts and Monographs in Computer Science. Springer, New York (2001). https://doi.org/10.1007/978-1-4757-3452-2

55. Cantone, D., Omodeo, E.G., Policriti, A.: Banishing ultrafilters from our consciousness. In: Omodeo, E.G., Policriti, A. (eds.) Martin Davis on Computability, Computational Logic, and Mathematical Foundations. OCL, vol. 10, pp. 255–283. Springer, Cham (2016). https://doi.org/10.1007/978-3-319-41842-1_10

56. Cantone, D., Omodeo, E.G., Schwartz, J.T., Ursino, P.: Notes from the logbook of a proof-checker's project*. In: Dershowitz, N. (ed.) Verification: Theory and Practice. LNCS, vol. 2772, pp. 182–207. Springer, Heidelberg (2003). https://doi.org/10.1007/978-3-540-39910-0_8

57. Cantone, D., Schwartz, J.T.: Decision procedures for elementary sublanguages of set theory. XI. Multilevel syllogistic extended by some elementary map constructs. J. Autom. Reason. **7**(2), 231–256 (1991)

58. Cantone, D., Ursino, P.: Formative processes with applications to the decision problem in set theory: II. Powerset and singleton operators, finiteness predicate. Inf. Comput. **237**, 215–242 (2014)

59. Cantone, D., Ursino, P.: An Introduction to the Technique of Formative Processes in Set Theory. Springer, Cham (2018). https://doi.org/10.1007/978-3-319-74778-1

60. Cantone, D., Ursino, P.: Decidability of the satisfiability problem for Boolean set theory with the unordered Cartesian product operator. ACM Trans. Comput. Log. **25**(1), 1–30 (2024). Article 5

61. Cantone, D., Ursino, P., Omodeo, E.G.: Formative processes with applications to the decision problem in set theory: I. Powerset and singleton operators. Inf. Comput. **172**(2), 165–201 (2002)

62. Cantone, D., Zarba, C.G.: A new fast tableau-based decision procedure for an unquantified fragment of set theory. In: Caferra, R., Salzer, G. (eds.) Automated Deduction in Classical and Non-Classical Logics, Selected Papers. LCNS, vol. 1761, pp. 126–136. Springer, Heidelberg (2000)

63. Cantone, D., Zarba, C.G.: A tableau-based decision procedure for a fragment of graph theory involving reachability and acyclicity. In: Beckert, B. (ed.) TABLEAUX 2005. LNCS (LNAI), vol. 3702, pp. 93–107. Springer, Heidelberg (2005). https://doi.org/10.1007/11554554_9

64. Cantone, D., Zarba, C.G.: A decision procedure for monotone functions over bounded and complete lattices. In: de Swart, H., Orłowska, E., Schmidt, G., Roubens, M. (eds.) Theory and Applications of Relational Structures as Knowledge Instruments II. LNCS (LNAI), vol. 4342, pp. 318–333. Springer, Heidelberg (2006). https://doi.org/10.1007/11964810_15

65. Casagrande, A., Di Cosmo, F., Omodeo, E.G.: On perfect matchings for some bipartite graphs. In: Cristiá et al. [70], pp. 38–51

66. Casagrande, A., Omodeo, E.G.: Reasoning about connectivity without paths. In: Bistarelli, S., Formisano, A. (eds.) Proceedings of the 15th Italian Conference on Theoretical Computer Science, Perugia, Italy, 17–19 September 2014, volume 1231 of CEUR Workshop Proc., pp. 93–108. CEUR-WS.org (2014)

67. Ceterchi, R., Omodeo, E.G., Tomescu, A.I.: The representation of Boolean alge-bras in the spotlight of a proof checker. In: Giordano, L., Gliozzi, V., Pozzato, G.L. (eds.) Proceedings 29th Italian Conference on Computational Logic, Torino, Italy, 16–18 June 2014, volume 1195 of CEUR Workshop Proc., pp. 287–301. CEUR-WS.org (2014)

68. Collins, G.E.: Quantifier elimination for real closed fields by cylindrical algebraic decomposition. In: Caviness, B.F., Johnson, J.R. (eds.) Quantifier Elimination and Cylindrical Algebraic Decomposition. LNCS, vol. 33, pp. 134–183. Springer, Vienna (1975). https://doi.org/10.1007/978-3-7091-9459-1_4

69. Cristiá, M., Delahaye, D., Dubois, C. (eds.) Proceedings of the 3rd International Workshop on Sets and Tools co-located with the 6th International ABZ Confer-ence, SETS@ABZ 2018, Southamptom, UK, 5 June 2018, volume 2199 of CEUR Workshop Proc. CEUR-WS.org (2018)

70. Cristiá, M., Rossi, G., Frydman, C.: *log* as a test case generator for the test template framework. In: Hierons, R.M., Merayo, M.G., Bravetti, M. (eds.) SEFM 2013. LNCS, vol. 8137, pp. 229–243. Springer, Heidelberg (2013). https://doi.org/10.1007/978-3-642-40561-7_16

71. Davis, M., Schonberg, E. (eds.): From Linear Operators to Computational Biol-ogy: Essays in Memory of Jacob T Schwartz. Springer, London (2013)

72. Davis, M., Schwartz, J.T.: Correct-program technology / Extensibility of verifiers - Two papers on Program Verification with Appendix of Edith Deak. Techni-cal Report No. NSO-12, Courant Institute of Mathematical Sciences, New York University, September 1977

73. Davis, M., Schwartz, J.T.: Metamathematical extensibility for theorem verifiers and proof-checkers. Comput. Math. Appl. **5**(3), 217–230 (1979)

74. Dewar, R.: SETL and the evolution of programming. In: Davis and Schonberg [72], pp. 39–46

75. Doberkat, E.-E., Fox, D.: Software Prototyping MIT SETL. Teubner, Stuttgart (1989). https://link.springer.com/book/10.1007/978-3-322-94710-9

76. Dovier, A., Formisano, A., Omodeo, E.G.: Decidability results for sets with atoms. ACM Trans. Comput. Log. **7**(2), 269–301 (2006)

77. Dovier, A., Omodeo, E.G., Pontelli, E., Rossi, G.: {log}: a logic programming language with finite sets. In: Furukawa, K. (ed.) Logic Programming, Proceedings of the Eighth International Conference, Paris, France, 24–28 June 1991, pp. 111–124. MIT Press (1991)

78. Dovier, A., Omodeo, E.G., Pontelli, E., Rossi, G.: {log}: a language for program-ming in logic with finite sets. J. Log. Program. **28**(1), 1–44 (1996)

79. Dovier, A., Piazza, C., Rossi, G.: A uniform approach to constraint-solving for lists, multisets, compact lists, and sets. ACM Trans. Comput. Log. **9**(3), 15:1-15:30 (2008)

80. Dovier, A., Pontelli, E., Rossi, G.: Set unification. Theory Pract. Log Program **6**(6), 645–701 (2006)

81. Dreben, B., Goldfarb, W.D.: The Decision Problem. Solvable Classes of Quantifi-cational Formulas. Advanced Book Program. Addison-Wesley Publishing Com-pany, Reading (1979)

82. Farmer, W.M., Guttman, J.D., Thayer, F.J.: IMPS: an interactive mathematical proof system. J. Autom. Reason. **11**, 213–248 (1993)

83. Ferro, A.: Decision procedures for some classes of unquantified set theoretic for-mulae. Ph.D. thesis, NYU / GSAS (1981)

84. Ferro, A.: A note on the decidability of MLS extended with the powerset operator. Commun. Pure Appl. Math. **38**(4), 367–374 (1985)

85. Ferro, A.: Decision procedures for elementary sublanguages of set theory: XII. Multilevel syllogistic extended with singleton and choice operators. J. Autom. Reason. **7**(2), 257–270 (1991)
86. Ferro, A., Omodeo, E.G.: An efficient validity test for formulae in extensional two-level syllogistic. Le Matematiche **XXXIII**, 130–137 (1978). In reality, this paper appeared in print in 1979 or 1980
87. Ferro, A., Omodeo, E.G.: Decision procedures for elementary sublanguages of set theory. VII. Validity in set theory when a choice operator is present. Comm. Pure Appl. Math. **40**(3), 265–280 (1987)
88. Ferro, A., Omodeo, E.G., Schwartz, J.T.: Decision procedures for elementary sublanguages of set theory. I. Multi-level syllogistic and some extensions. Comm. Pure Appl. Math. **33**(5), 599–608 (1980)
89. Ferro, A., Omodeo, E.G., Schwartz, J.T.: Decision procedures for some fragments of set theory. In: Bibel, W., Kowalski, R. (eds.) CADE 1980. LNCS, vol. 87, pp. 88–96. Springer, Heidelberg (1980). https://doi.org/10.1007/3-540-10009-1_8
90. Garey, M.R., Johnson, D.S.: Computers and Intractability; A Guide to the Theory of NP-Completeness. W. H. Freeman & Co., New York (1990)
91. Giordano, L., Policriti, A.: Adding the power-set to description logics. Theor. Comput. Sci. **813**, 155–174 (2020)
92. Gogol, D.: The $\forall_n\exists$-completeness of Zermelo-Fraenkel set theory. Zeitschr. f. math. Logik und Grundlagen d. Math. **24**(19–24):289–290 (1978)
93. Goguen, J.A., Malcolm, G.: Algebraic Semantics of Imperative Programs. MIT Press, Cambridge (1996)
94. Horrocks, I., Kutz, O., Sattler, U.: The even more irresistible SROIQ. In: Doherty, P., Mylopoulos, J., Welty, C.A. (eds.) Proceedings of the 10th International Conference on Princ. of Knowledge Representation and Reasoning, pp. 57–67. AAAI Press (2006)
95. Jech, T.: Set Theory. Springer Monographs in Mathematics, Third Millennium edition. Springer, Heidelberg (2003)
96. Keller, J.-P., Paige, R.: Program derivation with verified transformations - a case study. Comm. Pure Appl. Math. **48**(9), 1053–1113 (1995). Special issue https://onlinelibrary.wiley.com/toc/10970312/1995/48/9, in honor of J. T. Schwartz
97. Lévy, A.: A Hierarchy of Formulas in Set Theory. A Hierarchy of Formulas in Set Theory. American Mathematical Society (1965)
98. Lewis, H.R.: Unsolvable Classes of Quantificational Formulas. Advanced Book Program. Addison-Wesley Publishing Company, Reading (1979)
99. Lewis, H.R.: Complexity results for classes of quantificational formulas. J. Comput. Syst. Sci. **21**(3), 317–353 (1980)
100. Lücke, J.: Hilberticus - a tool deciding an elementary sublanguage of set theory. In: Goré, R., Leitsch, A., Nipkow, T. (eds.) IJCAR 2001. LNCS, vol. 2083, pp. 690–695. Springer, Heidelberg (2001). https://doi.org/10.1007/3-540-45744-5_57
101. Mancosu, P.: Between Russell and Hilbert: Behmann on the foundations of mathematics. Bull. Symbolic Logic **5**(3), 303–330 (1999)
102. Manin,Yu.I.: A Course in Mathematical Logic. Graduate Texts in Mathematics. Springer, New York (1977). https://doi.org/10.1007/978-1-4757-4385-2
103. Matiyasevich, Yu.V., Desyataya Problema Gilberta. Fizmatlit, Moscow (1993). English translation: Hilbert's Tenth Problem. The MIT Press, Cambridge (1993). French translation: Le dixième Problème de Hilbert: son indécidabilité, Masson, Paris Milan Barcelone (1995). http://logic.pdmi.ras.ru/~yumat/H10Pbook/
104. Omodeo, E.G.: Decidability and proof procedures for set theory with a choice operator. Ph.D. thesis, NYU/GSAS (1984)

105. Omodeo, E.G.: L'automazione della sillogistica. Le Scienze (Italian edition of Scientific American) **218**, 120–128, 143 (1986)
106. Omodeo, E.G.: The Ref proof-checker and its "common shared scenario". In: Davis and Schonberg [72], pp. 121–131
107. Omodeo, E.G., Parlamento, F., Policriti, A.: A derived algorithm for evaluating ε-expressions over abstract sets. J. Symb. Comput. **15**(5–6), 673–704 (1993). Special issue on Automatic Programming; A.W. Biermann and W. Bibel editors
108. Omodeo, E.G., Parlamento, F., Policriti, A.: Decidability of $\exists^*\forall$-sentences in membership theories. Math. Log. Quart. **42**(1), 41–58 (1996)
109. Omodeo, E.G., Policriti, A.: Solvable set/hyperset contexts: I. Some decision procedures for the pure, finite case. Comm. Pure Appl. Math. **48**(9), 1123–1155 (1995). Special issue in honor of Jacob T. Schwartz
110. Omodeo, E.G., Policriti, A.: The Bernays-Schönfinkel-Ramsey class for set theory: semidecidability. J. Symb. Logic **75**(2), 459–480 (2010)
111. Omodeo, E.G., Policriti, A.: The Bernays-Schönfinkel-Ramsey class for set theory: decidability. J. Symb. Logic **77**(3), 896–918 (2012)
112. Omodeo, E.G., Policriti, A., Tomescu, A.I.: Statements of ill-founded infinity in Set Theory. In: Cegielski, P. (ed.) Studies in Weak Arithmetics. CLSI Lecture Notes, vol. 196, pp. 173–199. CLSI, Stanford (2010)
113. Omodeo, E.G., Policriti, A., Tomescu, A.I.: Stating infinity in Set/Hyperset theory. Rend. Istit. Mat. Univ. Trieste **42**, 207–212 (2010)
114. Omodeo, E.G., Policriti, A., Tomescu, A.I.: Infinity, in short. J. Log. Comput. **22**(6), 1391–1403 (2012)
115. Omodeo, E.G., Policriti, A., Tomescu, A.I.: On Sets and Graphs: Perspectives on Logic and Combinatorics. Springer (2017). Foreword by Ernst-Erich Doberkat. https://link.springer.com/book/10.1007%2F978-3-319-54981-1
116. Omodeo, E.G., Policriti, A., Tomescu, A.I.: Set-syllogistics meet combinatorics. Math. Struct. Comput. Sci. **27**(2), 296–310 (2017)
117. Omodeo, E.G., Schwartz, J.T.: A 'Theory' mechanism for a proof-verifier based on first-order set theory. In: Kakas, A.C., Sadri, F. (eds.) Computational Logic: Logic Programming and Beyond, Part II. LNCS (LNAI), vol. 2408, pp. 214–230. Springer, Heidelberg (2002). https://doi.org/10.1007/3-540-45632-5_9
118. Omodeo, E.G., Tomescu, A.I.: Using ÆtnaNova to formally prove that the Davis-Putnam satisfiability test is correct. Le Matematiche LXIII, Fasc. **I**, 85–105 (2008)
119. Omodeo, E.G., Tomescu, A.I.: Appendix: claw-free graphs as sets. In: Davis and Schonberg [72], pp. 131–167
120. Omodeo, E.G., Tomescu, A.I.: Set graphs. III. Proof Pearl: claw-free graphs mirrored into transitive hereditarily finite sets. J. Autom. Reason. **52**(1), 1–29 (2014)
121. Omodeo, E.G., Tomescu, A.I.: Set Graphs. V. On representing graphs as membership digraphs. J. Log. Comput. **25**(3), 899–919 (2015)
122. Paige, R., Koenig, S.: Finite differencing of computable expressions. ACM Trans. Program. Lang. Syst. **4**(3), 402–454 (1982). https://doi.org/10.1145/357172.357177
123. Papadimitriou, C.H.: On the complexity of Integer Programming. J. ACM **28**(4), 765–768 (1981)
124. Parlamento, F., Policriti, A.: Decision procedures for elementary sublanguages of set theory. IX. Unsolvability of the decision problem for a restricted subclass of Δ_0-formulas in set theory. Comm. Pure Appl. Math. **41**, 221–251 (1988)
125. Parlamento, F., Policriti, A.: The logically simplest form of the infinity axiom. Proc. Amer. Math. Soc. **103**(1), 274–276 (1988)

126. Parlamento, F., Policriti, A.: Note on "The logically simplest form of the infinity axiom". Proc. Amer. Math. Soc. **108**(1), 285–286 (1990)
127. Parlamento, F., Policriti, A.: Expressing infinity without foundation. J. Symb. Logic **56**(4), 1230–1235 (1991)
128. Parlamento, F., Policriti, A.: Undecidability results for restricted universally quantified formulae of set theory. Comm. Pure Appl. Math. **46**(1), 57–73 (1993)
129. Pastre, D.: Automatic theorem proving in Set Theory. Artif. Intell. **10**(1), 1–27 (1978)
130. Pellegrini, M., Sepe, R.: SetLog, a tool for experimenting with new semantics. ACM SIGPLAN Not. **26**(2), 67–74 (1991)
131. Quine, W.V.: Methods of Logic. Henry Holt, New York (1950)
132. Robinson, R.M.: The theory of classes A modification of von Neumann's system. J. Symb. Logic **2**(1), 29–36 (1937)
133. Rogers, H.: Theory of Recursive Functions and Effective Computability. MIT Press, Cambridge (1987)
134. Schaub, T., Thielscher, M.: Skeptical query-answering in constrained default logic. In: Gabbay, D.M., Ohlbach, H.J. (eds.) FAPR 1996. LNCS, vol. 1085, pp. 567–581. Springer, Heidelberg (1996). https://doi.org/10.1007/3-540-61313-7_101
135. Schwartz, J.T.: Instantiation and decision procedures for certain classes of quantified set-theoretic formulae. Technical Report 78–10, Institute for Computer Applications in Science and Engineering, NASA Langley Research Center, Hampton, Virginia, August 1978
136. Schwartz, J.T., Cantone, D., Omodeo, E.G.: Computational Logic and Set Theory — Applying Formalized Logic to Analysis. Springer, London (2011). Foreword by Martin D. Davis
137. Schwartz, J.T., Dewar, R.B.K., Dubinsky, E., Schonberg, E.: Programming with sets: An introduction to SETL. Texts and Monographs in Computer Science, Springer, New York (1986)
138. Sigal, R.: Desiderata for logic programming with sets. In: Mello, P. (ed.) Atti del quarto convegno nazionale sulla programmazione logica (GULP 1989, Bologna), pp. 127–141 (1989)
139. Smullyan, R.M.: First-order Logic. Dover books on Advanced Math, Dover (1995)
140. Stevens, L.: Towards a verified prover for a ground fragment of set theory (2022). https://arxiv.org/abs/2209.14133
141. Stevens, L.: Towards a verified tableau prover for a quantifier-free fragment of set theory. In: Pientka, B., Tinelli, C. (eds.) Automated Deduction - CADE 29. LNCS, vol. 14132, pp. 491–508. Springer, Cham (2023). https://doi.org/10.1007/978-3-031-38499-8_28
142. Tarski, A.: A decision method for elementary algebra and geometry. Technical Report U.S. Air Force Project RAND R-109, the RAND Corporation, Santa Monica, Ca, iv+60 pp., 1948. Prepared for publication by J.C.C. McKinsey; a 2nd revised edition was published by the University of California Press, Berkeley and Los Angeles, iii+63 pp. (1951)
143. Vaught, R.L.: On a theorem of Cobham concerning undecidable theories. In: Nagel, E., Suppes, P., Tarski, A. (eds.) Proceedings of the 1960 International Congress on Logic, Methodology, and Philosophy of Science, pp. 14–25. Stanford University Press (1962)
144. World Wide Web Consortium (W3C): SWRL: A Semantic Web Rule Language Combining OWL and RuleML (2004)
145. World Wide Web Consortium: OWL 2 Web ontology language structural specification and functional-style syntax (second edition), December 2012

An Automatically Verified Prototype
of a Landing Gear System

Maximiliano Cristiá[1]([✉]) [iD] and Gianfranco Rossi[2] [iD]

[1] Universidad Nacional de Rosario and CIFASIS, Rosario, Argentina
cristia@cifasis-conicet.gov.ar
[2] Università di Parma, Parma, Italy
gianfranco.rossi@unipr.it

Abstract. In this paper we show how $\{log\}$ (read 'setlog'), a Constraint Logic Programming (CLP) language based on set theory, can be used as an automated verifier for B specifications. In particular we encode in $\{log\}$ an Event-B specification, developed by Mammar and Laleau, of the case study known as the Landing Gear System (LGS). Next we use $\{log\}$ to discharge all the proof obligations proposed in the Event-B specification by the Rodin platform. In this way, the $\{log\}$ program can be regarded as an automatically verified prototype of the LGS. We believe this case study provides empirical evidence on how CLP and set theory can be used in tandem as a vehicle for program verification.

1 Introduction

In the fourth edition of the ABZ Conference held in Toulouse (France) in 2014, ONERA's[1] Boniol and Wiels proposed a real-life, industrial-strength case study, known as the Landing Gear System (LGS) [5]. The initial objectives of the proposal were "to learn from experiences and know-how of the users of the ASM, Alloy, B, TLA, VDM and Z methods" and to disseminate the use of verification techniques based on those methods, in the aeronautic and space industries. The LGS was the core of an ABZ track of the same name. The track received a number of submissions of which eleven were published [5]. Six of the published papers approached the problem with methods and tools rooted in the B notation [2] (Event-B, ProB, Hybrid Event-B and Rodin). In this paper we consider the article by Mammar and Laleau [24] and a journal version [25] as the starting point for our work. In those articles, the authors use Event-B as the specification language and Rodin [4], ProB [23] and AnimB [1] as verification tools.

The B method was introduced by Abrial [2] after his work on the Z notation [29]. B is a formal notation based on state machines, set theory and first-order logic aimed at software specification and verification. Verification is approached by discharging proof obligations generated during specification refinement. That

---

[1] ONERA: Office National d'Etudes et Recherches Aérospatiales.

D. Cantone and A. Pulvirenti (Eds.): *From Computational Logic to Computational Biology*,
LNCS 14070, pp. 56–81, 2024.
https://doi.org/10.1007/978-3-031-55248-9_3

is, the engineer starts with a first, abstract specification and refines it into a second, less abstract specification. In order to ensure that the refinement step is correct a number of proof obligations must be discharged. This process is continued until an executable program is obtained. Given that all refinements have been proved correct, the executable program is correct by construction.

Discharging a proof obligation entails to perform a *formal proof* (most often) in the form of a *mechanized proof*. A formal proof is a proof of a mathematical theorem; a mechanized proof is a formal proof made by either an interactive theorem prover or an automated theorem prover[2]. That is, mechanized proofs are controlled, guided or verified by a program. Unless there are errors in the verification software, a mechanized proof is considered to be error-free because those programs are supposed to implement only sound proof steps. The mechanization of proofs is important for at least two reasons: errors cannot be tolerated in safety-critical systems and mechanization enables the possibility of proof automation which in turn reduces verification costs.

Our work starts by considering the Event-B specification of the LGS developed by Mammar and Laleau. Event-B [3] is a further development over B aimed at modeling and reasoning about discret-event systems. The basis of B are nonetheless present in Event-B: state machines, set theory, first-order logic, refinement and formal proof. The Event-B specification developed by Mammar and Laleau has an important property for us. They used the Rodin platform to write and verify the model. This implies that proof obligations were generated by Rodin according to a precise and complete algorithm [4].

In this paper we consider the LGS casy study and Mammar and Leleau's Event-B specification as a benchmark for {*log*} (read 'setlog') [17,18]. {*log*} is a Constraint Logic Programming (CLP) language and satisfiability solver based on set theory. As such, it can be used as a model animator and as an automated theorem prover. Being {*log*} based on set theory, it should also be a good candidate to encode Event-B (or classic B) specifications; since it implements several decision procedures for set theory, it should be a good candidate to automatically discharge proof obligations generated from refinement steps of Event-B (or classic B) specifications.

Therefore, we proceed as follows: *a)* Mammar and Leleau's Event-B specification is encoded as a {*log*} program; *b)* all the proof obligations generated by the Rodin platform are encoded as {*log*} queries; and *c)* {*log*} is used to automatically discharge all these queries. We say 'encode' and not 'implement' due to the similarities between the {*log*} language and the mathematical basis of the Event-B language; however, the encoding provides an implementation in the form of a prototype.

The contributions of this paper are the following:

– We provide empirical evidence on how CLP and set theory can be used in tandem as a vehicle for program verification. More specifically, {*log*} is shown to work well in practice.

---

[2] In this context the term 'automated theorem prover' includes tools such as satisfiability solvers.

– Given that the {*log*} prototype of the LGS has been (mechanically and automatically) proved to verify a number of properties, it can be regarded as
correct w.r.t. those properties.

The paper is structured as follows. In Sects. 2 and 3 we introduce {*log*}
by means of several examples. In particular, in Sect. 3 we shown the *formula-
program duality* enjoined by {*log*}. Section 4 presents the encoding of the Event-B
specification of the LGS in {*log*}. The encoding of the proof obligations generated
by the Rodin tool is introduced in Sect. 5. Finally, we discuss our approach in
Sect. 6 and give our conclusions in Sect. 7.

2   Overview of {*log*}

{*log*} is a publicly available satisfiability solver and a declarative set-based,
constraint-based programming language implemented in Prolog [26]. {*log*} is
deeply rooted in the work on Computable Set Theory [6], combined with the
ideas put forward by the set-based programming language SETL [28].

{*log*} implements various decision procedures for different theories on the
domain of finite sets and integer numbers. Specifically, {*log*} implements: a decision procedure for the theory of *hereditarily finite sets* ($\mathcal{SET}$), i.e., finitely nested
sets that are finite at each level of nesting [18]; a decision procedure for a very
expressive fragment of the theory of finite set relation algebras ($\mathcal{BR}$) [8,9]; a
decision procedure for the theory $\mathcal{SET}$ extended with restricted intensional sets
($\mathcal{RIS}$) [11]; a decision procedure for the theory $\mathcal{SET}$ extended with cardinality
constraints ($\mathcal{CARD}$) [15]; a decision procedure for the latter extended with integer intervals ($\mathcal{INTV}$) [13]; and integrates an existing decision procedure for the
theory of linear integer arithmetic ($\mathcal{LIA}$). All these procedures are integrated
into a single solver, implemented in Prolog, which constitutes the core part of
the {*log*} tool. Several in-depth empirical evaluations provide evidence that {*log*}
is able to solve non-trivial problems [8,9,11,16]; in particular as an automated
verifier of security properties [10,12].

2.1   The Theories

Figure 1 schematically describes the stack of the first-order theories supported
by {*log*}. The fact that theory T is over theory S means that T extends S. For
example, $\mathcal{CARD}$ extends both $\mathcal{LIA}$ and $\mathcal{SET}$.

$\mathcal{LIA}$ provides integer linear arithmetic constraints (e.g., $2*X+5 \leq 3*Y-2$)
by means of Prolog's CLP(Q) library [20]. $\mathcal{SET}$ [18] provides the Boolean algebra
of hereditarily finite sets; that is, it provides equality ($=$), union (un), disjointness
(disj), and membership (in). In turn, these operators enable the definition of
other operators, such as intersection and relative complement, as $\mathcal{SET}$ formulas.
In all set theories, set operators are encoded as atomic predicates, and are dealt
with as constraints. For example, un(A, B, C) is a constraint interpreted as $C =
A \cup B$. $\mathcal{CARD}$ [15] extends $\mathcal{SET}$ and $\mathcal{LIA}$ by providing the cardinality operator

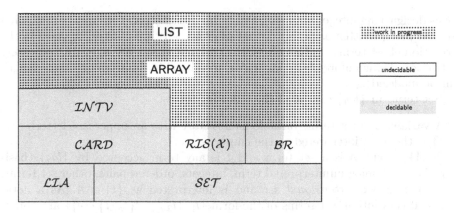

Fig. 1. The stack of theories dealt with by {*log*}

(size) which allows to link sets with integer constraints[3] (e.g., size(A, W) & $W - 1 \leq 2*U + 1$). $\mathcal{RIS}$ [11] extends $\mathcal{SET}$ by introducing the notion of restricted intensional set (RIS) into the Boolean algebra. RIS are finite sets defined by a property. For example, $x \in \{y : A \mid \phi(y)\}$, where A is a set and ϕ is a formula of a parameter theory $\mathcal{X}$. $\mathcal{BR}$ [8,9] extends $\mathcal{SET}$ by introducing ordered pairs, binary relations and Cartesian products (as sets of ordered pairs), and the operators of relation algebra lacking in $\mathcal{SET}$—identity (id), converse (inv) and composition (comp). For example[4], comp(R, S, $\{[X, Z]\}$)&inv(R, T)&$[X, Y]$in T. In turn, $\mathcal{BR}$ allows the definition of relational operators such as domain, range and domain restriction, as $\mathcal{BR}$ formulas. $\mathcal{INTV}$ [13] extends $\mathcal{CARD}$ by introducing finite integer intervals thus enabling the definition of several non-trivial set operators such as the minimum of a set and the partition of a set w.r.t. a given number. ARRAY and LIST are still work in progress. They should provide theories to (automatically) reason about arrays and lists from a set theoretic perspective. For example, A is an array of length n iff is a function with domain in the integer interval[5] int(1, n). Then, in ARRAY it is possible to define the predicate array(A, n) $\hat{=}$ pfun(A) & dom(A, int(1, n)), where pfun and dom are predicates definable in $\mathcal{BR}$. However, it is still necessary to study the decidability of such an extension to $\mathcal{BR} + \mathcal{INTV}$ as, at the bare minimum, it requires for dom to deal with integer intervals. The following subsections provide some clarifications on the syntax and semantics of the logic languages on which these theories are based on.

2.2 Set Terms

The integrated constraint language offered by {*log*} is a first-order predicate language with terms of three sorts: terms designating sets (i.e., *set terms*),

---

[3] '&' stands for conjunction ($\wedge$); see Sect. 2.4.

[4] $[x, y]$ stands for the ordered pair (x, y).

[5] int(a, b) stands for the integer interval $\{a, a + 1, \ldots, b\}$; see Sect. 2.2.

terms designating integer numbers, and terms designating ur-elements (including ordered pairs, written as $[x, y]$). Terms of either sort are allowed to enter in the formation of set terms (in this sense, the designated sets are hybrid), no nesting restrictions being enforced (in particular, membership chains of any finite length can be modeled).

Set terms in $\{log\}$ can be of the following forms:

- A variable is a set term; variable names start with an uppercase letter.
- $\{\}$ is the term interpreted as the empty set.
- $\{t/A\}$, where A is a set term and t is any term accepted by $\{log\}$ (basically, any Prolog uninterpreted term, integers, ordered pairs, other set terms, etc.), is called *extensional set* and is interpreted as $\{t\} \cup A$. As a notational convention, set terms of the form $\{t_1/\{t_2/ \cdots \{t_n/t\}\cdots\}\}$ are abbreviated as $\{t_1, t_2, \ldots, t_n/t\}$, while $\{t_1/\{t_2/ \cdots \{t_n/\{\}\}\cdots\}\}$ is abbreviated as $\{t_1, t_2, \ldots, t_n\}$.
- $\texttt{ris}(X \text{ in } A, \phi)$, where ϕ is any $\{log\}$ formula (see Sect. 2.4), A is a any set term among the first three, and X is a bound variable local to the $\texttt{ris}$ term, is called *restricted intensional set* (RIS) and is interpreted as $\{x : x \in A \wedge \phi\}$. Actually, RIS have a more complex and expressive structure [11].
- $\texttt{cp}(A, B)$, where A and B are any set term among the first three, is interpreted as $A \times B$, i.e., the Cartesian product between A and B.
- $\texttt{int}(m, n)$, where m and n are either integer constants or variables, is interpreted as $\{x \in \mathbb{Z} \mid m \le x \le n\}$.

Set terms can be combined in several ways: binary relations are hereditarily finite sets whose elements are ordered pairs and so set operators can take binary relations as arguments; RIS and integer intervals can be passed as arguments to $\mathcal{SET}$ operators and freely combined with extensional sets.

2.3 Set and Relational Operators

$\{log\}$ implements a wide range of set and relational operators covering most of those used in B. Some of the basic operators are provided as *primitive* constraints. For instance, $\texttt{pfun}(F)$ constrains F to be a (partial) function; $\texttt{dom}(F, D)$ corresponds to dom $F = D$; $\texttt{subset}(A, B)$ corresponds to $A \subseteq B$; $\texttt{comp}(R, S, T)$ is interpreted as $T = R \,\dfrac{\circ}{\circ}\, S$ (i.e., relational composition); and $\texttt{apply}(F, X, Y)$ is equivalent to $\texttt{pfun}(F) \,\&\, [X, Y] \text{ in } F$.

A number of other set, relational and integer operators (in the form of predicates) are defined as $\{log\}$ formulas, thus making it simpler for the user to write complex formulas. Dovier et al. [18] proved that the collection of predicate symbols implementing $\{=, \neq, \in, \notin, \cup, \|\}$ is sufficient to define constraints for the set operators $\cap$, $\subseteq$ and $\setminus$. This result has been extended to binary relations [9] by showing that adding to the previous collection the predicate symbols implementing $\{\text{id}, \frac{\circ}{\circ}, {}^{\smile}\}$ is sufficient to define constraints for most of the classical relational operators, such as dom, ran, $\lhd$, $\rhd$, etc. Similarly, $\{=, \neq, \le\}$ is sufficient to define $<$, $>$ and $\ge$. We call predicates defined in this way *derived constraints*.

$\{log\}$ provides also so-called *negated constraints*. For example, $\mathtt{nun}(A, B, C)$ is interpreted as $C \neq A \cup B$ and $\mathtt{nin}$ corresponds to $\notin$—in general, a constraint beginning with 'n' identifies a negated constraint. Most of these constraints are defined as derived constraints in terms of the existing primitive constraints; thus their introduction does not really require extending the constraint language.

2.4 $\{log\}$ Formulas

Formulas in $\{log\}$ are built in the usual way by using the propositional connectives (e.g., &, or) between atomic constraints. More precisely, $\{log\}$ formulas are defined according to the following grammar:

$$\mathcal{F} ::= \mathtt{true} \mid \mathtt{false} \mid \mathcal{C} \mid \mathcal{F} \mathbin{\&} \mathcal{F} \mid \mathcal{F} \mathbin{\mathtt{or}} \mathcal{F} \mid \mathcal{P} \mid \mathtt{neg}(\mathcal{F}) \mid \mathcal{F} \mathbin{\mathtt{implies}} \mathcal{F} \mid \mathcal{Q}(\mathcal{F})$$

where: $\mathcal{C}$ is any $\{log\}$ atomic constraint; $\mathcal{P}$ is any atomic predicate which is defined by a Horn clause of the form:

$$\mathcal{H} ::= \mathcal{P} \mid \mathcal{P} \mathbin{:\text{-}} \mathcal{F}$$

and the user-defined predicates possibly occurring in the body of the clause do not contain any direct or indirect recursive call to the predicate in $\mathcal{P}$; $\mathcal{Q}(\mathcal{F})$ is the quantified version of $\mathcal{F}$, using either restricted universal quantifiers (RUQ) or restricted existentially quantifiers (REQ). The definition and usage of RUQ and REQ in $\{log\}$ will be discussed in Sect. 4.4. It is worth noting that fresh variables occurring in the body of a clause but not in its head can be seen as implicitly existentially quantified inside the clause body.

$\mathtt{neg}$ computes the *propositional* negation of its argument. In particular, if $\mathcal{F}$ is an atomic constraint, $\mathtt{neg}(\mathcal{F})$ returns the corresponding negated constraint. For example, $\mathtt{neg}(x \mathbin{\mathtt{in}} A \mathbin{\&} z \mathbin{\mathtt{nin}} C)$ becomes $x \mathbin{\mathtt{nin}} A \mathbin{\mathtt{or}} z \mathbin{\mathtt{in}} C$. However, the result of $\mathtt{neg}$ is not always correct because, in general, the negated formula may involve existentially quantified variables, whose negation calls into play (general) universal quantification that $\{log\}$ cannot handle properly. Existentially quantified variables may appear in the bodies of clauses or in the formula part of RIS and RUQ. Hence, there are cases where $\{log\}$ users must manually compute the negation of some formulas.

The same may happen for some logical connectives, such as $\mathtt{implies}$, whose implementation uses the predicate $\mathtt{neg}$.[6]

2.5   Types in $\{log\}$

$\{log\}$ is an untyped formalism. This means that, for example, a set such as $\{(x, y), 1, \text{'}red\text{'}\}$ is accepted by the language. However, recently, a type system and a type checker have been defined and implemented in $\{log\}$. Typed as well as untyped formalisms have advantages and disadvantages [22]. For this reason, $\{log\}$ users can activate and deactivate the typechecker according to their needs.

---

[6] Indeed, $F \mathbin{\mathtt{implies}} G$ is implemented in $\{log\}$ as $\mathtt{neg}(F) \mathbin{\mathtt{or}} G$.

{ *log* } types are defined according to the following grammar:

$$\tau ::= \texttt{int} \mid \texttt{str} \mid Atom \mid \texttt{etype}([Atom, \ldots, Atom]) \mid [\tau, \ldots, \tau] \mid \texttt{stype}(\tau)$$

where *Atom* is any Prolog atom other than `int` and `str`. `int` corresponds to the type of integer numbers; `str` corresponds to the type of Prolog strings; if *atom* $\in$ *Atom*, then it defines the type given by the set $\{atom?t \mid t$ is a Prolog atom$\}$, where '?' is a functor of arity 2. $\texttt{etype}([c_1, \ldots, c_n])$, with $2 \leq n$, defines an enumerated type; $[T_1, \ldots, T_n]$ with $2 \leq n$ defines the Cartesian product of types $T_1, \ldots, T_n$; and $\texttt{stype}(T)$ defines the powerset type of type T. This type system is similar to B's.

When in typechecking mode, all variables and user-defined predicates must be declared to be of a precise type. Variables are declared by means of the $\texttt{dec}(V, \tau)$ constraint, meaning that variable V is of type τ. In this mode, every { *log* } atomic constraint has a polymorphic type much as in the B notation. For example, in $\texttt{comp}(R, S, T)$ the type of the arguments must be $\texttt{stype}([\tau_1, \tau_2])$, $\texttt{stype}([\tau_2, \tau_3])$ and $\texttt{stype}([\tau_1, \tau_3])$, for some types τ_1, τ_2, τ_3, respectively.

Concerning user-defined predicates, if the head of a predicate is $\texttt{p}(X_1, \ldots, X_n)$, then a predicate of the form $\texttt{:- dec\_p\_type}(\texttt{p}(\tau_1, \ldots, \tau_n))$, where $\tau_1, \ldots, \tau_n$ are types, must precede p's definition. This is interpreted by the typechecker as X_i is of type τ_i in p, for all $i \in 1 \ldots n$. See an example in Sect. 4.1.

More details on types in { *log* } can be found in the user's manual [27].

2.6 Constraint Solving

As concerns constraint solving, the { *log* } solver repeatedly applies specialized rewriting procedures to its input formula Φ and returns either **false** or a formula in a simplified form which is guaranteed to be satisfiable with respect to the intended interpretation. Each rewriting procedure applies a few non-deterministic rewrite rules which reduce the syntactic complexity of primitive constraints of one kind. At the core of these procedures is set unification [19]. The execution of the solver is iterated until a fixpoint is reached, i.e., the formula is irreducible.

The disjunction of formulas returned by the solver represents all the concrete (or ground) *solutions* of the input formula. Any returned formula is divided into two parts: the first part is a (possibly empty) list of equalities of the form $X = t$, where X is a variable occurring in the input formula and t is a term; and the second part is a (possibly empty) list of primitive constraints.

3 Using { *log* }

In this section we show examples on how { *log* } can be used as a programming language (3.1) and as an automated theorem prover (3.2). The main goal is to provide examples of the *formula-program duality* enjoined by { *log* } code.

3.1   {*log*} as a Programming Language

{*log*} is primarily a programming language at the intersection of declarative programming, set programming [28] and constraint programming. Specifically, {*log*} is an instance of the general CLP scheme [21]. As such, {*log*} programs are structured as a finite collection of *clauses*, whose bodies can contain both atomic constraints and user-defined predicates. The following example shows the program side of the formula-program duality of {*log*} code along with the notion of clause.

Example 1. The following clause corresponds to the Start_GearExtend event of the Doors machine of the Event-B model of the LGS. It takes four arguments and the body is a {*log*} formula.

start_GearExtend(*PositionsDG*, *Gear_ret_p*, *Door_open_p*, *Gear_ret_p_*) :-
   Po in *PositionsDG* &
   applyTo(*Gear_ret_p*, *Po*, *true*) &
   Door_open_p = cp(*PositionsDG*, {*true*}) &
   foplus(*Gear_ret_p*, *Po*, *false*, *Gear_ret_p_*).

Note that *Po* is an existential variable. applyTo and foplus are predicates defin-able in terms of the atomic constraints provided by $\mathcal{BR}$. For example, applyTo is as follows:

$$applyTo(F, X, Y) :- \hspace{4cm} (1)$$
$$F = \{[X, Y]/G\} \,\&\, [X, Y]\, \text{nin}\, G \,\&\, \text{comp}(\{[X, X]\}, G, \{\}).$$

Now we can call start_GearExtend by providing inputs and waiting for outputs:

Pos = {*front*, *right*, *left*} &
start_GearExtend(*Pos*, cp(*Pos*, {*true*}), cp(*Pos*, {*true*}), *Gear_ret_p_*).

returns:

$$Gear\_ret\_p\_ = \{[\mathit{front}, \mathit{false}]/\text{cp}(\{\mathit{right}, \mathit{left}\}, \{\mathit{true}\})\}$$

As a programming language, {*log*} can be used to implement set-based spec-ifications (e.g., B or Z specifications). As a matter of fact, an industrial-strength Z specification has been translated into {*log*} [12] and students of a course on Z taught both at Rosario (Argentina) and Parma (Italy) use {*log*} as the proto-typing language for their Z specifications [14]. This means that {*log*} can serve as a programming language in which a *prototype* of a set-based specification can be easily implemented. In a sense, the {*log*} implementation of a set-based specification can be seen as an *executable specification*.

Remark 1. A {*log*} implementation of a set-based specification is easy to get but usually it will not meet the typical performance requirements demanded by users. Hence, we see a {*log*} implementation of a set-based specification more as a *prototype* than as a final program. On the other hand, given the similarities between a specification and the corresponding {*log*} program, it is reasonable to think that the prototype is a *correct* implementation of the specification[7].

For example, in a set-based specification a table (in a database) with a primary key is usually modeled as a partial function, $t : X \nrightarrow Y$. Furthermore, one may specify the update of row r with data d by means of the *oplus* or *override* ($\oplus$) operator: $t' = t \oplus \{r \mapsto d\}$. All this can be easily and naturally translated into {*log*}: t is translated as variable T constrained to verify $\mathtt{pfun}(T)$, and the update specification is translated as $\mathtt{oplus}(T, \{[R, D]\}, T_-)$. However, the $\mathtt{oplus}$ constraint will perform poorly compared to the $\mathtt{update}$ command of SQL, given that $\mathtt{oplus}$'s implementation comprises the possibility to operate in a purely logical manner with it (e.g., it allows to compute $\mathtt{oplus}(T, \{[a, D]\}, \{[a, 1], [b, 3]\})$ while $\mathtt{update}$ does not). □

Then, we can use these prototypes to make an early validation of the requirements. Validating user requirements by means of prototypes entails executing the prototypes together with the users so they can agree or disagree with the behavior of the prototypes. This early validation will detect many errors, ambiguities and incompleteness present in the requirements and possible misunderstandings or misinterpretations generated by the software engineers. Without this validation many of these issues would be detected in later stages of the project thus increasing the project costs.

3.2   {*log*} as an Automated Theorem Prover

{*log*} is also a *satisfiability solver*. This means that {*log*} is a program that can decide if formulas of some theory are *satisfiable* or not. In this case the theory is the combination of the decidable (fragments of the) theories of Fig. 1.

Being a satisfiability solver, {*log*} can be used as an automated theorem prover and as a counterexample generator. To prove that formula ϕ is a theorem, {*log*} has to be called to prove that $\neg \phi$ is unsatisfiable.

Example 2. We can prove that set union is commutative by asking {*log*} to prove the following is unsatisfiable:

$$\mathtt{neg}(\mathtt{un}(A, B, C) \,\&\, \mathtt{un}(B, A, D) \,\mathtt{implies}\, C = D).$$

{*log*} first applies $\mathtt{neg}$ to the formula, returning:

$$\mathtt{un}(A, B, C) \,\&\, \mathtt{un}(B, A, D) \,\&\, C \,\mathtt{neq}\, D$$

As there are no sets satisfying this formula, {*log*} answers **no**. Note that the initial formula can also be written as: $\mathtt{neg}(\mathtt{un}(A, B, C) \,\mathtt{implies}\, \mathtt{un}(B, A, C))$.

---

[7] In fact, the translation process can be automated in many cases.

In this case, the result of applying **neg** uses the **nun** constraint, $\text{un}(A, B, C)$ & $\text{nun}(B, A, C)$. □

When $\{log\}$ fails to prove that a certain formula is unsatisfiable, it generates a counterexample. This is useful at early stages of the verification phase because it helps to find mismatches between the specification and its properties.

Example 3. If the following is run on $\{log\}$:

$$\text{neg}(\text{un}(A, BBBB, C) \ \& \ \text{un}(B, A, D) \ \texttt{implies} \ C = D).$$

the tool will provide the following as the first counterexample:

$$C = \{\_N3/\_N1\}, \_N3 \, \texttt{nin} \, D, \text{un}(A, \_N2, \_N1), \dots$$

More counterexamples can be obtained interactively. □

Remark 2. As we have mentioned in Sect. 2.4, there are cases where $\{log\}$'s **neg** predicate is not able to correctly compute the negation of its argument. For instance, if we want to compute the negation of `applyTo` as defined in Example 1 with **neg**, the result will be wrong—basically due to the presence of the existentially quantified variable G. In that case, the user has to compute and write down the negation manually.

Evaluating properties with $\{log\}$ helps to run correct simulations by checking that the starting state is correctly defined. It also helps to *test* whether or not certain properties are true of the specification or not. However, by exploiting the ability to use $\{log\}$ as a theorem prover, we can *prove* that these properties are true of the specification. In particular, $\{log\}$ can be used to automatically discharge verification conditions in the form of state invariants. Precisely, in order to prove that state transition T (from now on called *operation*) preserves state invariant I the following proof must be discharged:

$$I \wedge T \Rightarrow I' \tag{2}$$

where I' corresponds to the invariant in the next state. If we want to use $\{log\}$ to discharge (2) we have to ask $\{log\}$ to check if the negation of (2) is *unsatisfiable*. In fact, we need to execute the following $\{log\}$ *query*:

$$\text{neg}(I \ \& \ T \ \texttt{implies} \ I') \tag{3}$$

As we have pointed out in Example 2, $\{log\}$ rewrites (3) as:

$$I \ \& \ T \ \& \ \text{neg}(I') \tag{4}$$

Example 4. The following is the $\{log\}$ encoding of the state invariant labeled inv2 in machine Doors of the LGS:

```
doors_inv2(PositionsDG, Gear_ext_p, Gear_ret_p, Door_open_p) :-
    exists(Po in PositionsDG,
        applyTo(Gear_ext_p, Po, false) & applyTo(Gear_ret_p, Po, false)
    ) implies   Door_open_p = cp(PositionsDG, {true}).
```

Besides, Rodin generates a proof obligation as (2) for `start_GearExtend` and `doors_inv2`. Then, we can discharge that proof obligation by calling $\{log\}$ on its negation:

neg(doors_inv2($PosDG, Gear\_ext\_p, Gear\_ret\_p, Door\_open\_p$) &

      start_GearExtend($PosDG, Gear\_ret\_p, Door\_open\_p, Gear\_ret\_p\_$)

      implies doors_inv2($PosDG, Gear\_ext\_p, Gear\_ret\_p\_, Door\_open\_p$))

The consequent corresponds to the invariant evaluated in the next state due to the presence of $Gear\_ret\_p\_$ instead of $Gear\_ret\_p$ in the third argument—the other arguments are not changed by `start_GearExtend`.   □

Examples 1 and 4 show that $\{log\}$ is a programming and proof platform exploiting the program-formula duality within the theories of Fig. 1. Indeed, in Example 1 `start_GearExtend` is treated as a program (because it is executed) while in Example 4 is treated as a formula (because it participates in a theorem).

4 Encoding the Event-B Specification of the LGS in $\{log\}$

We say 'encoding' and not 'implementing' the Landing Gear System (LGS) because the resulting $\{log\}$ code: *a)* is a prototype rather than a production program; and *b)* is a formula as the LGS specification is. In particular, $\{log\}$ provides all the logical, set and relational operators used in the LGS specification. Furthermore, these operators are not mere imperative implementations but real mathematical definitions enjoying the formula-program duality discussed in Sect. 3.

Since the $\{log\}$ program verifies the properties proposed by Rodin (see Sect. 5), we claim the prototype is a faithful encoding of the Event-B specification.

Due to space considerations we are not going to explain in detail the Event-B development of Mammar and Laleau. Instead, we will provide some examples of how we have encoded it in $\{log\}$. The interested reader can first take a quick read to the problem description [5], and then download the Event-B development[8] and the $\{log\}$ program[9] in order to make a thorough comparison. The Event-B development of the LGS consists of eleven models organized in a refinement pipeline [25]. Each model specifies the behavior and state invariants of increasingly complex and detailed versions of the LGS. For example, the first and simplest model specifies the gears of the LGS; the second one adds the doors that allow the gears to get out of the aircraft; the third one adds the hydraulic cylinders that either open (extend) or close (retract) the doors (gears); and so on and so forth. In this development each Event-B model consists of one Event-B machine.

In the following subsections we show how the main features of the Event-B model of the LGS are encoded in $\{log\}$.

---

[8] http://deploy-eprints.ecs.soton.ac.uk/467.

[9] http://www.clpset.unipr.it/SETLOG/APPLICATIONS/lgs.zip.

4.1   Encoding Event-B Machines

Figure 2 depicts at the left the Event-B machine named Gears and at the right the corresponding {*log*} encoding. From here on, {*log*} code is written in `typewriter` font. We tried to align as much as possible the {*log*} code w.r.t. the corresponding Event-B code so the reader can compare both descriptions line by line. The {*log*} code corresponding to a given Event-B machine is saved in a file with the name of the machine (e.g., `gears.pl`).

As can be seen in the figure, each invariant, the initialization predicate and each event are encoded as {*log*} clauses. In Event-B the identifiers for each of these constructs can be any word but in {*log*} clause predicates must begin with a lowercase letter. For instance, event Make_GearExtended corresponds to the {*log*} clause predicate named `make_GearExtended`[10].

Each clause predicate receives as many arguments as variables and constants are used by the corresponding Event-B construct. For example, `inv1` waits for two arguments: `PositionsDG`, corresponding to a set declared in the Event-B context named PositionsDoorsGears (not shown); and `Gear_ext_p`, corresponding to the state variable declared in the machine. When an event changes the state of the machine, the corresponding {*log*} clause predicate contains as many more arguments as variables the event modifies. For example, Make_GearExtended modifies variable $gear\_ext\_p$, so `make_GearExtended` has `Gear_ext_p_` as the third argument. `Gear_ext_p_` corresponds to the value of `Gear_ext_p` in the next state; i.e., the value of $gear\_ext\_p$ after considering the assignment $gear\_ext\_p(po) := TRUE$. Observe that in {*log*} variables must begin with an uppercase letter and constants with a lowercase letter, while such restrictions do not apply in Event-B. For example, in the Event-B model we have the Boolean constant $TRUE$ which in our encoding is written as `true`. Likewise, in Event-B we have variable $gear\_ext\_p$ which is encoded as `Gear_ext_p`.

In Fig. 2 we have omitted type declarations for brevity. We will include them only when strictly necessary. For instance, the {*log*} clause `inv1` is actually preceded by its type declaration:

```
:- dec_p_type(inv1(stype(positionsdg),stype([positionsdg,bool]))).
```

where `positionsdg` is a synonym for `etype([front,right,left])` and `bool` is for `etype([true,false])`.

Guards and actions are encoded as {*log*} predicates. The identifiers associated to guards and actions can be provided as comments. {*log*} does not provide language constructs to label formulas. As an alternative, each guard and action can be encoded as a clause named with the Event-B identifier. These clauses are then assembled together in clauses encoding events. More on the encoding of actions in Sect. 4.3.

---

[10] Actually, the name of each clause predicate is prefixed with the name of the machine. Then, is `gears_make_GearExtended` rather than `make_GearExtended`. We omit the prefix whenever it is clear from context.

MACHINE Gears
SEES PositionsDoorsGears
VARIABLES gear_ext_p
INVARIANTS
  INV1: $gear\_ext\_p \in PositionsDG \rightarrow BOOL$
EVENTS
Initialisation
  begin
    ACT1: $gear\_ext\_p := PositionsDG \times \{TRUE\}$
  end
Event Make_GearExtended ≙
any
  po
where
  GRD1: $po \in PositionsDG \land gear\_ext\_p(po) = FALSE$
then
  ACT1: $gear\_ext\_p(po) := TRUE$
end
Event Start_GearRetract ≙
any
  po
where
  GRD1: $po \in PositionsDG \land gear\_ext\_p(po) = TRUE$
then
  ACT1: $gear\_ext\_p(po) := FALSE$
end
END

```
inv1(PositionsDG,Gear_ext_p) :-
  pfun(Gear_ext_p) & dom(Gear_ext_p,PositionsDG).

init(PositionsDG,Gear_ext_p) :-
  Gear_ext_p = cp(PositionsDG,{true}).

make_GearExtended(PositionsDG,Gear_ext_p,Gear_ext_p_) :-
  Po in PositionsDG &
  applyTo(Gear_ext_p,Po,false) &
  foplus(Gear_ext_p,Po,true,Gear_ext_p_).

start_GearRetract(PositionsDG,Gear_ext_p,Gear_ext_p_) :-
  Po in PositionsDG &
  applyTo(Gear_ext_p,Po,true) &
  foplus(Gear_ext_p,Po,false,Gear_ext_p_).
```

Fig. 2. The Event-B Gears machine at the left and its $\{log\}$ encoding at the right

4.2   Encoding (Partial) Functions

Functions play a central role in Event-B specifications. In Event-B functions are sets of ordered pairs; i.e., a function is a particular kind of binary relation. {*log*} supports functions as sets of ordered pairs and supports all the related operators. Here we show how to encode functions and their operators.

Function Definition. In order to encode functions, we use a combination of types and constraints. In general, we try to encode as much as possible with types, as typechecking performs better than constraint solving. Hence, to encode $f \in X \nrightarrow Y$, we declare it as dec(F,stype([X,Y])) and then we assert pfun(F). If f is a total function, then we conjoin dom(F,D), where D is the set representation of the domain type.

For instance, predicate INV1 in Gears asserts $gear\_ext\_p \in PositionsDG \rightarrow BOOL$ (Fig. 2). In {*log*} we declare Gear_ext_p to be of type stype([positionsdg,bool]) (see the dec_p_type predicate in Sect. 4.1). This type declaration only ensures $gear\_ext\_p \in PositionsDG \leftrightarrow BOOL$—i.e., a binary relation between *PositionsDG* and *BOOL*. The type assertion is complemented by a constraint assertion: pfun(Gear_ext_p) & dom(Gear_ext_p,PositionsDG), as explained above.

Function Application. Function application is encoded by means of the applyTo predicate defined in (1). The encoding is a little bit more general than Event-B's notion of function application. For instance, in Make_GearExtended (Fig. 2):

$$gear\_ext\_p(po) = FALSE \qquad (5)$$

might be undefined because Event-B's type system can only ensure $gear\_ext\_p \in PositionsDG \leftrightarrow BOOL$ and $po \in PositionsDG$. Actually, the following well-definedness proof obligation is required by Event-B:

$$po \in \mathrm{dom}\, gear\_ext\_p \wedge gear\_ext\_p \in PositionsDG \nrightarrow BOOL$$

As said, we encode (5) as applyTo(Gear_ext_p,Po,false). applyTo cannot be undefined but it can fail because: *a)* Po does not belong to the domain of Gear_ext_p; *b)* Gear_ext_p contains more than one pair whose first component is Po; or *c)* false is not the image of Po in Gear_ext_p. Then, by discharging the proof obligation required by Event-B we are sure that applyTo will not fail due to *a)* and *b)*. Actually, $gear\_ext\_p \in PositionsDG \nrightarrow BOOL$ is unnecessarily strong for function application. As applyTo suggests, $f(x)$ is meaningful when f is *locally functional* on x. For example, $\{x \mapsto 1, y \mapsto 2, y \mapsto 3\}(x)$ is well-defined in spite that the set is not a function.

Encoding Membership to Dom. In different parts of the Event-B specification we find predicates such as $po \in$ dom $gear\_ext\_p$. There is a natural way of encoding this in $\{log\}$: dom(Gear_ext_p,D) & Po in D. However, we use an encoding based on the ncomp constraint because it turns out to be more efficient in $\{log\}$: ncomp({[Po,Po]},Gear_ext_p,{}). ncomp is the negation of the comp constraint (composition of binary relations). Then, ncomp({[X,X]},F,{}) states that $\{(X, X)\} \, \mathring{,}\, F \neq \emptyset$ which can only hold if there is a pair in F of the form $(X, \_)$. Then, X is in the domain of F.

4.3 Encoding Action Predicates

Action predicates describe the state change performed due to an event or the value of the initial state. Next, we show the encoding of two forms of action predicates: simple assignment and functional override. In events, the next state variable is implicitly given by the variable at the left of the action predicate. In $\{log\}$ we must make these variables explicit.

Encoding Simple Assignments. Consider a simple assignment $x := E$. If this is part of the initialization, we interpret it as $x = E$; if it is part of an event, we interpret it as $x' = E$. As we have said in Sect. 4.1, x' is encoded as X_. Hence, simple assignments are basically encoded as equalities to before-state variables in the case of initialization and to after-state variables in the case of events.

As an example, part of the initialization of the Gears machine is:

$$\text{ACT1} : gear\_ext\_p := PositionsDG \times \{TRUE\} \tag{6}$$

This is simply encoded in $\{log\}$ as follows:

Gear_ext_p = cp(PositionsDG,{true})

An example of a simple assignment in an event, can be the following (event ReadDoors, machine Sensors):

ACT1: $door\_open\_ind :=$
$\{front \mapsto door\_open\_sensor\_valueF,$
$left \mapsto door\_open\_sensor\_valueL,$
$right \mapsto door\_open\_sensor\_valueR\}$

Then, we encode it as follows:

Door_open_ind_ =
  {[front,Door_open_sensor_valueF],
   [left,Door_open_sensor_valueL],
   [right,Door_open_sensor_valueR]}

Functional Override. The functional override $f(x) := E$ is interpreted as $f' = f \oplus \{x \mapsto E\}$, which in turn is equivalent to $f' = f \setminus \{x \mapsto y \mid y \in \text{ran} f\} \cup \{x \mapsto E\}$. This equality is encoded by means of the `foplus` constraint:

```
foplus(F,X,Y,G) :-
  F = {[X,Z]/H} & [X,Z] nin H & comp({[X,X]},H,{}) & G = {[X,Y]/H}
  or    comp({[X,X]},F,{}) & G = {[X,Y]/F}.
```

That is, if there is more than one image of X through F, `foplus` fails. Then, `foplus(F,X,Y,G)` is equivalent to:

$$G = F \setminus \{(X, Z)\} \cup \{(X, Y)\}, \text{ for some } Z$$

This is a slight difference w.r.t. the Event-B semantics but it is correct in a context where F is intended to be a function (i.e., $\text{pfun}(F)$ is an invariant).

As an example, consider the action part of Make_GearExtended (Fig. 2): $gear\_ext\_p(po) := TRUE$. As can be seen in the same figure, the encoding is: `foplus(Gear_ext_p,Po,true,Gear_ext_p_).`

The following is a more complex example appearing in event Start_ GearExtend of machine TimedAspects:

$$\text{ACT2: } deadline GearsRetractingExtending(po) :=$$
$$\{front \mapsto cT + 12, left \mapsto cT + 16, right \mapsto 16\}(po)$$

The encoding is the following:

```
foplus(DeadlineGearsRetractingExtending,
       Po,M1,DeadlineGearsRetractingExtending_) &
applyTo({[front,M2],[left,M3],[right,16]},Po,M1) &
M2 is CT + 12 & M3 is CT + 16
```

Given that in $\{log\}$ function application is a constraint, we cannot put in the third argument of `foplus` an expression encoding $\{front \mapsto cT + 12, \dots\}(po)$. Instead, we have to capture the result of that function application in a new variable (M1) which is used as the third argument. Along the same lines, it would be wrong to write CT + 12 in place of M2 in `applyTo({[front,M2],...},Po,M1)` because $\{log\}$ (as Prolog) does not interpret integer expressions unless explicitly indicated. Precisely, the `is` constraint forces the evaluation of the integer expression at the right-hand side.

4.4 Encoding Quantifiers

The best way of encoding quantifiers in $\{log\}$ is by means of the `foreach` and `exists` constraints, which implement RUQ and REQ. Unrestricted existential quantification is also supported. Then, if an Event-B universally quantified formula cannot be expressed as a RUQ formula, it cannot be expressed in $\{log\}$.

Universal Quantifiers. In its simplest form, the RUQ formula $\forall\, x \in A : \phi$ is written in $\{log\}$ as foreach(X in A, ϕ) [11, Sect. 5.1][11]. However, foreach can also receive four arguments: foreach(X in A, [vars], ϕ, ψ), where ψ is a conjunction of *functional predicates*. Functional predicates play the role of LET expressions; that is, they permit to define a name for an expression [11, Sect. 6.2]. In turn, that name must be one of the variables listed in *vars*. These variables are implicitly existentially quantified inside the foreach. A typical functional predicate is applyTo(F,X,Y) because there is only one Y for given F and X. Functional predicates can be part of ϕ but in that case its negation will not always correct.

As an example, consider the second invariant of machine GearsIntermediate:

$$\text{INV2} : \forall\, po \cdot (po \in PositionsDG$$
$$\Rightarrow \neg\, (gear\_ext\_p(po) = TRUE \wedge gear\_ret\_p(po) = TRUE))$$

This is equivalent to a restricted universal quantified formula:

$$\text{INV2} : \forall\, po \in PositionsDG \cdot$$
$$\neg\, (gear\_ext\_p(po) = TRUE \wedge gear\_ret\_p(po) = TRUE) \quad (7)$$

Then, we encode (7) as follows:

```
inv2(PositionsDG, Gear_ext_p, Gear_ret_p) :-
  foreach(Po in PositionsDG,[M1,M2],
    neg(M1 = true & M2 = true),
    applyTo(Gear_ext_p,Po,M1) & applyTo(Gear_ret_p,Po,M2)).
```

In some quantified formulas of Mammar and Laleau's Event-B project the restricted quantification is not as explicit as in (7). For instance, INV3 of the Cylinders machine is:

$$\forall\, po \cdot door\_cylinder\_locked\_p(po) = TRUE \Rightarrow door\_closed\_p(po) = TRUE$$

However, given that *po* must be in the domain of *door_closed_p* then it must belong to *PositionsDG*.

Existential Quantifiers. Like universal quantifiers, also existential quantifiers are encoded by first rewriting them as REQ, and then by using $\{log\}$'s exists constraint. However, $\{log\}$ supports also general (i.e., unrestricted) existential quantification: in fact, all variables occurring in the body of a clause but not in its head are implicitly existentially quantified. Thus, the Event-B construct **any** is not explicitly encoded. For instance, in the encoding of event Make_GearExtended we just state Po in PositionsDG because this declares Po as an existential variable.

---

[11] In turn, the foreach constraint is defined in $\{log\}$ by using RIS and the $\subseteq$ constraint, by exploiting the equivalence $\forall\, x \in D : \mathcal{F}(x) \Leftrightarrow D \subseteq \{x \in D \mid \mathcal{F}(x)\}$.

4.5   Encoding Types

In Event-B type information is sometimes given as membership constraints. Besides, type information sometimes becomes what can be called *type invariants*. That is, state invariants that convey what normally is typing information. As we have said, we try to encode as much as possible with $\{log\}$ types rather than with constraints. Some type invariants can be enforced by the typechecker. Nevertheless, for the purpose of using this project as a benchmark for $\{log\}$, we have encoded all type invariants as regular invariants so we can run all the proof obligations involving them.

Type Guards and Type Invariants. For example, in event ReadHandleSwitchCircuit (machine Failures) we can find the guard:

$Circuit\_pressurized\_sensor\_value \in BOOL$

Guards like this are encoded as actual type declarations and not as constraints:

```
dec(Circuit_pressurized_sensor_value,bool)
```

In this way, the typechecker will reject any attempt to bind this variable to something different from `true` and `false`.

As an example of a type invariant we have the following one in machine HandleSwitchShockAbsorber:

INV5: $Intermediate1 \in BOOL$

This is an invariant that can be enforced solely by the typechecker. However, for the purpose of the empirical comparison, we encoded it as a state invariant:

```
:- dec_p_type(inv5(bool)).
inv5(Intermediate1) :- Intermediate1 in {true,false}.
```

Encoding the Set of Natural Numbers ($\mathbb{N}$). The Event-B model includes type invariants involving the set of natural numbers. This requires to be treated with care because $\{log\}$ can only deal with finite sets so there is no set representing $\mathbb{N}$. As these invariants cannot be enforced solely by the typechecker we need to combine types and constraints. Then, an invariant such as:

INV1: $currentTime \in \mathbb{N}$

is encoded by typing the variable as an integer and then asserting that it is non-negative:

```
:- dec_p_type(inv1(int)).
inv1(CurrentTime) :- 0 =< CurrentTime.
```

There are, however, more complex invariants involving $\mathbb{N}$ such as:

$$deadline\,UnlockLockDoorsCylinders \in PositionsDG \rightarrow \mathbb{N}$$

In this case $\{log\}$'s type system can only ensure:

$$deadline\,UnlockLockDoorsCylinders \in PositionsDG \leftrightarrow \mathbb{Z} \qquad (8)$$

Then, we need to use constraints to state: *a)* *deadline UnlockLockDoorsCylinders* is a function; *b)* the domain is *PositionsDG*; and *c)* the range is a subset of $\mathbb{N}$. The first two points are explained in Sect. 4.2. The last one is encoded with the foreach constraint:

```
foreach([X,Y] in DeadlineUnlockLockDoorsCylinders, 0 =< Y)
```

As can be seen, the control term of the foreach predicate can be an ordered pair [11, Sect. 6.1]. Then, the complete encoding of (8) is the following:

```
:- dec_p_type(inv5(stype(positionsdg),stype([positionsdg,int]))).
inv5(PositionsDG,DeadlineUnlockLockDoorsCylinders) :-
  pfun(DeadlineUnlockLockDoorsCylinders) &
  dom(DeadlineUnlockLockDoorsCylinders,PositionsDG) &
  foreach([X,Y] in DeadlineUnlockLockDoorsCylinders, 0 =< Y).
```

5   Encoding Proof Obligations in $\{log\}$

Rodin is one of the tools used by Mammar and Leleau in the project. Rodin automatically generates a set of proof obligations for each model according to the Event-B verification rules [4]. Among the many kinds of proof obligations defined in Event-B, in the case of the LGS project Rodin generates proof obligations of the following three kinds:

- Well-definedness (WD). A WD condition is a predicate describing when an expression or predicate can be safely evaluated. For instance, the WD condition for x div y is $y \neq 0$. Then, if x div y appears in some part of the specification, Rodin will generate a proof obligation asking for $y \neq 0$ to be proved in a certain context.
- Invariant initialization (INIT). Let I be an invariant depending on variable x. Let $x := V$ be the initialization of variable x. Then, Rodin generates the following proof obligation[12]:

$$\vdash I[V/x] \qquad (9)$$

- Invariant preservation (INV). Let I be as above. Let $E \mathrel{\widehat{=}} G \wedge A$ be an event where G are the preconditions (called *guards* in Event-B) and A are the postconditions (called *actions* in Event-B). Say E changes the value of x;

---

[12] This is a simplification of the real situation which, nonetheless, captures the essence of the problem. All the technical details can be found in the Event-B literature [4].

e.g., there is an assignment such as $x := V$ in E. Then, Rodin generates the following proof obligation:

$$\mathcal{I} \wedge I \wedge G \wedge A \vdash I[V/x] \tag{10}$$

where $\mathcal{I} \triangleq \bigwedge_{j \in J} I_j$ is a conjunction of invariants in scope other than I.

Dealing with INV proof obligations deserves some attention. When confronted with a *mechanized* proof we can think on two general strategies concerning the hypothesis for that proof:

- The interactive strategy. When proving a theorem interactively, the more the hypothesis are available, the easiest the proof. In other words, users of an interactive theorem prover will be happy to have as many hypothesis as possible.
- The automated strategy. When using an automated theorem prover, some hypothesis can be harmful. Given that automated theorem provers do not have the intelligence to chose (only) the right hypothesis in each proof step, they might chose hypothesis that do not lead to the conclusion or that produce a long proof path. Therefore, a possible working strategy is to run the proof with just the necessary hypothesis.

Our approach is to encode proof obligations by following the 'automated strategy'. For example, in the case of INV proofs, we first try (10) *without* $\mathcal{I}$. If the proof fails, the counterexample returned by $\{log\}$ is analyzed, just the necessary I_j are added, and the proof is attempted once more. Furthermore, in extreme cases where a proof is taking too long, parts of G and A are dropped to speed up the prover. As we further discuss it in Sect. 6, this proof strategy considerably reduces the need for truly interactive proofs.

The combination between typechecking and constraint solving also helps in improving proof automation. As we have explained in Sect. 4, the encoding of INV1 in Fig. 2 does not include constraints to state that the range of $gear\_ext\_p$ is a subset of $BOOL$. This is so because this fact is enforced by the typechecker. Hence, the $\{log\}$ encoding of INV1 is simpler than INV1 itself. As a consequence, proof obligations involving INV1 will be simpler, too. For example, the INV proof for INV1 and event Make_GearExtended (of machine Gears) is the following:

$$gear\_ext\_p \in PositionsDG \rightarrow BOOL \wedge \text{Make\_GearExtended}$$
$$\Rightarrow gear\_ext\_p' \in PositionsDG \rightarrow BOOL \tag{11}$$

However, the $\{log\}$ encoding of the negation of (11) is the following[13]:

```
neg(pfun(Gear_ext_p) & dom(Gear_ext_p, PositionsDG) &
    make_GearExtended(PositionsDG, Gear_ext_p, Gear_ext_p_)
    implies pfun(Gear_ext_p_) & dom(Gear_ext_p_, PositionsDG)).
```

That is, $\{log\}$ will not have to prove that:

---

[13] Predicate inv1 of Fig. 2 is expanded to make the point more evident.

```
ran(Gear_ext_p,M) & subset(M,{true,false})
```

is an invariant because this is ensured by the typechecker.

PositionsDG is a set declared in the context named PositionsDoorsGears where it is bound to the set $\{front, right, left\}$. *PositionsDG* is used by the vast majority of events and thus it participates in the vast majority of proof obligations. In the $\{log\}$ encoding, instead of binding *PositionsDG* to that value we leave it free unless it is strictly necessary for a particular proof. For example, the proof obligation named ReadDoors/grd/WD (machine Sensors) cannot be discharged if *left* is not an element of *PositionsDG*. Hence, we encoded that proof obligation as follows:

```
ReadDoors_grd2 :-
  neg(PositionsDG = {left / M} &
      doors_inv1(PositionsDG, Door_open_p) &
      Door_open_sensor_valueL = true
      implies
        ncomp({[left,left]},Door_open_p,{}) & pfun(Door_open_p)).
```

Note that: we state the membership of only `left` to `PositionsDG`; INV1 of machine Doors is needed as an hypothesis; the encoding based on `ncomp` is used to state membership to dom; and `pfun` ensures that `Door_open_p` is a function while the type system ensures its domain and range are correct (not shown).

Furthermore, binding *PositionsDG* to $\{front, right, left\}$ is crucial in a handful of proof obligations because otherwise the encoding would fall outside the decision procedures implemented in $\{log\}$. One example is passing-Time/grd7/WD (machine TimedAspects):

$$
\begin{aligned}
&\text{ran } deadlineOpenCloseDoors \neq \{0\} \\
&\Rightarrow \text{ran}(deadlineOpenCloseDoors \rhd \{0\}) \neq \emptyset \\
&\quad \wedge (\exists\, b \in \text{ran}(deadlineOpenCloseDoors \rhd \{0\}) \cdot \\
&\qquad (\forall\, x\, \text{ran}(deadlineOpenCloseDoors \rhd \{0\}) \cdot b \leq x))
\end{aligned}
\tag{12}
$$

As can be seen, the quantification domain of both the REQ and RUQ is the same, the REQ is before the RUQ and the REQ is at the consequent of an implication. This kind of formulas lies outside the decision procedures implemented in $\{log\}$ unless the quantification domain is a closed set—such as $\{front, right, left\}$.

6   Discussion and Comparison

The LGS Event-B project of Mammar and Leleau comprises 4.8 KLOC[14] of LATEX code which amounts to 213 Kb. The $\{log\}$ encoding is 7.8 KLOC long weighting 216 Kb. Although $\{log\}$'s LOC are quite more than the encoding of the specification in LATEX, we would say both encodings are similar in size—we tend to use very short lines. Beyond these numbers, it is worth noting that

---

[14] LOC stands for Lines Of Code.

several key state variables of the Event-B specification are Boolean functions. We wonder why the authors used them instead of sets because this choice would have implied less proof obligations—for instance, many WD proofs would not be generated simply because function application would be absent.

Table 1 summarizes the verification process carried out with $\{log\}^{15}$. Each row shows the proof obligations generated by Rodin and discharged by $\{log\}$ for each refinement level (machine). The meaning of the columns is as follows: PO stands for the total number of proof obligations; INIT, WD and INV is the number of each kind of proof obligations; and TIME shows the computing time (in seconds) needed to discharge those proof obligations (ϵ means the time is less than one second). As can be seen, all the 465 proofs are discharged, roughly, in 290 s, meaning 0.6 s in average. Mammar and Leleau [25, Sect. 10.1] refer that there are 285 proof obligations, of which 72% were automatically discharged. We believe that this number corresponds to the INV proofs—i.e., those concerning with invariant preservation. Then, Mammar and Leleau had to work out 80 proof obligations interactively in spite of Roding using external provers such as Atelier B and SMT solvers.

Table 1. Summary of the verification process

MACHINE	PO	INIT	WD	INV	TIME
Gears	5	1	2	2	ϵ
GearsIntermediate	13	2	5	6	ϵ
Doors	10	2	2	6	ϵ
DoorsIntermediate	13	2	5	6	ϵ
Cylinders	37	5	14	18	ϵ
HandleSwitchShockAbsorber	29	5	4	20	ϵ
ValvesLights	12	2	2	8	ϵ
Sensors	52	12	17	23	55 s
TimedAspects	98	14	34	50	4 s
Failures	38	9	9	20	3 s
PropertyVerification	158	27	1	130	228 s
TOTALS	465	81	95	289	290 s

The figures obtained with $\{log\}$ for the LGS are aligned with previous results concerning the verification of a $\{log\}$ prototype of the Tokeneer ID Station written from a Z specification [12] and the verification of the Bell-LaPadula security model [10].

---

[15] The verification process was run on a Dell Latitude E7470 with a 4 core Intel(R) Core^{TM} i7-6600U CPU at 2.60GHz with 8 GB of main memory, running Linux Ubuntu 18.04.5, SWI-Prolog 7.6.4 and $\{log\}$ 4.9.8-10i.

There is, however, a proof obligation that, as far as we understand, cannot be discharged. This proof is HandleFromIntermediate2ToIntermediate1/INV2 of the TimedAspects machine. This lemma would prove that event Handle-FromIntermediate2ToIntermediate1 preserves $0 \leq deadlineSwitch$. However, the assignment $deadlineSwitch := currentTime + (8 - (2/3) * (deadlineSwitch - currentTime))$, present in that event, implies that the invariant is preserved iff $deadlineSwitch \leq (5/2) * currentTime + 12$. But there is no invariant implying that inequality. Both $deadlineSwitch$ and $currentTime$ are first declared in the TimedAspects machine which states only two invariants concerning these variables: both of them must be non-negative integers.

Mammar and Leleau use ProB and AnimB besides Rodin during the verification process. They use these other tools in the first refinement steps to try to find obvious errors before attempting any serious proofs. For instance, they use ProB to check that invariants are not trivially falsified. In this sense, the tool is used to find counterexamples for invalid invariants. In general, ProB cannot prove that an invariant holds, it can only prove that it does not hold—as far as we know ProB does not implement a decision procedure for a significant fragment of the set theory underlying Event-B. $\{log\}$ could potentially be used as a back-end system to run the checks carried out by Mammar and Leleau. It should produce more accurate and reliable results as it implements several decision procedures as stated in Sect. 2.

Nevertheless, the main point we would like to discuss is our approach to automated proof. That is, the automated strategy mentioned in Sect. 5 plus some details of the $\{log\}$ encoding of the LGS. Our approach is based on the idea expressed as *specify for automated proof*. In other words, when confronted with the choice between two or more ways of specifying a given requirement, we try to chose the one that improves the chances for automated proof. Sooner or later, automated proof hits a computational complexity wall that makes progress extremely difficult. However, there are language constructs that move that wall further away than others.

For example, the encoding of function application by means of `applyTo`, the encoding of functional override by means of `foplus`, and the encoding of membership to dom by means of `ncomp`, considerably simplify automated proofs because these are significantly simpler than encodings based on other constraints.

However, in our experience, the so-called automated strategy provides the greatest gains regarding automated proof. As we have said, we first try to discharge a proof such as (10) without $\mathcal{I}$. If the proof fails, we analyze the counterexample returned by $\{log\}$ and add a suitable hypothesis—i.e., we pick the right $I_j \in \mathcal{I}$. This process is iterated until the proof succeeds. Clearly, this proof strategy requires some degree of interactivity during the verification process. The question is, then, whether or not this approach is better than attempting to prove (10) as it is and if it does not succeed, an interactive proof assistant is called in. Is it simpler and faster our strategy than a truly interactive proof?

We still do not have strong evidence to give a conclusive answer, although we can provide some data. We have developed a prototype of an interactive proof tool based on the automated strategy [7]. Those results provide evidence for a positive answer to that question. Now, the data on the LGS project further contributes in the same direction. In this project, $\{log\}$ discharges roughly 60% of the proofs of Table 1 in the first attempt. In the vast majority of the remaining proofs only *one* evident $I_j \in \mathcal{I}$ is needed. For example, in many proofs we had to add the invariant $0 \leq currentTime$ as an hypothesis; and in many others an invariant stating that some variable is a function.

In part, our approach is feasible because it is clear what $\{log\}$ can prove and what it cannot, because it implements decision procedures. As a matter of fact, the proof of (12) is a good example about the value of working with decision procedures: users can foresee the behavior of the tool and, if possible, they can take steps to avoid undesired behaviors.

7   Final Remarks

This paper provides evidence that many B specifications can be easily translated into $\{log\}$. This means that $\{log\}$ can serve as a *programming language* in which *prototypes* of those specifications can be immediately implemented. Then, $\{log\}$ itself can be used to automatically prove or disprove that the specifications verify the proof obligations generated by tools such as Rodin. In the case study presented in this paper, $\{log\}$ was able to discharge all such proof obligations.

In turn, this provides evidence that CLP and set theory are valuable tools concerning formal specification, formal verification and prototyping. Indeed $\{log\}$, as a CLP instance, enjoys properties that are hard to find elsewhere. In particular, $\{log\}$ code can be seen as a program but also as a set formula. This duality allows to use $\{log\}$ as both, a programming language and an automated verifier for its own programs. In $\{log\}$, users do not need to switch back and forth between programs and specifications: programs *are* specifications and specifications *are* programs.

References

1. AnimB: B model animator. http://www.animb.org/
2. Abrial, J.R.: The B-Book: Assigning Programs to Meanings. Cambridge University Press, New York (1996)
3. Abrial, J.R.: Modeling in Event-B: System and Software Engineering, 1st edn. Cambridge University Press, New York (2010)
4. Abrial, J., Butler, M.J., Hallerstede, S., Hoang, T.S., Mehta, F., Voisin, L.: Rodin: an open toolset for modelling and reasoning in Event-B. Int. J. Softw. Tools Technol. Transf. **12**(6), 447–466 (2010). https://doi.org/10.1007/s10009-010-0145-y
5. Boniol, F., Wiels, V.: The landing gear system case study. In: Boniol, F., Wiels, V., Ameur, Y.A., Schewe, K. (eds.) ABZ 2014. CCIS, vol. 433, pp. 1–18. Springer, Cham (2014). https://doi.org/10.1007/978-3-319-07512-9_1
6. Cantone, D., Ferro, A., Omodeo, E.: Computable Set Theory. Clarendon Press, USA (1989)

7. Cristiá, M., Katz, R.D., Rossi, G.: Proof automation in the theory of finite sets and finite set relation algebra. Comput. J. **65**(7), 1891–1903 (2022). https://doi. org/10.1093/comjnl/bxab030
8. Cristiá, M., Rossi, G.: A set solver for finite set relation algebra. In: Desharnais, J., Guttmann, W., Joosten, S. (eds.) RAMiCS 2018. LNCS, vol. 11194, pp. 333–349. Springer, Cham (2018). https://doi.org/10.1007/978-3-030-02149-8_20
9. Cristiá, M., Rossi, G.: Solving quantifier-free first-order constraints over finite sets and binary relations. J. Autom. Reason. **64**(2), 295–330 (2020). https://doi.org/ 10.1007/s10817-019-09520-4
10. Cristiá, M., Rossi, G.: Automated proof of Bell-LaPadula security properties. J. Autom. Reason. **65**(4), 463–478 (2021). https://doi.org/10.1007/s10817-020-09577-6
11. Cristiá, M., Rossi, G.: Automated reasoning with restricted intensional sets. J. Autom. Reason. **65**(6), 809–890 (2021). https://doi.org/10.1007/s10817-021-09589-w
12. Cristiá, M., Rossi, G.: An automatically verified prototype of the Tokeneer ID station specification. J. Autom. Reason. **65**(8), 1125–1151 (2021). https://doi.org/ 10.1007/s10817-021-09602-2
13. Cristiá, M., Rossi, G.: A decision procedure for a theory of finite sets with finite integer intervals. CoRR abs/2105.03005 (2021). https://arxiv.org/abs/2105.03005
14. Cristiá, M., Rossi, G.: {*log*}: applications to software specification, prototyping and verification. CoRR abs/2103.14933 (2021). https://arxiv.org/abs/2103.14933
15. Cristiá, M., Rossi, G.: Integrating cardinality constraints into constraint logic programming with sets. Theory Pract. Log. Program. **23**(2), 468–502 (2023). https:// doi.org/10.1017/S1471068421000521
16. Cristiá, M., Rossi, G., Frydman, C.S.: {*log*} as a test case generator for the Test Template Framework. In: Hierons, R.M., Merayo, M.G., Bravetti, M. (eds.) SEFM 2013. LNCS, vol. 8137, pp. 229–243. Springer, Heidelberg (2013). https://doi.org/ 10.1007/978-3-642-40561-7_16
17. Dovier, A., Omodeo, E.G., Pontelli, E., Rossi, G.: A language for programming in logic with finite sets. J. Log. Program. **28**(1), 1–44 (1996). https://doi.org/10. 1016/0743-1066(95)00147-6
18. Dovier, A., Piazza, C., Pontelli, E., Rossi, G.: Sets and constraint logic programming. ACM Trans. Program. Lang. Syst. **22**(5), 861–931 (2000)
19. Dovier, A., Pontelli, E., Rossi, G.: Set unification. Theory Pract. Log Program. **6**(6), 645–701 (2006). https://doi.org/10.1017/S1471068406002730
20. Holzbaur, C., Menezes, F., Barahona, P.: Defeasibility in CLP(Q) through generalized slack variables. In: Freuder, E.C. (ed.) CP 1996. LNCS, vol. 1118, pp. 209–223. Springer, Heidelberg (1996). https://doi.org/10.1007/3-540-61551-2_76
21. Jaffar, J., Maher, M.J., Marriott, K., Stuckey, P.J.: The semantics of constraint logic programs. J. Log. Program. **37**(1–3), 1–46 (1998). https://doi.org/10.1016/ S0743-1066(98)10002-X
22. Lamport, L., Paulson, L.C.: Should your specification language be typed? ACM Trans. Program. Lang. Syst. **21**(3), 502–526 (1999). https://doi.org/10.1145/ 319301.319317
23. Leuschel, M., Butler, M.: ProB: a model checker for B. In: Keijiro, A., Gnesi, S., Mandrioli, D. (eds.) FME 2003. LNCS, vol. 2805, pp. 855–874. Springer, Heidelberg (2003). https://doi.org/10.1007/978-3-540-45236-2_46
24. Mammar, A., Laleau, R.: Modeling a landing gear system in Event-B. In: Boniol, F., Wiels, V., Ameur, Y.A., Schewe, K. (eds.) ABZ 2014. CCIS, vol. 433, pp. 80–94. Springer, Heidelberg (2014). https://doi.org/10.1007/978-3-319-07512-9_6

25. Mammar, A., Laleau, R.: Modeling a landing gear system in Event-B. Int. J. Softw. Tools Technol. Transf. **19**(2), 167–186 (2017). https://doi.org/10.1007/s10009-015-0391-0
26. Rossi, G.: {*log*} (2008). http://www.clpset.unipr.it/setlog.Home.html. Accessed 2022
27. Rossi, G., Cristiá, M.: {*log*} user's manual. Technical report, Dipartimento di Matematica, Universitá di Parma (2020). http://www.clpset.unipr.it/SETLOG/setlog-man.pdf
28. Schwartz, J.T., Dewar, R.B.K., Dubinsky, E., Schonberg, E.: Programming with Sets - An Introduction to SETL. Texts and Monographs in Computer Science, Springer, New York (1986). https://doi.org/10.1007/978-1-4613-9575-1
29. Spivey, J.M.: The Z notation: a reference manual. Prentice Hall International (UK) Ltd., Hertfordshire (1992)

A Sound and Complete Validity Test for Formulas in Extensional Multi-level Syllogistic

Alfredo Ferro[1] and Eugenio G. Omodeo[2(✉)]

[1] University of Catania, Catania, Italy
alfredo.ferro@unict.it
[2] University of Trieste, Trieste, Italy
eomodeo@units.it

1 Introduction

In this paper we introduce both the syntax and the semantics of a formal language which has been designed to express multi-level syllogistic. This syllogistic (which could also be called ω-level syllogistic) is a 0th order set theory and differs from two-level syllogistic in the following respect:

> whereas, in the latter, the values allowed for variables are either individuals (i.e. memberless entities which are non-sets) or sets of individuals, in the former the value of each variable can be: an individual, a set of individuals, a set whose members are individuals or sets of individuals, etc.

We shall describe a validity test for formulas in multi-level syllogistic; the same method will enable us to exhibit a counter-example for each formula which is not valid. The correctness of the method will be proved, and its scope will be broadened by characterizing an effective validity-preserving translation of two-level syllogistic into multi-level syllogistic.

A procedure similar to the decision method described here was implemented in the programming language SETL, before the correctness proofs were available. The implemented algorithm, which demostrated to be quite efficient in practice, was developed by modifying a pre-existing program due to prof. J. T. Schwartz and based on Behmann's techniques [1,2], which determined the validity of quantifier-free boolean formulas. It is interesting to note that some of the heuristic ideas underlying our initial, tentative implementations, were partially incorrect, as emerged from certain wrong answers given by the tester. Thus, in the discovery of the validity test we are about to present, the computer played a significant role.

Unlike two-level syllogistic, which has been studied in depth since Aristotle's time, multi-level syllogistic occupies little or no room in the logical literature. However, we feel that the decidability of multi-level syllogistic has both an intrinsic philosophical interest and (possibly relevant) practical applications in the fields of automated deduction and program-correctness verification.

D. Cantone and A. Pulvirenti (Eds.): *From Computational Logic to Computational Biology*,
LNCS 14070, pp. 82–95, 2024.
https://doi.org/10.1007/978-3-031-55248-9_4

We deal with extensional syllogistic systems and have made no attempt to strengthen our results so as to include modal syllogistic systems, although this seems to be an interesting direction for future research. Another syllogistic system which seems to deserve some study is the one whose variables are allowed to range over classes (in Morse's sense) and whose language contains, in addition to all features to be found in our language, a unary operator designating the complementation operation over classes.

2   Entities: Individuals and Sets

We will consider two kinds of ENTITIES: individuals and sets.

If e is an individual, then it is different from any set (in particular, e differs from the empty set $\emptyset$): furthermore, the entities $e \cup x$, $x \cup e$, $e \cap x$, $x \cap e$, $e \setminus x$, $x \setminus e$, and the proposition $x \in e$ are undefined for every (individual or set) x.

We assume the following mild version of the axiom of regularity: The membership relation has no finite cycles $x_0 \in x_1 \in \cdots \in x_{n+1} = x_0$. Equivalently, every finite set is partially ordered by the transitive closure of $\in$. In particular, $x \notin x$ for any set x.

We postulate that for every set S there exists an injective function from S into the collection of individuals.

The sets $\emptyset$ and $\{\emptyset\}$ will be identified with falsehood and truth respectively. They are also identified with the natural numbers 0 and 1 respectively; more generally, the natural number n is identified with the set $\{0, \ldots, n-1\}$.

3   Definition of the Language

SYMBOLS:
  variables $\nu_0, \nu_1, \nu_2, \ldots$ (*ad inf.*);
  a *constant* 0;
  closed parenthesis) and *open parenthesis* (;
  binary *operators* $\setminus$ and $\cup$,
  connectives $\neg$ (unary) and $\vee$ (binary);
  binary *relators* $\in$ and $=$.

EXPRESSIONS: finite strings of symbols.

TERMS: members of the smallest set, T, of expressions which has 0 and all variables among its members and is closed under the property:
$$\text{if } \tau, \theta \in T, \text{ then } (\tau \cup \theta), (\tau \setminus \theta) \in T.$$

ATOMIC FORMULAS: expressions of the form $(\tau \in \theta)$ or $(\tau = \theta)$, where $\tau, \theta \in T$.

LITERALS: expressions of the form α or $\neg \alpha$, where α is an atomic formula.

FORMULAS: members of the smallest set, F, of expressions which has all of the atomic formulas among its members and is closed under the property:
$$\text{if } \varphi, \psi \in F, \text{ then } \neg \varphi, (\varphi \vee \psi) \in F.$$

4   Interpretations

Definition 1. *An* INTERPRETATION *is a function I from a (possibly empty) set of variables into the collection of all entities.*   ⊣

Every interpretation I has a natural extension $\hat{I}$, which is recursively defined by the conditions below:

(1) The domain of $\hat{I}$ consists of terms and formulas.
(2) $\hat{I}0 = \emptyset$; $\hat{I}\xi = I\xi$ for every variable ξ such that $I\xi$ is defined.
   (2′) $\hat{I}\xi$ is undefined whenever ξ is a variable and $I\xi$ is undefined.
(3) $\hat{I}$ maps every formula φ that belongs to its domain to one of the two values:
   $\emptyset$ (falsehood), $\{\emptyset\}$ (truth).
      Whenever it is defined, $\hat{I}\neg\varphi = \{\emptyset\} \setminus \hat{I}\varphi$;
      whenever it is defined, $\hat{I}(\varphi \vee \psi) = \hat{I}\varphi \cup \hat{I}\psi$;
      if σ is $\in$ or $=$ and $\hat{I}(\tau\,\sigma\,\theta)$ is defined, then
   $$\hat{I}(\tau\,\sigma\,\theta) = \{\emptyset\} \text{ if } \hat{I}\tau\,\sigma\,\hat{I}\theta\,, \text{ and } \hat{I}(\tau\,\sigma\,\theta) = \emptyset \text{ otherwise.}$$
(4) If $\hat{I}(\tau\,\sigma\,\theta)$ is defined, where σ is $\cup$ or $\setminus$, then $\hat{I}(\tau\,\sigma\,\theta) = \hat{I}\tau\,\sigma\,\hat{I}\theta$.
(5) $\hat{I}(\tau\,\sigma\,\theta)$ is undefined, where σ is one of the symbols $=, \in, \setminus, \cup$, if and only if one of the following holds:
      either $\hat{I}\tau$ is undefined or $\hat{I}\theta$ is undefined;
      $\hat{I}\theta$ is an individual and σ is $\in$, $\cup$, or $\setminus$;
      $\hat{I}\tau$ is an individual and σ is either $\cup$ or $\setminus$.
(6) $\hat{I}\neg\varphi$ is undefined if and only if $\hat{I}\varphi$ is undefined.
(7) $\hat{I}(\varphi \vee \psi)$ is undefined if and only if either one of $\hat{I}\varphi$, $\hat{I}\psi$ is undefined.

5   Classification of Interpretation

An interpretation I is said to be:

1. a *2-interpretation* if $I\xi \in \{0,1\}$ for every variable ξ (hence $\hat{I}\tau \in \{0,1\}$ for every term τ);
2. *local* if there are only a finite number n (possibly $n = 0$) of variables ξ such that $I\xi$ is defined and $I\xi \neq 0$;
3. *everywhere defined* if the domain of $\hat{I}$ is $\{\,\text{terms}\,\} \cup \{\,\text{formulas}\,\}$.

Clearly, if I is an everywhere defined interpretation, then $\hat{I}\tau$ is a set for every term τ. Conversely, if $I\xi$ is a set for every variable ξ, then I is everywhere defined.

6   Abbreviations, Truth-Value Assignments

$$(\varphi_0 \vee \ldots \vee \varphi_n) \text{ stands for } \begin{cases} \varphi_0 & \text{if } n = 0\,, \\ (\varphi_0 \vee \ldots \vee \varphi_{n-1}) \vee \varphi_n & \text{if } n > 0\,. \end{cases}$$

$(\varphi_0 \,\&\, \ldots \,\&\, \varphi_n)$ stands for $\neg(\neg\varphi_0 \vee \ldots \vee \neg\varphi_n)$ and is called a *conjunction*.
$(\varphi \Rightarrow \psi)$ stands for $(\neg\varphi \vee \psi)$; $(\varphi \Leftrightarrow \psi)$ stands for $((\varphi \Rightarrow \psi) \,\&\, (\psi \Rightarrow \varphi))$.
$(\tau \notin \theta)$ stands for $\neg(\tau \in \theta)$; $(\tau \neq \theta)$ stands for $\neg(\tau = \theta)$.
$(\tau \cap \theta)$ stands for $(\tau \setminus (\tau \setminus \theta))$; $(\tau \bigtriangleup \theta)$ stands for $((\tau \cup \theta) \setminus (\tau \cap \theta))$.
$(\tau \subseteq \theta)$ stands for $((\tau \setminus \theta) = 0)$; $(\tau \subset \theta)$ stands for $((\tau \subseteq \theta) \,\&\, (\tau \neq \theta))$.

A formula $(\varphi_0 \vee \ldots \vee \varphi_n)$ is said to be *in disjunctive normal form* if each φ_i is a conjunction of literals.

7   Models, Satisfiability, and Validity

An interpretation I is said to be a MODEL for a formula φ if $\hat{I}\varphi$ is defined and $\hat{I}\varphi = 1$. A model for φ which is a 2-interpretation is called a 2-MODEL for φ. A formula which has a model is said to be SATISFIABLE, and a formula φ for which $\hat{I}\varphi = 1$ under every interpretation I such that $\hat{I}\varphi$ is defined is said to be VALID. A formula ψ is said to be a LOGICAL CONSEQUENCE of the formula φ if $(\varphi \Rightarrow \psi)$ is valid.

Remark 1. If a 2-model for a formula φ exists, then a local 2-model for φ can be effectively found. This is an easy consequence of the decidability of sentential calculus.   ⊣

The following two lemmas are easily proved:

Lemma 1. *Let $\mathcal{N}$ be a class of interpretations, fulfilling the following condition $a(\mathcal{N})$: every satisfiable formula has a model $I \in \mathcal{N}$. Then $\mathcal{N}$ also meets the condition: every formula φ such that $\hat{I}\varphi = 1$ whenever $I \in \mathcal{N}$ and $\hat{I}\varphi$ is defined, is valid.*   ⊣

Lemma 2. *Let $\mathcal{N}$ and $\mathcal{N}'$ be classes of interpretations. If $a(\mathcal{N})$ holds, and for every $I \in \mathcal{N}$ and every formula φ in the domain of $\hat{I}$ there exists an $I^\varphi \in \mathcal{N}'$ under which $\widehat{I^\varphi}\varphi = \hat{I}\varphi$, then $a(\mathcal{N}')$ is also true.*   ⊣

The first lemma tells us that if $a(\mathcal{N})$ is true, then $\mathcal{N}$ is an adequate substitute for the notion of interpretation as far as one is concerned with the satisfiability or validity of isolated formulas. The second lemma supplies us with a criterion for proving $a(\mathcal{N}')$. This criterion in particular allows us, from now on till the end of Sect. 17, to confine our consideration to local interpretations. The notion of interpretation will be further restricted, using the same criterion, at the end of Sect. 9.

8   Superstructures

Let $U = U_0$ be a (possibly empty) set of individuals. For $i = 0, 1, 2, \ldots$ (*ad inf.*), we define $U_{i+1} = U_i \cup \{ x : x \subseteq U_i \}$. The set $U_\infty = \bigcup_{i=0}^\infty U_i$ is called the SUPERSTRUCTURE BASED ON U.

Remarks 1.(1) Every superstructure is closed under the binary operations: union, intersection, and difference of sets.

(2) For each $x \in U_\infty$, define lev (x) to be the smallest i for which $x \in U_i$. Since $x \in y$ implies lev $(x) <$ lev (y) for all $x, y \in U_\infty$, it follows (with no use of the regularity axiom) that the proposition $x_0 \in x_1 \in \cdots \in x_{n+1} = x_0$ is either undefined or false for every $n \geqslant 0$ and for all $x_0, x_1, \ldots, x_{n+1} \in U_\infty$.

(3) If U_0 is finite, then every member of $U_\infty \setminus U_0$ is a finite set.

(4) $0, 1, 2, \ldots$ belong to every superstructure, and lev $(i) = i + 1$ holds for $i = 0, 1, 2, \ldots$. Thus $U_{3+i} \setminus U_{2+i}$ has at least $i + 1$ distinct members, namely $\{0, 1 + i\}, \ldots, \{i, 1 + i\}$, for every $i = 0, 1, 2, \ldots$.   ⊣

Let I be an interpretation such that $I\xi \in (U_\infty \setminus U_0)$ for every variable ξ. Then we say that I is AN INTERPRETATION OVER U_∞. Note that an interpretation over U_∞ is everywhere defined and sends every term or formula to a member of U_∞.

9   Narrowing down the Notion of Interpretation

Lemma 3. *Let* $S = \{s_1, \ldots, s_n\}$. *Then the smallest superset,* $\tilde{S}$, *of* $\{\emptyset\} \cup S$ *which is closed with respect to finite unions and differences of sets is finite.*

Proof. Without loss of generality we assume that every member of S is a set. If the s_i's are pairwise disjoint, then the statement is obvious. Otherwise we observe that $\tilde{S} = \tilde{S'}$, where $S' = \{(\cap J) \setminus (\cup(S \setminus J)) : J \subseteq S\}$ (here, when $J = \emptyset$, by $\cap J$ we refer to $\cup S$).   ⊣

Lemma 4. *Let* $S = \{s_1, \ldots, s_n\}$. *Then there exists a superstructure,* U_∞, *and a function,* λ, *from* $\tilde{S}$ *into* $U_\infty \setminus U_0$ *which fulfills the conditions (where* $x, y \in S$ *):* $\lambda\emptyset = \emptyset$, $\lambda x = \lambda y$ *implies* $x = y$; $\lambda x \in \lambda y$ *if and only if* $x \in y$; *if* x *and* y *are sets, then* $\lambda(x \cup y) = \lambda x \cup \lambda y$ *and* $\lambda(x \setminus y) = \lambda x \setminus \lambda y$.

Proof. Let σ be an injective function from $S \cup \cup\tilde{S}$, where $\cup\tilde{S}$ indicates $\cup\{s : s \in \tilde{S} \text{ and } s \text{ is a set}\}$, into the collection of individuals. For every $s \in \tilde{S}$, define: $\lambda s = \{\sigma s\}$ if s is an individual; $\lambda s = \{\sigma z : z \in (s \setminus \tilde{S})\} \cup \{\lambda z : z \in (s \cap \tilde{S})\}$ otherwise. Since $\tilde{S}$ is finite (and therefore $\tilde{S}$ is partially ordered by the transitive closure of $\in$), this definition makes sense. The base, U_0, of U_∞ be $\{\sigma z : z \in (S \cup \cup\tilde{S})\}$. The injectivity of λ is proved by induction on the maximum of the heights of x, y in $\tilde{S}$.   ⊣

Corollary 1. *If* I *is a local interpretation, then there exist a superstructure* U_∞ *and a local interpretation,* I', *over* U_∞ *such that* $\hat{I'}\varphi = \hat{I}\varphi$ *for every formula* φ *in the domain of* $\hat{I}$.   ⊣

In view of the above corollary, from now on till the end of Sect. 17 interpretation will mean: local interpretation over a superstructure.

10   First Simplifications of the Decision Problem for Validity

We are given a formula φ and want to determine whether it is valid or not. Our method will effectively find a model for $\neg\varphi$ if any exists. If ψ is a disjunctive normal form of $\neg\varphi$, the models for $\neg\varphi$ and for ψ are the same. ψ has the form $(\psi_0 \vee \ldots \vee \psi_n)$, where each ψ_i is a conjunction of literals; therefore, the models for ψ are exactly those interpretations that are models for at least one of the ψ_i's and the validity problem gets reduced to the following: given a conjunction χ of literals, determine whether χ has a model or not. Without loss of generality, we can assume that χ is $(\chi_0 \mathrel{\&} \chi_1 \mathrel{\&} \ldots \mathrel{\&} \chi_h)$, where:

χ_0 is $(\zeta = 0)$ for some variable ζ;

$\chi_1, \ldots, \chi_k$ are of the forms $(\xi_0 = (\xi_1 \cup \xi_2))$, $(\xi_0 = (\xi_1 \setminus \xi_2))$, $(\xi_0 = \xi_1)$;

$\chi_{k+1}, \ldots, \chi_h$ are of the forms $(\xi_0 \neq \xi_1)$, $(\xi_0 \in \xi_1)$, $(\xi_0 \notin \xi_1)$.

Here ξ_0, ξ_1, ξ_2 stand for variables and $0 \leqslant k \leqslant h$. That we can make such an assumption about the form of the χ_i's is an easy corollary of the following easy lemma:

Lemma 5. *Let χ, τ, and ξ be, in this order, a formula, a term, and a variable occurring neither in χ nor in τ. Obtain χ' by substituting ξ for zero or more occurrences of τ within χ. Then:*

- *every model for $\left(\chi' \mathbin{\&} (\xi = \tau)\right)$ is also a model for χ;*
- *to every model M for χ there corresponds a model M' for $\left(\chi' \mathbin{\&} (\xi = \tau)\right)$ in such a way that if the base of the superstructure of M is finite, then M' can be effectively obtained from M.* ⊣

11   Further Simplifications of the Validity Problem

We keep using the same notation used in the preceding section.

Lemma 6. *If η_0, η_1 are distinct variables and $(\chi_0 \mathbin{\&} \ldots \mathbin{\&} \chi_k \mathbin{\&} (\eta_0 \neq \eta_1))$ has a model M, then it has a 2-model M'.*

Proof. Let e be a member of $M\eta_i \setminus M\eta_{1-i}$ where $i = 0$ or $i = 1$. Define:

$M'\xi = 1$ for every variable ξ such that $e \in M\xi$;

$M'\xi = 0$ for every variable ξ such that $e \notin M\xi$. ⊣

Corollary 2. *$(\chi_0 \mathbin{\&} \ldots \mathbin{\&} \chi_k)$ has only finitely many logical consequences of the form $(\eta_0 = \eta_1)$ where η_0 and η_1 are distinct variables, and these can be effectively found.*

Proof. Both η_0 and η_1 must occur in $(\chi_0 \mathbin{\&} \ldots \mathbin{\&} \chi_k)$. (Note that the interpretation sending every variable to 0 is a model for $(\chi_0 \mathbin{\&} \ldots \mathbin{\&} \chi_k)$. Cf. Appendix A for more details. ⊣

Thanks to this corollary and to Lemma 5—cf. the preceding section–, with no loss of generality we may from now on assume that $(\chi_0 \mathbin{\&} \ldots \mathbin{\&} \chi_k)$ has no logical consequences of the form $(\eta_0 = \eta_1)$ where η_1 and η_0 are distinct variables. In particular, no formula of this form is one of the χ_i's. Finally, if any χ_i has the form $(\eta = \eta)$, then it can simply be dropped; if any χ_i has the form $(\eta \neq \eta)$, the answer to the particular instance of the decision problem is trivial (we will therefore assume that this is not the case); χ_0, which is of the form $(\zeta = 0)$, can be replaced by $\left(\zeta = (\zeta \setminus \zeta)\right)$.

12   Decidability of the Validity Problem

In this section and the subsequent ones, we will indicate:

by χ, a conjunction of literals which has the normal form that we have been describing in the preceding two sections;

by χ_i, one of the literals whose conjunction makes χ;

by $\chi_=$, the conjunction of all the atomic formulas that equal some of the χ_i's and in which the symbol $=$ appears;

by $\chi_{\neq}$, the set of all the χ_i's which have the form $(\xi \neq \eta)$;

by L, the set of those variables that occur immediately to the left of either $\notin$ or $\in$ in χ;

by V, the set

$L \cup \{$ variables that occur immediately to the right of either $\in$ or $\notin$ in $\chi\}$.

We claim here—and shall prove in the forthcoming four sections—that the problem of finding a model for χ amounts to the (obviously decidable) one of finding a binary relation, Π, which fulfills the conditions below:

(1) the transitive closure of Π is a (strict) partial ordering of the members of V;

(2a) if $(\xi \notin \eta)$ is one of the χ_i's, then $\langle \xi, \eta \rangle \notin \Pi$;

(2b) if $(\xi \in \eta)$ is one of the χ_i's, then $\langle \xi, \eta \rangle \in \Pi$;

(3) for every $\xi \in L$:

if $\{\eta_0, \ldots, \eta_p\} = \{\eta : \langle \xi, \eta \rangle \in \Pi\}$, with $p \geqslant 0$ and $\{\xi, \xi_1, \ldots, \xi_q\} = V \setminus \{\eta_0, \ldots, \eta_p\}$, then there exists a 2-model for

$$(\quad \chi_= \quad \& \ (\eta_0 \neq 0) \ \& \ \ldots \ \& \ (\eta_p \neq 0) \ \& $$
$$(\xi = 0) \ \& \ (\xi_1 = 0) \ \& \ \ldots \ \& \ (\xi_q = 0) \quad);$$

(4) the members of $V \setminus L$ are maximal in Π, i.e., $\xi \in V \setminus L$ and $\eta \in V$ implies $\langle \xi, \eta \rangle \notin \Pi$.

Intuitively speaking, $\langle \xi, \eta \rangle \in \Pi$ means that $\xi \in L$, $\eta \in V$, and $I\xi \in I\eta$ in the model I for χ we are after.

13   Completeness Theorem

Theorem 1. *If χ has a model, then there is a function $\mathcal{M}$, defined on L, that associates with every $\sigma \in L$ a 2-model for $\chi_=$ so that:*

(1) *for every $\xi \in L$, $(\mathcal{M}\xi)\xi = 0$;*

(2a) *for every $\xi \in L$ and every variable η such that $(\xi \notin \eta)$ is one of the χ_i's, $(\mathcal{M}\xi)\eta = 0$;*

(2b) *for every $\xi \in L$ and every variable η such that $(\xi \in \eta)$ is one of the χ_i's, $(\mathcal{M}\xi)\eta = 1$;*

(3) *the transitive closure of the binary relation*

$$\{\langle \xi, \eta \rangle : \xi \in L, \eta \in L \text{ and } (\mathcal{M}\xi)\eta = 1\}$$

on L is a partial ordering of the elements of L.

Proof. Let I be a model for χ. For every $\sigma \in L$ and every variable ξ, define:

$$(\mathcal{M}\sigma)\xi = \begin{cases} 0 \text{ if } I\sigma \notin I\xi, \\ 1 \text{ if } I\sigma \in I\xi. \end{cases}$$

$\dashv$

Remarks 2. A function $\mathcal{M}$ as in the above theorem exists if and only if there exists a relation Π like the one characterized in the preceding Sect. 12. In fact, given $\mathcal{M}$, one can define

$$\Pi = \{\langle \xi, \eta \rangle : \xi \in L, \eta \in V \text{ and } (\mathcal{M}\xi)\eta = 1\}.$$

Conversely, given Π, one can define $\mathcal{M}\xi$ for every $\xi \in L$, as being the 2-model whose existence is asserted by (3) in Sect. 12.

In Appendix B we describe an algorithm which ascertains the existence of the above function $\mathcal{M}$, thereby deciding—by virtue of the completeness theorem just proved and of the soundness theorem to be proved in Sect. 16 the satisfiability of χ.

$\dashv$

14 Soundness Theorem: 1st Version

Theorem 2. *Let U be a finite set of individuals, $U \supseteq (L \cup \chi_{\neq})$. Let Ξ be the set of all the variables in the language that are members of U. Let $\mathcal{M}$ be a function which associates with every $\sigma \in U$ a 2-model, $\mathcal{M}\sigma$, for $\chi_=$ so that:*

(0) *for every $\upsilon \in \chi_{\neq}$, $\widehat{\mathcal{M}\upsilon\upsilon} = 1$;*
(1) *for every $\xi \in \Xi$, $(\mathcal{M}\xi)\xi = 0$;*
(2a) *for every $\xi \in L$ and every variable η such that $(\xi \notin \eta)$ is one of the χ_i's, $(\mathcal{M}\xi)\eta = 0$;*
(2b) *for every $\xi \in L$ and every variable η such that $(\xi \in \eta)$ is one of the χ_i's, $(\mathcal{M}\xi)\eta = 1$;*
(3) *the transitive closure of the binary relation*

$$\{\langle \xi, \eta \rangle : \xi \in \Xi, \eta \in \Xi \text{ and } (\mathcal{M}\xi)\eta = 1\}$$

on Ξ is a partial ordering of the elements of Ξ;
(4) *if we define*

$$I\xi = \{\sigma : \sigma \in U \text{ and } (\mathcal{M}\sigma)\xi = 1\}$$

for every variable ξ, then I has the property that for every pair $\xi, \eta \in \Xi$, if there is a variable σ for which $(\mathcal{M}\xi)\sigma \neq (\mathcal{M}\eta)\sigma$, then

$$\left([(I\xi \cup I\eta) \setminus (I\xi \cap I\eta)] \setminus \Xi \right) \neq \emptyset.$$

Under these hypotheses, χ has a model I^* which can be effectively determined from $\mathcal{M}$.

$\dashv$

15   Proof of the Soundness Theorem, 1st Version

The variables ξ for which $I\xi \neq 0$ are the members of the set

$$\Xi' = \bigcup_{\sigma \in U} \{\xi : (\mathcal{M}\sigma)\xi \neq 0\} \, ,$$

which is finite, because U is finite and each $\mathcal{M}\sigma$ is a local model. It follows from (3) that the transitive closure of

$$\{\langle \xi, \eta \rangle : \xi \in \Xi, \, \eta \in \Xi', \, (\mathcal{M}\xi)\eta \neq 0\}$$

is a partial ordering of the elements of $\Xi \cup \Xi'$.

For each variable $\xi \notin (\Xi \cup \Xi')$, define $I^*\xi = 0$; for each variable $\xi \in (\Xi \cup \Xi')$, by means of recursion on the height of ξ, define

$$I^*\xi = (I\xi \setminus \Xi) \cup \{I^*\eta : \eta \in (\Xi \cap I\xi)\} \, ,$$

so that in particular $I^*\xi = I\xi$ whenever ξ has height 0. Clearly every $I^*\xi$ is a member of the superstructure based on $U \setminus \Xi$ and meets the conditions:

for every $\sigma \in (U \setminus \Xi)$, $\sigma \in I^*\xi$ if and only if $\sigma \in I\xi$;
for every $\sigma \in \Xi$, $I^*\sigma \in I^*\xi$ if and only if $\sigma \in I\xi$.

(In proving that $\sigma \in \Xi$, $\sigma \notin I\xi$ together imply $I^*\sigma \notin I^*\xi$, use is made of (4)).

From the conditions above, it follows that $I^*\xi_2 = \widehat{I}^*(\xi_0 \cup \xi_1)$ [respectively: $I^*\xi_2 = \widehat{I}^*(\xi_0 \setminus \xi_1)$] whenever $(\xi_2 = (\xi_0 \cup \xi_1))$ [resp.: $(\xi_2 = (\xi_0 \setminus \xi_1))$] is one of the χ_i's.

By (2a), $I^*\xi_0 \notin I^*\xi_1$ whenever $(\xi_0 \notin \xi_1)$ is one of the χ_i's.
By (2b), $I^*\xi_0 \in I^*\xi_1$ whenever $(\xi_0 \in \xi_1)$ is one of the χ_i's.
By (0), $I^*\xi_0 \neq I^*\xi_1$ whenever $(\xi_0 \neq \xi_1)$ is one of the χ_i's.

16   Soundness Theorem: 2nd Version

Theorem 3. *Let U be a finite set of individuals, $U \supseteq L$. Let Ξ be the set of all variables that are members of U, and let $\mathcal{M}$ be a function which associates with every $\sigma \in U$ a 2-model, $\mathcal{M}\sigma$, for $\chi_=$ so that the condition*

(0') *for every $\varphi \in (\chi_{\neq} \cap U)$, it holds that $\widehat{\mathcal{M}\varphi}\varphi = 1$*

and the conditions (1), (2), (3) of the first version of the soundness theorem are satisfied. Then χ has a model, which can be effectively determined from $\mathcal{M}$.

Proof. Define $I\xi = \{\sigma : \sigma \in U \text{ and } (\mathcal{M}\sigma)\xi = 1\}$, for every variable ξ. We have assumed that no equality of the form $(\eta = \zeta)$, where η and ζ are distinct variables, is a logical consequence of $\chi_=$. Therefore we can find a 2-model $\mathcal{M}'(\eta \neq \zeta)$ for $(\chi_= \& (\eta \neq \zeta))$, for every pair $\langle \eta, \zeta \rangle$ of variables such that either $(\eta \neq \zeta) \in (\chi_{\neq} \setminus U)$ or $\eta \in I\sigma$, $\zeta \notin I\sigma$ holds for some variable σ. Taking the union of $\mathcal{M}$ and $\mathcal{M}'$, we get a new function, $\mathcal{M}''$, which meets the first version of the soundness theorem. ⊣

Remark 2. For $U = L$, the above theorem is the converse of the completeness theorem. ⊣

17 Getting Rid of Individuals

In proving the decidability of the validity problem, we have incidentally seen that if a formula is satisfiable, then it has a model over a superstructure whose base is finite.

Let us now strengthen this result and show that if a formula is satisfiable, then it has a model over the superstructure, $\varnothing_\infty$, whose base is empty.

Lemma 7. *Let $S = \{s_1, \ldots, s_n\} \subseteq (U_\infty \backslash U_0)$, where U_0 is finite. Then there is a function, λ, from $\tilde{S}$ into $\varnothing_\infty$ such that for all $x, y \in \tilde{S}$: $\lambda\emptyset = \emptyset$; $\lambda x = \lambda y$ implies $x = y$; $\lambda x \in \lambda y$ if and only if $x \in y$; $\lambda(x \cup y) = \lambda x \cup \lambda y$; $\lambda(x \backslash y) = \lambda x \backslash \lambda y$.*

Proof. Let σ be an injective function from $\bigcup \tilde{S} \backslash \tilde{S}$ into $\varnothing_\ell \backslash \varnothing_{\ell-1}$ for a suitable $\ell > \mathsf{lev}\left(\tilde{S} \cap \varnothing_\infty\right)$. For every $s \in \tilde{S}$, define

$$\lambda s = \left\{ \sigma z : z \in (s \backslash \tilde{S}) \right\} \cup \left\{ \lambda z : z \in (s \cap \tilde{S}) \right\} .$$

By induction on the height of s in $\tilde{S}$ (with respect to the transitive closure of $\in$), one proves that: either $\lambda s = s$ (hence $\mathsf{lev}(\lambda s) < \ell$), or $\mathsf{lev}(\lambda s) > \ell$. The injectivity of λ then follows easily, by induction on the maximum of the heights of x, y in $\tilde{S}$. ⊣

Corollary 3. *If I is an interpretation over U_∞, where U_0 is finite, then an interpretation, I', over $\varnothing_\infty$ can be effectively determined from I, so that $\hat{I'}\varphi = \hat{I}\varphi$ holds for every formula φ.*

Proof. Let $S = \{I\xi_1, \ldots, I\xi_n\}$, where $\xi_1, \ldots, \xi_n$ are all of the variables ξ for which $I\xi \neq \emptyset$. Define $I'\xi = \lambda I\xi$ for every variable ξ. Then $\hat{I'}\tau = \lambda\hat{I}\tau$ holds for every term τ. ⊣

18 A Language for 2-Level Syllogistic

Let us informally describe a suitable language for 2-level syllogistic. The set of symbols of this language includes a new constant, 1, and two infinite sequences of variables. The constant 1 designates a nonempty set of individuals, U; the variables of one sort, which are called INDIVIDUAL VARIABLES, designate the members of U; the variables of the other sort, which are called SET VARIABLES, designate subsets of U; compound terms, which are built up in the usual manner using the constants $0, 1$ and the set variables, also designate subsets of U. Having these semantics in mind, one regards as ill-formed any expression of the form $(\tau \in \theta)$ where either θ is an individual variable, or τ is a set variable, or τ is a compound term. An expression $(\tau_0 = \tau_1)$ where τ_i is an individual variable whereas τ_{1-i} is not (i standing for either 0 or 1) is also regarded as an ill-formed formula.

Rather than characterizing in a precise fashion the language for 2-level syllogistic, we give the following definitions, which refer to the multi-level language which we have been considering so far:

A 2-LEVEL FORMULA is one in which only three kinds of atomic formulas may occur, namely $(\nu_{2i+1} = \nu_{2j+1})$, $(\tau = \theta)$, and $(\nu_{2i+1} \in \tau)$, where τ and θ are terms containing no occurrences of any variable ν_{2j+1}.

A 2-LEVEL INTERPRETATION is an interpretation (in the original sense of the expression, as defined in Sect. 4) which sends:

ν_0 to a nonempty set, U, of individuals;
ν_{2n} to a subset of U, for every $n > 0$;
ν_{2n+1} to a member of U, for every $n \geqslant 0$.

Clearly, 2-level formulas can be identified with the well-formed formulas of the language for 2-level syllogistic that we have described above. In this identification, ν_0 acts as the constant 1 and the variables ν_{2i+1} [respectively: ν_{2i}, with $i > 0$] play the role of individual [resp.: set] variables.

We note that $\hat{I}\varphi$ is defined whenever I is a 2-level interpretation and φ is a 2-level formula.

19   A Connection Between 2-Level Syllogistic and Multi-level Syllogistic

Let φ be a 2-level formula in which no variables ν_j with $j > 2n$ occur. We indicate by φ' the 2-level formula:

$$\Big(\big((\nu_0 \neq 0) \,\&\, (\nu_1 \in \nu_0) \,\&\, (\nu_2 \subseteq \nu_0) \,\&\, \cdots \,\&\, (\nu_{2n-1} \in \nu_0) \,\&\, (\nu_{2n} \subseteq \nu_0)\big) \Rightarrow \varphi\Big).$$

Clearly, if φ' is valid, then $\widehat{I'}\varphi = 1$ under every 2-level interpretation I'. In fact: $\widehat{I'}\varphi' = 1$ follows from the hypothesis, and $\widehat{I'}\big((\nu_0 \neq 0)\,\&\cdots\&(\nu_{2n} \subseteq \nu_0)\big) = 1$ since I' is 2-level.

Conversely: If $\widehat{I'}\varphi = 1$ under every 2-level interpretation I', then φ' is valid. In fact, if there is any interpretation I such that $\hat{I}\varphi' = 0$, then we can establish a one-one correspondence, f, between $I\nu_0$ and a nonempty set of individuals, U. If, for every i such that $0 \leqslant i \leqslant n$, we set

$$I'\nu_{2i} = \{\, fx : x \in I\nu_{2i}\,\}, \qquad I'\nu_{2i+1} = f\,I\nu_{2i+1},$$

and, for every $i > n$, we set:

$$I'\nu_{2i} = 0, \qquad I'\nu_{2i+1} = \text{a fixed member of } U,$$

then I' is a 2-level interpretation such that

$$\widehat{I'}\tau = \Big\{\, fx : x \in \hat{I}\tau \Big\} \ [\text{respectively: } \widehat{I'}\psi = \hat{I}\psi]$$

for every term τ [resp.: 2-level formula ψ] that involves at most the variables $\nu_0, \nu_2, \nu_4, \ldots, \nu_{2n}$ [resp.: $\nu_0, \nu_1, \nu_2, \ldots, \nu_{2n}$]. In particular, $\widehat{I'}\varphi = \hat{I}\varphi = 0$.

Acknowledgement. We wish to particularly thank prof. Jacob T. Schwartz for his guidance and many suggestions throughout the development of this and related work.

A   Appendix 1

In this appendix, we describe an algorithm which determines all logical consequences of μ that have the form $(\xi = \eta)$, where ξ and η are distinct variables. We assume that μ is a conjunction of literals, each of which has one of the forms:

$$(\zeta = 0), \; (\xi_0 = \xi_1), \; (\xi_0 = (\xi_1 \cup \xi_2)), \; (\xi_0 = (\xi_1 \setminus \xi_2)),,$$

where ξ_0, ξ_1, ξ_2, and ζ are variables.

Our algorithm calls a one-argument procedure, named '2-model', which is supposed to return a 2-model for the formula that is passed as a value for its parameter.

The names 'poteq' and 'neceq' indicate those equalities that potentially [respectively: necessarily] follow from μ. The names 'posmod' and 'negmod' indicate those variables that have been assigned the value true [resp.: false] in a 2-model.

```
procedure yieldedeq( μ );
    poteq ← neceq ← posmod ← negmod ← ∅;
    model ← 2_model( μ );
    (for every variable ξ occurring in μ)
        if ξ is true in model then
            poteq ← ( poteq ∪ { (ξ = η) : η ∈ posmod } );
            posmod ← ( posmod ∪ {ξ} );
        else
            poteq ← ( poteq ∪ { (ξ = η) : η ∈ negmod } );
            negmod ← ( negmod ∪ {ξ} );
        end if;
    end for;
    (while ∃ε ∈ poteq)
        poteq ← ( poteq \ {ε} );
        if ∃ model ← 2_model( (μ & ¬ ε) ) then
            ( poteq ← poteq \ { (ξ = η) : ξ and η have different
                                          truth-values in model } );
        else
            neceq ← neceq ∪ {ε};
        end if;
    end while;
    return neceq;
end procedure yieldedeq;
```

B   Appendix 2

In this appendix, we give a semi-formal specification of an algorithm which decides the existence of a model for χ, on the basis of the completeness and soundness theorems that have been stated and proved in Sects. 13 and 16 respectively.

Our notation here will be consistent with the one adopted in Sect. 12; further-more, we are assuming that the χ_i's are global entities known to every subroutine of the main program.

We omit the specification of the procedure '2-model', which is called by the recursive procedure 'buildmodels' below: this is supposed to return a 2-model for a given formula of our language and can easily be implemented as a variation of the well-known Davis-Putnam-Logemann-Loveland procedure for deciding satisfiability in sentential calculus.

Our algorithm follows:

```
unvisited ← L ;
(while ∃ ξ ∈ unvisited)
    visited ← buildmodels( ξ , χ= , unvisited ) ;
    if visited = ∅ then
        print('there exist no models'); stop;
    else
        unvisited ← (unvisited \ visited) ;
    end if;
end while;
```

procedure buildmodels(ξ , conjunction, unvisited) ;

$$\text{conj} \leftarrow \left(\text{conjunction} \ \& \ \underset{\substack{(\xi \in \eta) \text{ is one} \\ \text{of the } \chi_i\text{'s}}}{\&} (\eta \neq 0) \ \& \right.$$

$$\left. \underset{\substack{(\xi \notin \eta) \text{ is one} \\ \text{of the } \chi_i\text{'s}}}{\&} (\eta = 0) \right) \ \& \ (\xi = 0) ;$$

```
( while ∃ model ← 2_model( conj ) )
    visited ← { ξ } ;
    newunvisited ← ( unvisited \ { ξ } ) ;
    successors ← { η : η ∈ unvisited and (η ≠ 0) is true in model } ;
    (while ∃ ξ' ∈ successors )
        newvisited ← buildmodels( ξ' , (conjunction  &  (ξ = 0)),
                                                         newunvisited ) ;
        if newvisited = ∅ then go to quitwhile ;
        else
            successors ← ( successors \ ({ξ'} ∪ newvisited) ) ;
            visited ← ( visited ∪ newvisited ) ;
            newunvisited ← ( newunvisited \ newvisited ) ;
        end if;
    end while ;
    quitwhile : if successors = ∅ then return visited;
                  else conj ← ( conj  &  (ξ' = 0) ) ;        end if;
end while ;
return ∅ ;
end procedure buildmodels ;
```

References

1. Behmann, H.: Beiträge zur Algebra der Logik insbesondere zum Entscheidungsproblem. Math. Ann. **86**, 163–220 (1922)
2. Schwartz, J.T.: Instantiation and decision procedures for certain classes of quantified set-theoretic formulae. Technical report 78-10, Institute for Computer Applications in Science and Engineering, NASA Langley Research Center, Hampton, Virginia (1978)

Courant Institute of Mathematical Sciences, New York University, July 1979.

Computational Biology and Complex Systems

Advances in Network-Based Drug Repositioning

Marco Pellegrini[✉][iD]

CNR-IIT, Area della Ricerca, Via Giuseppe Moruzzi, 1, 56124 Pisa, Italy
marco.pellegrini@iit.cnr.it

Abstract. Network-based drug repositioning is a fast-growing approach to computational drug repositioning that has had a powerful impulse at the onset of the COVID-19 pandemic. In this review, we collect several recent results in this area of research to highlight the current trends and to clarify the open challenges for further improving the state-of-the-art.

1 Introduction

Network biology [4] is based on the intuition that the quantitative algorithmic tools of network theory offer new possibilities to understand, model, and simulate the cell's internal organization and evolution, fundamentally altering our view of cell biology. As network biology has been gaining ground and recognition in the last 20 years, the scope of its application has moved steadily from biology to medicine, including modeling of diseases, and applications in drug design and drug action prediction [32,50,53,55].

Drug repositioning (DR), aka drug repurposing, defined as the activity of predicting the beneficial effect of existing drugs for novel diseases or for existing diseases different from those the drugs were initially designed for, has become a hot topic since the beginning of the COVID-19 pandemic in early 2020 [61], with concerns about the scientific validity of many initial reports not backed up by rigorous methodologies, validation protocols, and peer-review standards [37].

Galindez et al. [23] describe and critically assess the computational strategies for COVID-19 drug repurposing in the wider context of clinical trials to increase their chances of success by tightly coupling the computational and the experimental aspects.

Computational drug repositioning is a burgeoning field of research (see. e.g. [37,39] [1,17,23,32,41,42,55,73]) receiving much attention for its high potential impact on coping with present and future pandemic events by novel pathogens [50,76].

Network-based drug repositioning [4] is a specific approach that leverages building and analyzing various types of biological networks integrating several layers of 'omic' data [56].

As many excellent surveys on this topic have appeared up to 2019, and network-based approaches are gaining momentum with many algorithms proposed in the last couple of years, we focus on the latest developments in the area of bio-networks applied to drug repositioning.

© The Author(s), under exclusive license to Springer Nature Switzerland AG 2024
D. Cantone and A. Pulvirenti (Eds.): *From Computational Logic to Computational Biology*,
LNCS 14070, pp. 99–114, 2024.
https://doi.org/10.1007/978-3-031-55248-9_5

2   Recent Review Articles

In this section, we recapitulate the main literature for computational and network-oriented drug repositioning by referencing main surveys and review articles on this subject.

Csermely et al. [16] provide a comprehensive treatment of the use of molecular networks data representation in drug discovery, which includes both the design of new chemical compounds and the repositioning of existing chemical compounds. They summarize the major network types, the network analysis types, the drug design areas helped by network studies and give examples in four key areas of drug design. Their key message on the future directions of research points at two basic strategies: (1) the "central hit strategy" that selectively targets central nodes/edges of the bio-networks of the infectious agents or cancer cells with the aim of killing the infectious agent, or the cancer cell; and (2) the "network influence strategy" that may work against other types of diseases, where an efficient reconfiguration of the bio-networks needs to be achieved by targeting the neighbors of central nodes/edges.

Mernea et al. [47] report on the application of deep learning techniques, up to December 2020, for a better understanding of the druggable targets for COVID-19, covering, among other aspects, AI-based Biological Network analysis and Gene-Expression-based approaches.

Liu et al. [40] provide a comprehensive overview of the existing computational methods to discover potential inhibitors of the SARS-CoV-2 virus. They categorize and describe computational approaches by the basic algorithms employed and by the related biological datasets used in such predictions.

Alaimo et al. [1] describe advances in the field of network-based drug repositioning up to 2017 (included), covering several strategies for computational drug repositioning, network-based computational techniques, drug-related resources, and bio-network types. A similar scope is covered by Lotfi Shahreza et al. in [41] and Xue et al. in [73].

Li et al. [39] provide a comprehensive survey of advances in the critical areas of computational drug repositioning updated up to 2015, including available data sources, computational repositioning strategies, commonly used computational techniques, validation strategies for repositioning studies, including both computational and experimental methods, and a brief discussion of the challenges in computational drug repositioning.

Dotolo et al. [17] survey network-based drug repurposing methods broadly categorizing them into (i) network-based models, (ii) structure-based approaches, and (iii) artificial intelligence (AI) approaches. Network-based approaches are further divided into two categories: network-based clustering approaches and network-based propagation approaches, both mostly applied to multi-level heterogeneous biological graphs.

Zhou et al. [76] focus on the role of AI-based techniques in drug repositioning, advocating a convergence of AI with network precision medicine as key to further advances in this field.

A survey focusing specifically on anti-cancer drugs repurposed for COVID-19 is in [14].

3   Eighteen Network-Based Results

3.1   Taguchi et al. Tensor Decomposition

Taguchi et al. [65,66] apply an unsupervised method based on a 4D tensor decomposition to perform feature extraction of gene expression profiles in multiple lung cancer cell lines infected with severe acute respiratory syndrome coronavirus 2 [7]. They thus identified, using EnrichR [35], drug candidates that significantly altered the expression of the 163 genes selected in the previous phase. Drug perturbation data is collected from GEO, DrugMatrix, L1000, and other repositories. A key property of the method is that it does not rely on prior guessing of drug-target relationships.

3.2   Ruiz et al. Multiscale-Interactome

Ruiz et al. [57,58] build a multi-layer disease-gene-drug-pathway network and develop a random-walk-based score that captures the indirect effects of drugs on diseases via commonly affected pathways. They validate their approach in a leave-one-out validation on a golden standard of 6000 drug-disease pairs commonly used in clinical practice. The paper does not report on the specific application of this approach to COVID-19 but has a generalist approach. They observe that "...a drug's effectiveness can often be attributed to targeting genes that are distinct from disease-associated genes but that affect the same functional pathways.", in contrast with "... existing approaches assume that, for a drug to treat a disease, the proteins targeted by the drug need to be close to or even need to coincide with the disease-perturbed proteins." Interestingly here the relationship between the proteins that are affected by the disease and those that are modulated by the drugs is mediated by their co-occurrence in validated biological pathways (e.g. categories recorded in Gene Ontology). In [44], a similar effect is attained by clustering the bio-network topology.

3.3   Fiscon et al. SaveRUNNER

Fiscon et al. [21,22] use an approach based on computing distances between regions of a large human interactome network affected by a disease (disease targets) and those affected by drugs (drug perturbations). This work relies on disease target similarities between a group of diseases including COVID-19, SARS, MERS, and frequent co-morbidities of COVID-19. The final assessment of drug candidates for COVID-19 is done with the C-map database [36].

3.4   Gysi et al. a Multimodal Approach

Gysi et al. [28, 29] define several proximity-based distance functions between COVID-19 human target proteins (as listed in [27]) and drug target proteins as listed in DrugBank [71], in order to prioritize repurposable drugs. Several pipelines are developed and then their results are merged to form a consensus drug ranking. Refinements taking into account tissue specificity, drug action on gene expression levels (using data from [7]), comorbidities, and drug toxicity lead to a final list of 81 repurposable drugs for COVID-19.

A measure of performance chosen is the AUC of the predicted list versus the list of drugs currently employed in clinical trials for the treatment of COVID-19 as listed in https://www.clinicaltrials.gov/.

The final drug list is furthermore compared to infected cell viability data for 918 drugs experimentally screened in VeroE6 cells, showing that 22 out of the top 100 drugs in the listing have a measurable (strong or weak) effect on the VeroE6 cells. The hit rate is then further increased by applying expert knowledge and curation.

Interestingly the large majority of the drugs in this final listing do not bind directly to human proteins targeted by the viral proteome, but rather to protein in their vicinity (in a network topology), thus indicating that the indirect approaches may be more suitable than the direct target-oriented ones.

3.5   Lucchetta et al. DrugMerge

Lucchetta et al. [43, 44] build modular disease active subnetwork (DAS) of a large gene co-expression network to capture disease-enriched pathways [43] and to rank drugs according to their contrary effect on the DAS [44]. The methodology called DrugMerge is tested on four benchmark diseases (Colorectal Cancer, Prostate Cancer, Asthma, and Rheumatoid Arthritis) and it is able to find in top-ranking positions drugs in clinical use for the specific disease. DrugMerge applied to COVID-19 data guesses in top-ranking positions drugs currently in clinical trials. The performance evaluation in DrugMerge is similar to the one in [29] since drugs in clinical trials are used as the golden standard for COVID-19.

Thus DrugMerge's evaluation methodology measures how well an automated ranking system comes close to the collective wisdom of human experts that shortlisted existing drugs for repurposing on COVID-19 during 2020 based on a large variety of considerations such as available clinical and pre-clinical data, pharmacological background, and putative mechanism of action.

3.6   Zhou et al. a Network Proximity Measure

Zhou et al. [75] also use proximity-based measures in biological networks to find a list of 16 repurposable drugs. We notice that Zhou et al. use data collected from the family of human coronaviruses (HCoVs) to build the network, thus relying heavily on the evolutionary conservation of relevant coding parts of the viral genome across the species in this family, and phylogenetic considerations.

3.7   Sadegh et al. CoVex

Sadegh et al. [60] use the algorithm KeyPathwayMiner [2] to define the active subnetwork in an integrated COVID-19 resource network from CoronaVirus Explorer CoVex resource (https://exbio.wzw.tum.de/covex/). The main contribution of [60] is a resource (CoVex) that can be used interactively in many different scenarios and modalities to explore disease-gene-drug relationships for COVID-19.

3.8   Mall et al. Drug Autoencoder

Mall et al. [45] use deep learning autoencoder-induced vector embedding of drugs and viral proteins to build a representation model. The representation model is then used to build an ensemble framework of ML methods to rank approved drugs based on their ability to inhibit the three main protease enzymes of the SARS-CoV-2 virus. This pipeline produces a shortlist of 19 drugs potentially useful against COVID-19.

3.9   Maria et al. 2021: PHENSIM

Maria et al. 2021 [46] introduce a systems biology tool (called PHENotype SIMulator: PHENSIM) that leverages transcriptomic and proteomic data to simulate in silico SARS-CoV-2 infection of host cells. This tool produces infection signatures for the host cells which are used to highlight potential target proteins and the drugs affecting these targets. Drugs with an effect signature anti-correlated with the disease signatures are considered good candidates for repositioning. Validation is performed by comparing the results of simulations with the proteomic measurements of SARS-CoV-2 infections in [10] and the transcriptomic measures in [7], and a good match is found.

3.10   Ge et al. NeoDTI Applied to COVID-19

Ge et al. [25] propose a pipeline for identifying putative drug targets in COVID-19. Initially, a tripartite knowledge graph is built, having as nodes: drugs, human proteins, and viral proteins, where the relational edges are derived from a variety of sources and databases, including the literature. Next, this graph is analyzed via a variant of the NeoDTI method [70] that uses information passing and aggregation techniques generalizing convolution neural networks so to accurately predict drug-target interactions. This initial pool of candidates is then refined by AI-based large-scale literature search, and by ranking with the connectivity map approach [36] using gene expression data from patients and cell lines infected by SARS-CoV and SARS-CoV-2 viruses. In vitro validation is finally performed, highlighting CVL218, a poly-ADP-ribose polymerase 1 (PARP1) inhibitor, as having an effective inhibitory activity against SARS-CoV-2 replication, as well as reduced toxicity.

3.11   Saberian et al. Learning the Similarity Metric

Saberian et al. [59], perform a data fusion of three main sources: (i) large-scale gene expression profiles corresponding to human cell lines treated with small molecules, (ii) gene expression profile of a given human disease and (iii) the known relationship between Food and Drug Administration (FDA)-approved drugs and diseases. Based on these data they propose a method that learns a similarity metric by mapping the metric identification problem as a convex optimization problem, through a supervised machine learning-based algorithm. The optimal metric is such that any disease and its associated FDA-approved drugs have a smaller distance than the other disease-drug pairs.

3.12   Peyvandipour et al. DrugDiseaseNet

Peyvandipour et al. [54] build a drug-disease network by considering all interactions between drug targets and disease-related genes in the context of all known signaling pathways. Many methods are built on the hypothesis that if the gene expression signature of a particular drug is opposite to the gene expression signature of a disease, then this drug may have a potential therapeutic effect on the disease. This paper refines this approach by taking the specific disease context into account. in particular, the method accounts for upstream and downstream effects on the genes expressed in a gene regulatory network.

3.13   Draghici et al. iPathwayGuide

Draghici et al. [19] tackle the drug repositioning problem for COVID-19 by a pipeline that involves (i) identification of biological processes significantly affected by SARS-CoV-2, taking care to correct for multiple comparisons (ii) pathway analysis [18,51,52] is then used to identify the pathways affected, including the direction and type of every signal connecting the genes; (iii) an upstream analysis is applied to identify key regulators and key target proteins; finally, drug-target analysis homes in the repurposable drugs. The promising FDA-approved drugs that resulted from the process above were then validated in additional independent datasets and in clinical studies ('in vivo' testing).

3.14   Stolfi et al. DIAMOND Applied to COVID-19

Stolfi et al. [63] define new network proximity functions in human PPI Networks to measure the 'distance' between the COVID-19-affected proteins (including 338 proteins listed in [27], 47 additional human proteins considered critical for the virus's entry into the host cell, and 115 proteins known to interact with other viruses of the coronaviruses family.

The proposed methodology uses the "DIAMOND" algorithm [26] and the "VarElect scores" [62] along with rank aggregation by stochastic optimization to rank the target genes in order of their likely effectiveness as drug targets. A final list of 18 potentially effective pharmacological targets with associated approved drugs, is produced.

3.15   Muratov et al. Viribus Unitis

Muratov et al. [49] combine data-mining, text-mining, and knowledge-mining within a model aiming at the identification of drug combinations against SARS-CoV-2. As a result, they identified 281 combinations of 38 antiviral drugs that may serve as a potential treatment for COVID-19; of these twenty binary combinations were shortlisted for in vitro experimental testing.

The main tool used is ROBOKOP [48], a data-mining tool developed within Biomedical Data Translator Initiative (https://ncats.nih.gov/translator) to efficiently store, query, rank, and explore sub-graphs of a complex knowledge graph (KG). ROBOKOP provides a unique query mechanism based on meta-graphs and hypothesis ranking algorithms. Drug-Drug Interactions (DDI) are modeled using the QSAR models [74] and the Random Forest ML predictor. In vitro testing of selected drug combinations shortlisted in [49] is reported in Broboswki et al. [8,9].

3.16   Cheng et al. Complementary Exposure

Cheng et al. [13] explore the network configurations arising from drug-drug-disease interactions in the context of hypertension. Analysis of drug pairs known to increase (adverse pairs) or decrease (synergic pairs) hypertension reveals that a specific configuration is associated with favorable hypertension decrease. In particular, this configuration termed *Complementary Exposure* has drug-related subnetworks with small or null overlap among themselves, while both drug-related subnetworks overlap significantly the disease subnetwork.

This principle is turned into a scoring function and a ranking algorithm that is used by [75] to prioritize pairs of drugs with potential synergic effects on SARS-CoV-2.

3.17   Zitnik et al. Decagon

Zitnik et al. [77] tackle the problem of side effects prediction and control in polypharmacy. Here the problem is somewhat dual to that of drug combinations: i.e. predict and minimize the complications due to adverse side effects of drugs administered in conjunction.

Interestingly, the approach relies on the use of well-annotated databases of side effects records for single drugs and pairs of drugs.

Initially, a multimodal graph of protein-protein interactions, drug-protein target interactions, and explicit annotated polypharmacy side effects is built. Next, this graph is used as input for a new graph convolutional neural network algorithm for multi-relational link prediction. The link prediction is in turn interpreted as the prediction of the clinically relevant side effects for a query pair of drugs.

3.18   Gordon et al. a Direct Proteomic Approach

A more direct experimental approach has been described by [27] and [69]. Gordon et al. [27] performed a direct proteomic assay to uncover 332 human proteins potentially interacting with 26 SARS-CoV-2 proteins. Sixty-nine drugs targeting these proteins are then shortlisted as repurposable drugs for COVID-19, using a mix of chemo-informatic searches and expert advice. Note that in this setting a priority ranking is not explicitly provided.

4   Discussion

4.1   Data Availability

In the initial stages of the COVID-19 pandemic, few disease-specific data sets were available for building 'omic' bionetworks, thus researchers did use extensive data collected during outbreaks of other coronaviruses (such as SARS-CoV and MERS-CoV). With time as specific SARS-CoV-2 data is accumulated, reliance on data from other coronaviruses is becoming a matter of design choice, rather than a necessity. One can notice that most of the results cited in this survey rely on a small pool of 'omic' data sets (e.g. [27] for proteomic data and [7] for transcriptomic data), while efforts in extensive data collection and organization into standardized form (e.g. covidgraph.org, the NCATS repository [11] and CoVex) are not yet interfaced effectively with the algorithmic pipelines. Easy integration of data and algorithms is a key factor in the effective development of network-based methods for DR, thus we expect that efforts in this direction might be beneficial for the research community. A notable example of effective integration of data and algorithms via a variety of APIs is given by The Cancer Genome Atlas (TCGA) project in the area of cancer studies. Importantly, the use of a standardized data repository and API will make it possible to quickly apply network-based DR pipelines to new data, and cope with the effects of new viral variants.

4.2   Relationship to Clinical Trials

Most network-based methods for DR make predictions on drugs whose validity is mainly at a pre-clinical level. In other words, the 'in silico'. prediction of the drug efficacy is aimed at shortlisting a drug for the next step of 'in vitro' validation. This is a necessary intermediate step, but it does not in itself guarantee eventual efficacy in 'in vivo' and in clinical trials. It is well known that the vast majority of drugs that give good results 'in vitro' and in model species, eventually fail on humans. Incorporating prior knowledge into biological graph form to improve the chances of final success in a clinical trial is a key challenge for future research. Identification of the drug's mechanism of action and drug target proteins may help in this regard, however, these data are not often known for a new disease or a new drug, thus there is ample scope for progress in this area.

4.3   Network Construction Vs Network Usage

Most of the methods listed in this survey have as a common theme a two-phased approach, in the first phase relevant 'omics' data from the literature and databases are collected to form a searchable network, in the second phase an exploration mechanism to extract useful candidate repurposable drugs is set up (e.g. a new metric, a label diffusion algorithm, a matching rule, etc.), eventually followed by filtering, ranking, and validating finishes based on additional data and considerations (including human expert knowledge). As the final result may depend critically on this sequential approach, it remains often unresolved the issue whether a different way of building the initial network may alter the final outcome. Most methods using PPIN as the main graph input do not take advantage of the fact that several such networks exist each built with different and often complementary criteria (e.g. supplementary material in [23] reports 32 PPIN repositories and DB). This issue is important since at the moment no 'true' large-scale biological network exists (and indeed it is debatable if one definitive global network will ever be built) and the existing networks can be considered just an approximation subject to noise, errors, and vast under-explored areas due, for example, to publication bias. Having algorithms that are robust to network perturbations and bias is a way to increase confidence that 'incomplete' and 'noisy' network data are not detrimental to the final result. A second mitigation strategy for this problem is to use an ensemble of different networks, instead of just one network as input, and merge the final drug lists by their resilience to variations in the input .

4.4   End-to-end and Intermediate Validations

While the final hallmark of success is the identification of a repurposed drug that can pass successfully all stages of the drug approval process for clinical use (see e.g. [20]), it is important to be able to perform *in silico* validation on intermediate results. A recent study by Brown and Patel [12] gives a critical assessment of the options for *in silico* drug validations, listing the weak and strong points of each strategy.

COVID-19 is a recent phenomenon, and, to date, almost no drug has performed to satisfaction, in particular for acute cases[1]. This fact impacts on the validation of drug repurposing pipelines for which almost no 'true positive' drug exists to date for end-to-end validation of the prediction. Thus surrogate criteria must be used. One method is to consider a prediction successful if it predicts as repurposable a drug currently under clinical trial for COVID-19. Usually, drugs administered in a clinical trial must show evidence of effectiveness in 'in vitro'

---

[1] As of now (May 2023) the antiviral drug Veklury (remdesivir) has been approved by FDA for adult use and for certain pediatric conditions. The FDA has also approved the immune modulators Olumiant (baricitinib) and Actemra (tocilizumab) for certain hospitalized adults with COVID-19. Other drugs, such as monoclonal antibody treatments, are allowed by FDA under Emergency Use Authorization (EUA) even if their efficacy is not yet conclusively proven.

experiments, thus this validation criterion is akin to predicting effectiveness in 'in vitro' assays (this criterion is used for example in [29] and [44]). A second method uses actual 'in vitro' COVID-19-infected cell viability assays on a subset of the drugs as golden truth (see e.g. [28]). This is then considered as a test of the reliability of 'in silico' predictions for the drugs not yet tested 'in vitro'. Other methods rely on validating the performance of the pipeline in different pathologies for which effective drugs are known, as a hint of its effectiveness also on COVID-19.

Benchmarking the 'intermediate' steps is important in any pipeline, but it becomes a particularly difficult task in our context. As a general rule, to avoid over-optimistic results and information leaking, data used for the validation of a prediction should not be used to generate the prediction itself. However, since not much proteomic or transcriptomic data have been collected and made public on human samples, the decision as to whether use such data for computation or validation is a delicate one. Thus often validation is based on 'indirect' evidence or the use of data not specific to COVID-19. For example, known effectiveness on SARS-CoV and MERS-Cov, or known effectiveness against symptoms associated with COVID-19, may be used as a hint of success in guessing a COVID-19 repurposable drug.

4.5   Efficacy vs Adverse Effects

In most pipelines, it is often assumed that the drugs under consideration are FDA-approved. This is justified by the rationale that for such drugs their adverse effect profile is known since collecting such data is part of the FDA approval protocol, and, moreover, data on adverse effects is routinely collected, updated, and recorded for drugs that are prescribed to patients by practitioners. This collection of data on adverse effects is usually used to offset some initial stages of a standard clinical trial process, whose aim is to assess drug safety. The downside is that many drug repositioning methods do not model the adverse effects of a drug at the same level as the model for the drug efficacy, which is usually the main criterion for ranking repurposable drugs. Methods that rely on knowing the human (or viral) proteins that bind a specific drug may overlook the fact that often the repurposing effect of a drug is obtained via its off-target binding, in other words, often a side effect of a drug for disease A is exactly what triggers efficacy of the same drug for disease B. Thus for network-based DR, it is important in the long run to be able to model adverse effects without complete reliance on previously recorded knowledge and to be able to address the off-target mechanisms of action, as well as the on-target ones.

4.6   New and Legacy Software Tools

From a bioinformatics point of view, it is interesting to notice that most methods reported in this survey are novel and have been developed specifically to cope with the COVID-19 contextual data and limitations. This is a symptom that network-based drug repositioning is still in its infancy and the current pressure to deliver timely results to cope with the COVID-19 pandemic is an opportunity to mature the field beyond its current limitations. Often off-the-shelf software tools are used to solve particular steps of the pipeline, but, in general, all the surveyed approaches need to be creative in algorithmic design and insight in order to be successful.

4.7   AI and Network-Based Methods

Artificial Intelligence (AI) and in particular Deep Neural Networks (DNN) techniques [76] have been advocated as key to producing the next generation of computational drug repurposing systems, through convergence with network-oriented graph-based approaches. Given the success of DNN in uncovering hidden patterns in several applied areas, ranging from Image Recognition to Protein Folding and Natural Language Translation (to name a few), this line of research is certainly promising. However, a few hurdles must be taken into consideration. First of all, training DNN to attain high success rates requires a vast amount of data, which, in the specifics of a new pandemic, may not be readily available, or come at a high cost. The problem of data paucity can be mitigated by the careful application of *transfer learning* techniques [38], which make use of data derived from similar domains. Second, and maybe more fundamentally, is the issue of result explanation. For example, often auto-encoders are applied to the initial data to perform dimensionality reduction, to eliminate noise, and to produce a compact vector representation of the data, hopefully giving also clues to the hidden rules and semantics. However, in the process of producing such internal representations, often the identity of single 'omic' factors is blurred and hard to trace. This makes it harder to uncover the biological rationale for a certain outcome of the computation. On the positive side, the application of the Graph Neural Network techniques may indeed retain sufficiently the native identity of the network nodes and edges, helping in the final result interpretation (see e.g. [24,77]).

4.8   Pharmaco-Chemical Kinetics

Important aspects of drug development/repurposing such as the mechanism of delivery, the effect of concentrations, the timing of effects, and the drug's kinetics are as yet not well addressed in network-oriented DR. While these aspects are important, and they should be addressed at least in their main aspects, there are well-established pharmaco-chemical protocols and techniques [68] that can be used in subsequent steps of the drug repurposing pipeline for fine-tuning the results of 'in silico' assays.

4.9   Precision Medicine and Biomarkers

Due to the paucity of 'omics' data, initial efforts in computational DR for COVID-19 did not address explicitly in model-building the relevant sub-populations that are at higher risk of mortality or long-term effects (long covid syndrome). For example, it is known that age is one of the key risk factors. However, it is still debatable if this is due mostly to a generalized decay of the immune system in old age [5] or there are specific aging-related pathways that interact with the viral molecular fingerprint and thus can become druggable targets [6]. More generally, age and co-morbidities are obvious phenotypic traits easy to ascertain, with an associated important impact on human health at a molecular level, but it is possible that more subtle molecular characterizations not obviously linked to easy-to-measure phenotypic traits or conditions may have an impact on the disease progression and on drug efficacy [31]. Thus incorporating sub-population-specific molecular biomarkers in the search for druggable targets and repurposable drugs is the next level of model complexity that network-oriented DR should achieve. By its systemic nature, network-oriented DR approaches are uniquely well-posed to incorporate these aspects within their scope.

4.10   Drug Combinations

Finding repurposed drugs and ranking them is the main focus of this review. A main related theme is finding repurposed drug combinations [49, 64]. Often drug combinations offer lower toxicity and more effective disease treatment. Most drug combinations currently under consideration for COVID-19 result from ad-hoc considerations, as screening drug combinations automatically incurs easily a combinatorial explosion of cases to be considered. A few network-based principled approaches have been proposed within the network-oriented DR approach: [13, 33, 75] and [77]. This area however is rather under-researched for general diseases, although, for example, in cancer studies, the availability of large data repositories helped in developing several 'in silico' predictors of drug combination effect [30, 34, 67]. The extension of single drug network-oriented DR to drug combination is an interesting source of challenging questions [3].

4.11   Comparative Assessment of Network-Based Methods

Due to the rapid development of this field, little work has been made in terms of algorithmic consolidation, interface standardization, and comparative performance evaluation. In particular, comparing quantitatively the different proposed tools is problematic due to the variety of assumptions on the input data type, on the underlying biological knowledge, and on the relevant output quality measures. One promising way to move forward in this respect is to expand the crowd-sourced competitions advocated by the DREAM Challenges (Dialogue for Reverse Engineering Assessment and Methods) (https://dreamchallenges. org) [3, 15, 72].

5   Conclusion

The area of network-based drug repositioning is vast and continuously evolving. For example, we do not touch areas such as literature-based networks, methods aiming mainly at target-protein detection, and drug repositioning methods not aimed at a specific disease. We gave preference to disease-specific 'omic'-based approaches as these have a better chance to be applicable to new diseases with little prior knowledge.

The methods and approaches selected in this survey are certainly incomplete in number and reflect the inclinations of the author. However, this bird-view of the field has led us to spot interesting trends and relevant open challenges.

Acknowledgment. This work has been done within the research project DIT.AD021. 153 "Network Bioscience and Applications" funded by CNR.

References

1. Alaimo, S., Pulvirenti, A.: Network-based drug repositioning: approaches, resources, and research directions. In: Vanhaelen, Q. (ed.) Computational Methods for Drug Repurposing. Methods in Molecular Biology, vol. 1903, pp. 97–113. Humana Press, New York (2019). https://doi.org/10.1007/978-1-4939-8955-3_6
2. Alcaraz, N., Kücük, H., Weile, J., Wipat, A., Baumbach, J.: KeyPathwayMiner: detecting case-specific biological pathways using expression data. Internet Math. **7**(4), 299–313 (2011)
3. Bansal, M., et al.: A community computational challenge to predict the activity of pairs of compounds. Nat. Biotechnol. **32**(12), 1213–1222 (2014)
4. Barabasi, A.L., Oltvai, Z.N.: Network biology: understanding the cell's functional organization. Nat. Rev. Genet. **5**(2), 101–113 (2004)
5. Bartleson, J.M., Radenkovic, D., Covarrubias, A.J., Furman, D., Winer, D.A., Verdin, E.: SARS-CoV-2, COVID-19 and the aging immune system. Nat. Aging **1**(9), 769–782 (2021)
6. Belyaeva, A., et al.: Causal network models of SARS-CoV-2 expression and aging to identify candidates for drug repurposing. Nat. Commun. **12**(1), 1–13 (2021)
7. Blanco-Melo, D., et al.: SARS-CoV-2 launches a unique transcriptional signature from in vitro, ex vivo, and in vivo systems. BioRxiv (2020)
8. Bobrowski, T., et al.: Discovery of synergistic and antagonistic drug combinations against SARS-CoV-2 in vitro. BiorXiv (2020). https://doi.org/10.1101/2020.06.29.178889
9. Bobrowski, T., et al.: Synergistic and antagonistic drug combinations against SARS-CoV-2. Mol. Therapy **29**(2), 873–885 (2021)
10. Bojkova, D., et al.: Proteomics of SARS-CoV-2-infected host cells reveals therapy targets. Nature **583**(7816), 469–472 (2020)
11. Brimacombe, K.R., et al.: An OpenData portal to share COVID-19 drug repurposing data in real time. BioRxiv (2020)
12. Brown, A.S., Patel, C.J.: A review of validation strategies for computational drug repositioning. Briefings Bioinform. **19**(1), 174–177 (2018)
13. Cheng, F., Kovács, I.A., Barabási, A.-L.: Network-based prediction of drug combinations. Nat. Commun. **10**(1), 1–11 (2019)
14. Ciliberto, G., Mancini, R., Paggi, M.G.: Drug repurposing against COVID-19: focus on anticancer agents. J. Exp. Clin. Cancer Res. **39**, 1–9 (2020)
15. Costello, J.C., et al.: A community effort to assess and improve drug sensitivity prediction algorithms. Nat. Biotechnol. **32**(12), 1202–1212 (2014)

16. Csermely, P., Korcsmaros, T., Kiss, H.J., London, G., Nussinov, R.: Structure and dynamics of molecular networks: a novel paradigm of drug discovery: a comprehensive review. Pharmacol. Ther. **138**(3), 333–408 (2013)
17. Dotolo, S., Marabotti, A., Facchiano, A., Tagliaferri, R.: A review on drug repurposing applicable to COVID-19. Briefings Bioinform. **22**, 726–741 (2020)
18. Draghici, S., et al.: A systems biology approach for pathway level analysis. Genome Res. **17**(10), 1537–1545 (2007)
19. Draghici, S., et al.: COVID-19: disease pathways and gene expression changes predict methylprednisolone can improve outcome in severe cases. Bioinformatics (Oxford, England) **37**(17), 2691–2698 (2021)
20. Edwards, A.: What are the odds of finding a COVID-19 drug from a lab repurposing screen? J. Chem. Inf. Model. **60**, 5727–5729 (2020)
21. Fiscon, G., Conte, F., Farina, L., Paci, P.: SAveRUNNER: a network-based algorithm for drug repurposing and its application to COVID-19. arXiv preprint: arXiv:2006.03110 (2020)
22. Fiscon, G., Conte, F., Farina, L., Paci, P.: SAveRUNNER: a network-based algorithm for drug repurposing and its application to COVID-19. PLoS Comput. Biol. **17**(2), e1008686 (2021)
23. Galindez, G., et al.: Lessons from the COVID-19 pandemic for advancing computational drug repurposing strategies. Nat. Comput. Sci. **1**(1), 33–41 (2021)
24. Gaudelet, T., et al.: Utilizing graph machine learning within drug discovery and development. Briefings in Bioinform. **22**(6), bbab159 (2021)
25. Ge, Y., et al.: An integrative drug repositioning framework discovered a potential therapeutic agent targeting COVID-19. Signal Transduct. Target. Ther. **6**(1), 1–16 (2021)
26. Ghiassian, S.D., Menche, J., Barabasi, A.L.: A disease module detection (diamond) algorithm derived from a systematic analysis of connectivity patterns of disease proteins in the human interactome. PLoS Comput. Biol. **11**(4), e1004120 (2015)
27. Gordon, D. E., et al.: A SARS-CoV-2 protein interaction map reveals targets for drug repurposing. Nature, 1–13 (2020)
28. Gysi, D.M., et al.: Network medicine framework for identifying drug-repurposing opportunities for COVID-19. Proc. Nat. Acad. Sci. **118**(19), e2025581118 (2021)
29. Morselli Gysi, D.: Network medicine framework for identifying drug repurposing opportunities for COVID-19 (2020)
30. He, L., et al.: Network-guided identification of cancer-selective combinatorial therapies in ovarian cancer. Briefings Bioinform. **22**(6), bbab272 (2021)
31. COVID-19 Host Genetics Initiative et al. Mapping the human genetic architecture of COVID-19. Nature (2021)
32. Jarada, T.N., Rokne, J.G., Alhajj, R.: A review of computational drug repositioning: strategies, approaches, opportunities, challenges, and directions. J. Cheminform. **12**(1), 1–23 (2020)
33. Jin, W., Barzilay, R., Jaakkola, T.: Modeling drug combinations based on molecular structures and biological targets. arXiv preprint: arXiv:2011.04651 (2020)
34. Julkunen, H., et al.: Leveraging multi-way interactions for systematic prediction of pre-clinical drug combination effects. Nat. Commun. **11**(1), 1–11 (2020)
35. Kuleshov, M.V., et al.: Enrichr: a comprehensive gene set enrichment analysis web server 2016 update. Nucleic Acids Res. **44**(W1), W90–W97 (2016)
36. Lamb, J., et al.: The connectivity map: using gene-expression signatures to connect small molecules, genes, and disease. Science **313**(5795), 1929–1935 (2006)
37. Levin, J.M., et al.: Artificial intelligence, drug repurposing and peer review. Nat. Biotechnol. **38**(10), 1127–1131 (2020)

38. Li, H., et al.: A compact review of progress and prospects of deep learning in drug discovery. J. Mol. Model. **29**(4), 117 (2023)
39. Li, J., Zheng, S., Chen, B., Butte, A.J., Swamidass, S.J., Lu, Z.: A survey of current trends in computational drug repositioning. Briefings Bioinform. **17**(1), 2–12 (2016)
40. Liu, Q., Wan, J., Wang, G.: A survey on computational methods in discovering protein inhibitors of SARS-CoV-2. Briefings Bioinform. **23**, bbab416 (2021)
41. Lotfi Shahreza, M., Ghadiri, N., Mousavi, S.R., Varshosaz, J., Green, J.R.: A review of network-based approaches to drug repositioning. Briefings Bioinfo. **19**(5), 878–892 (2018)
42. Low, Z.Y., Farouk, I.A., Lal, S.K.: Drug repositioning: new approaches and future prospects for life-debilitating diseases and the COVID-19 pandemic outbreak. Viruses **12**(9), 1058 (2020)
43. Lucchetta, M., Pellegrini, M.: Finding disease modules for cancer and COVID-19 in gene co-expression networks with the core&peel method. Sci. Rep. **10**, 17628 (2020). https://doi.org/10.1038/s41598-020-74705-6
44. Lucchetta, M., Pellegrini, M.: Drug repositioning by merging active subnetworks validated in cancer and COVID-19. Sci. Rep. **11**(1), 19839 (2021)
45. Mall, R., Elbasir, A., Al Meer, H., Chawla, S., Ullah, E.: Data-driven drug repurposing for covid-19. ChemRxiv chemrxiv 12661103, v1 (2020)
46. Maria, N., et al.: Rapid identification of druggable targets and the power of the phenotype simulator for effective drug repurposing in COVID-19. Res. Square., rs-3 (2021)
47. Mernea, M., Martin, E., Petrescu, A.-J., Avram, S.: Deep learning in the quest for compound nomination for fighting COVID-19. Curr. Med. Chem. **28**, 5699–5732 (2021)
48. Morton, K., et al.: ROBOKOP: an abstraction layer and user interface for knowledge graphs to support question answering. Bioinformatics **35**(24), 5382–5384 (2019)
49. Muratov, E., Zakharov, A.: Viribus unitis: drug combinations as a treatment against COVID-19. ChemRxiv (2020)
50. Nabirotchkin, S., Peluffo, A.E., Rinaudo, P., Yu, J., Hajj, R., Cohen, D.: Next-generation drug repurposing using human genetics and network biology. Curr. Opin. Pharmacol. **11**, 1–15 (2020)
51. Nguyen, T., Mitrea, C., Draghici, S.: Network-based approaches for pathway level analysis. Curr. Protoc. Bioinformatics **61**(1), 8–25 (2018)
52. Nguyen, T., Shafi, A., Nguyen, T.M., Schissler, A.G., Draghici, S.: NBIA: a network-based integrative analysis framework-applied to pathway analysis. Sci. Rep. **10**(1), 1–11 (2020)
53. Pellegrini, M., Antoniotti, M., Mishra, B. (eds.) Network Bioscience, second edition. Frontiers Media SA, Lousanne (2020)
54. Peyvandipour, A., Saberian, N., Shafi, A., Donato, M., Draghici, S.: A novel computational approach for drug repurposing using systems biology. Bioinformatics **34**(16), 2817–2825 (2018)
55. Pushpakom, S., et al.: Drug repurposing: progress, challenges and recommendations. Nat. Rev. Drug Disc. **18**(1), 41–58 (2019)
56. Re, M., Valentini, G.: Network-based drug ranking and repositioning with respect to DrugBank therapeutic categories. IEEE/ACM Trans. Comput. Biol. Bioinf. **10**(6), 1359–1371 (2013)
57. Ruiz, C., Zitnik, M., Leskovec, J.: Discovery of disease treatment mechanisms through the multiscale interactome. BioRxiv (2020)

58. Ruiz, C., Zitnik, M., Leskovec, J.: Identification of disease treatment mechanisms through the multiscale interactome. Nat. Commun. **12**(1), 1–15 (2021)
59. Saberian, N., Peyvandipour, A., Donato, M., Ansari, S., Draghici, S.: A new computational drug repurposing method using established disease-drug pair knowledge. Bioinformatics **35**(19), 3672–3678 (2019)
60. Sadegh, S., et al.: Exploring the SARS-CoV-2 virus-host-drug interactome for drug repurposing. Nat. Commun. **11**(1), 3518 (2020)
61. Singh, T.U., Parida, S., Lingaraju, M.C., Kesavan, M., Kumar, D., Singh, R.K.: Drug repurposing approach to fight COVID-19. Pharmacol. Rep., 1–30 (2020)
62. Stelzer, G., et al.: VarElect: the phenotype-based variation prioritizer of the genecards suite. BMC Genomics **17**(2), 195–206 (2016)
63. Stolfi, P., Manni, L., Soligo, M., Vergni, D., Tieri, P.: Designing a network proximity-based drug repurposing strategy for COVID-19. Front. Cell Dev. Biol. **8**, 545089 (2020)
64. Sun, W., Sanderson, P.E., Zheng, W.: Drug combination therapy increases successful drug repositioning. Drug Discov. Today **21**(7), 1189–1195 (2016)
65. Taguchi, Y.H., Turki, T.: A new advanced in silico drug discovery method for novel coronavirus (SARS-CoV-2) with tensor decomposition-based unsupervised feature extraction. PLoS ONE **15**(9), e0238907 (2020)
66. YH Taguchi and Turki Turki: A new advanced in silico drug discovery method for novel coronavirus (SARS-COV-2) with tensor decomposition-based unsupervised feature extraction. PLoS ONE **15**(9), e0238907 (2020)
67. Tanoli, Z., Vaha-Koskela, M., Aittokallio, T.: Artificial intelligence, machine learning, and drug repurposing in cancer. Expert Opin. Drug Disc., 1–13 (2021)
68. Tonge, P.J.: Drug-target kinetics in drug discovery. ACS Chem. Neurosci. **9**(1), 29–39 (2018)
69. Tutuncuoglu, B., et al.: The landscape of human cancer proteins targeted by SARS-CoV-2. Cancer Disc. **10**(7), 916–921 (2020)
70. Wan, F., Hong, L., Xiao, A., Jiang, T., Zeng, J.: NeoDTI: neural integration of neighbor information from a heterogeneous network for discovering new drug-target interactions. Bioinformatics **35**(1), 104–111 (2019)
71. Wishart, D.S., et al.: DrugBank 5.0: a major update to the DrugBank database for 2018. Nucleic Acids Res. **46**(D1), D1074–D1082 (2018)
72. Xiong, Z., et al.: Crowdsourced identification of multi-target kinase inhibitors for RET- and TAU- based disease: the multi-targeting drug dream challenge. PLoS Comput. Biol. **17**(9), 1–19 (2021)
73. Xue, H., Li, J., Xie, H., Wang, Y.: Review of drug repositioning approaches and resources. Int. J. Biol. Sci. **14**(10), 1232 (2018)
74. Zakharov, A.V., et al.: QSAR modeling and prediction of drug-drug interactions. Mol. Pharm. **13**(2), 545–556 (2016)
75. Zhou, Y., Hou, Y., Shen, J., Huang, Y., Martin, W., Cheng, F.: Network-based drug repurposing for novel coronavirus 2019-NCOV/SARS-CoV-2. Cell Disc. **6**(1), 1–18 (2020)
76. Zhou, Y., Wang, F., Tang, J., Nussinov, R., Cheng, F.: Artificial intelligence in COVID-19 drug repurposing. Lancet Digit. Health **2**, e667–e676 (2020)
77. Zitnik, M., Agrawal, M., Leskovec, J.: Modeling polypharmacy side effects with graph convolutional networks. Bioinformatics **34**(13), i457–i466 (2018)

Integer Programming Based Algorithms for Overlapping Correlation Clustering

Barel I. Mashiach[ID] and Roded Sharan[✉][ID]

Blavatnik School of Computer Science, Tel Aviv University, 69978 Tel Aviv, Israel
roded@tauex.tau.ac.il

Abstract. Clustering is a fundamental problem in data science with diverse applications in biology. The problem has many combinatorial and statistical variants, yet few allow clusters to overlap which is common in the biological domain. Recently, Bonchi et al. defined a new variant of the clustering problem, termed overlapping correlation clustering, which calls for multi-label cluster assignments that correlate with an input similarity between elements as much as possible. This variant is NP-hard and was solved by Bonchi et al. using a local search heuristic. We revisit this heuristic and develop exact integer-programming based variants for it. We show that these variants perform well across several datasets and evaluation measures.

1 Introduction

Clustering is a fundamental problem in data science. While most clustering methods look for disjoint clusters [2,11], in many real-world application, and specifically in biology, an element might belong to more than one cluster. For example, a gene may have multiple functions, and a protein may belong to multiple protein complexes. One framework for dealing with such multi-label scenarios is *Overlapping Correlation Clustering (OCC)*, which was introduced by Bonchi et al. [3] and generalized a non-overlapping variant called *Correlation Clustering (CC)* [2] or *Cluster Editing* [14]. In *OCC*, one seeks an assignment of clusters to elements that maximizes the correlation with a given pairwise similarity.

Bonchi et al. [3] tackled this problem by iteratively finding the optimal labeling for one element given the labels of all other elements, using efficient, yet heuristic techniques (see Sect. 3.2). Another iterative algorithm for the problem was developed by Andrade et al. [1] which uses *Biased Random-Key Genetic Algorithm (BRKGA)*. A third method used a weighted Lovász theta function for node embedding and subsequently clustering [10]. Finally, Gartzman et al. [8] developed an exact solution by defining an integer-linear-programming formulation for the Jaccard variant of the problem called *ECLIP* (see Sect. 3.1).

Here we revisit previous local methods and develop exact integer-programming based algorithms for them that borrow ideas from both [3,8]. We show a simple algorithm solving each iterative step of Bonchi's algorithm by checking all possible labelings, assuming the number of labels is bounded. Our main contribution is a new iterative method for *OCC* that optimally assigns a

D. Cantone and A. Pulvirenti (Eds.): *From Computational Logic to Computational Biology*,
LNCS 14070, pp. 115–127, 2024.
https://doi.org/10.1007/978-3-031-55248-9_6

specific label given the assignments to all other labels using an integer linear program (ILP) formulation.

2   Preliminaries and Problem Definition

Consider a set of n elements V over which we define pairwise similarity function $s(u, v) \in [0, 1]$. Moreover let K be a set of k clusters $K = \{1, \ldots k\}$. In CC the goal is to construct a mapping $\mathcal{L} : V \to K$ that minimizes the cost:

$$C_{CC}(V, \mathcal{L}) = \sum_{\substack{(u,v) \in V \times V \\ \mathcal{L}(u) = \mathcal{L}(v)}} (1 - s(u, v)) + \sum_{\substack{(u,v) \in V \times V \\ \mathcal{L}(u) \neq \mathcal{L}(v)}} s(u, v) \tag{1}$$

In OCC, a multi-labeling mapping $\mathcal{L} : V \to 2^K$ is sought. Let $H(\mathcal{L}(u), \mathcal{L}(v))$ denote a similarity between two cluster assignments of elements u and v according to the mapping $\mathcal{L}$. Following Bonchi et al., we consider two different variants of H: the Jaccard coefficient J and the set-intersection indicator I, defined as follows:

$$J(\mathcal{L}(u), \mathcal{L}(v)) = \frac{|\mathcal{L}(u) \cap \mathcal{L}(v)|}{|\mathcal{L}(u) \cup \mathcal{L}(v)|} \tag{2}$$

$$I(\mathcal{L}(u), \mathcal{L}(v)) = \begin{cases} 1 & \text{if } |\mathcal{L}(u) \cap \mathcal{L}(v)| \neq \emptyset \\ 0 & \text{o.w} \end{cases} \tag{3}$$

Using these definitions the OCC objective is the entry-wise cost function:

$$C_{OCC}(V, \mathcal{L}) = \frac{1}{2} \sum_{u \in V} \sum_{v \in V \setminus \{u\}} |H(\mathcal{L}(u), \mathcal{L}(v)) - s(u, v)| \tag{4}$$

The original problem also addresses an additional constraint p on the mapping $\mathcal{L}$ which limits the number of clusters a single element can be assigned to, and in some cases we also require $p \geq 1$. One should notice that the OCC generalizes CC, which corresponds to the case of $p = 1$, and hence its hardness follows from the fact that CC is NP-hard [6].

3   Methods

3.1   ILP-Based Algorithms

As mentioned above, Gartzman et al. [8] gave a formulation of OCC as an integer-linear program (ILP) for the Jaccard case. The main drawback of this method is that it is not scalable for large datasets and limited due to significant long running time on very small subsets of up to 40 elements (see Fig. 1 below). For consistency with the methods before, we also formulate an ILP for the set-intersection case. The formulation described by Algorithm 1 is simple - we want to find a binary matrix $L \in \{0, 1\}^{n \times k}$ which represents assignments of each of the n elements to the k labels, denoting $L(i, l)$ as whether element i is assigned to label l. We further define $w_{i,j}^l$ as $L(i, l) \cdot L(j, l)$ representing whether both i and j share the label l. Using these variables we can compute the total number of shared labels between i and j as $y_{i,j}$ by summing $w_{i,j}^l$ over all possible labels and use $b_{i,j}$ as an indicator telling whether i and j share at least one label by setting $b_{i,j}$ to 1 if $y_{i,j} > 0$ and to 0 otherwise. From these definitions we can create the following minimization problem, claiming that $b_{i,j} + s(v_i, v_j) - 2b_{i,j} \cdot s(v_i, v_j)$ is exactly the expression $|H(\mathcal{L}(u), \mathcal{L}(v)) - s(u, v)|$, therefore achieving the objective of the original OCC problem.

Algorithm 1. ILP for OCC with set-intersection (ISEC-ILP)

$\min_L \sum_{i=1}^n \sum_{j=i+1}^n b_{i,j} + s(v_i, v_j) - 2b_{i,j} \cdot s(v_i, v_j)$ ▷ Equivalent to Eq. 4 (isec)

s.t. $\forall 1 \leq i < j \leq n, 1 \leq l \leq k$
  $w_{i,j}^l, b_{i,j} \in \{0, 1\}, y_{i,j} \in \mathbb{Z}^+$

  $w_{i,j}^l \leq L(i, l)$ ▷ Setting $w_{i,j}^l = L(i, l) \cdot L(j, l)$
  $w_{i,j}^l \leq L(j, l)$
  $L(i, l) + L(j, l) - 1 \leq w_{i,j}^l$

  $y_{i,j} = \sum_{l'=1}^k w_{i,j}^{l'}$ ▷ Setting $y_{i,j}$ as total number of shared labels

  $y_{i,j} - (k+1)b_{i,j} \geq -k$ ▷ Setting $b_{i,j} = 1$ if $y_{i,j} > 0$ and to 0 o.w
  $y_{i,j} - (k+1)b_{i,j} \leq 0$

  $1 \leq \sum_{l'=1}^k L(i, l') \leq p$ ▷ Setting total assignments of an element up to p labels

return L

3.2   Row-Based Clustering

The method of Bonchi et al. [3] for solving the OCC objective iteratively finds the optimal labeling vector for an element v given fixed labeling vectors of all other elements. In each step we aim to find labeling $\mathcal{L}^{t+1}(v)$ which minimizes the cost produced by v with all other elements:

$$\min_{\mathcal{L}^{t+1}(v)} C_{v,p}(\mathcal{L}^{t+1}(v) | \mathcal{L}^t) = \min_{\mathcal{L}^{t+1}(v)} \sum_{u \in V \setminus \{v\}} |H(\mathcal{L}^t(u), \mathcal{L}^t(v)) - s(u, v)| \quad (5)$$

This iterative step requires solving non-trivial optimization subproblems:

- For $H = J$, the problem is related to the *Jaccard-Triangulation* problem [3] which generalizes the NP-hard *Jaccard-Median* problem [5].
- For $H = I$, the problem is related to the *Hit-N-Miss* problem [3] which is isomorphic to the NP-hard *positive-negative partial set-cover* problem [12].

Therefore, Bonchi et al. employ heuristic approaches to tackle the resulting problems. Here we show that one can derive exact yet practical solutions for these problems using fixed parameter and integer programming techniques.

If we denote the number of clusters by a parameter k, then we can use a naive fixed parameter algorithm for precisely solving the local step optimization for every element in $O(2^k)$ time by enumerating all possible labeling vectors for v. Moreover when we know a bound $p \leq k$ on how many clusters a single element can be assigned to, then the number of possible options reduces to $\binom{k}{p} = O(k^p)$.

3.3   Column-Based Clustering

Our main contribution is a new optimization approach which looks in every iteration on each column (label) separately, aiming to assign the label's elements in a way that will minimize the overall clustering objective. Let L^t be a binary matrix of size $n \times k$ which represents the labeling assignment by $\mathcal{L}^t$. We denote $L^t(\cdot, l)$ as the l column representing the assignment for label l at iteration t. Algorithm 2 shows the main flow of our new local search method which iteratively updates the elements assignments L^{t+1} to every label according to previous assignments $\bar{L}^t$ (which for simplicity of notations would be referred just as L^t) to all other labels.

Algorithm 2. Label-by-Label Local Search (COL-ILP)

1: Initialize L^0 to a valid labeling
2: $t \leftarrow 1$
3: **while** $C_{\mathrm{OCC}}(V, L^t)$ decreases **do**
4:     Set $\bar{L}^t \leftarrow L^t$
5:     **for** $l \in [k]$ **do**
6:         Find the changes x in label l which maximizes the cost difference (ILP)
7:         Set $\bar{L}^t(\cdot, l) \leftarrow \bar{L}^t(\cdot, l) \cdot (1 - x) + (1 - \bar{L}^t(\cdot, l)) \cdot x$
8:         Set $L^{t+1} \leftarrow \bar{L}^t$
9:     **end for**
10:     $t \leftarrow t + 1$
11: **end while**
12: **return** L^t

In order to improve the assignment of a label l we need to calculate the difference in cost of every pair of elements v_i, v_j with respect to the proposed changes in the assignment. There are four cases to consider regarding a specific element v_i:

1. Element v_i is added to label l, i.e., $L^t(i, l) = 0$ and $L^{t+1}(i, l) = 1$.
2. Element v_i remains unlabeled with l, i.e., $L^{t+1}(i, l) = L^t(i, l) = 0$.
3. Element v_i is removed from label l, i.e., $L^t(i, l) = 1$ and $L^{t+1}(i, l) = 0$.
4. Element v_i remains labeled with l, i.e., $L^{t+1}(i, l) = L^t(i, l) = 1$.

These four cases induce 10 types of element pair combinations. To enumerate them, we define $e_i = L^t(i, l)$, and let the variable x_i be 1 if there was a change in v_i with respect to label l and 0 otherwise. By calculating cost differences for each of the 10 pair types we can define an optimization criterion for maximizing the total difference in cost over all pairs:

$$\max_x \sum_i \sum_{j>i} \begin{bmatrix} x_i \cdot x_j \cdot \left[(1-e_i)\cdot(1-e_j)\cdot \hat{C}_{i,j}^{1,1} + e_i \cdot e_j \cdot \hat{C}_{i,j}^{3,3} + (1-e_i)\cdot e_j \cdot \hat{C}_{i,j}^{1,3} + e_i \cdot(1-e_j)\cdot \hat{C}_{i,j}^{3,1}\right] \\ +x_i \cdot(1-x_j)\cdot\left[(1-e_i)\cdot(1-e_j)\cdot \hat{C}_{i,j}^{1,2} + e_i \cdot(1-e_j)\cdot \hat{C}_{i,j}^{3,2} + (1-e_i)\cdot e_j \cdot \hat{C}_{i,j}^{1,4} + e_i \cdot e_j \cdot \hat{C}_{i,j}^{3,4}\right] \\ +(1-x_i)\cdot x_j \cdot\left[(1-e_i)\cdot(1-e_j)\cdot \hat{C}_{i,j}^{2,1} + (1-e_i)\cdot e_j \cdot \hat{C}_{i,j}^{2,3} + e_i \cdot(1-e_j)\cdot \hat{C}_{i,j}^{4,1} + e_i \cdot e_j \cdot \hat{C}_{i,j}^{4,3}\right] \end{bmatrix} \quad (6)$$

where $\hat{C}_{i,j}^{r,q}$ is the cost difference, for element pairs involving one element i with a change of type r and the other element j with a change of type q, for $r, q \in \{1, 2, 3, 4\}$. By definition, $\hat{C}_{i,j}^{r,q} = \hat{C}_{j,i}^{q,r}$, hence from now on we assume that $r \leq q$.

Note that this objective defines a quadratically constrained quadratic program (QCQP) which we can transform into an integer linear program by defining the variables $w_{i,j}$ for every $i < j \in [n]$ to reflect $x_i \cdot x_j$ and adding the constrains: $x_i + x_j - 1 \leq w_{i,j} \leq x_i, x_j$.

Jaccard Label Fix. Let $J_{i,j}^t$ be the Jaccard coefficient of elements v_i and v_j at iteration t, with $|\mathcal{L}^t(v_i) \cap \mathcal{L}^t(v_j)| = n_{i,j}$ and $|\mathcal{L}^t(v_i) \cup \mathcal{L}^t(v_j)| = d_{i,j}$ (i.e., $J_{i,j}^t = \frac{n_{i,j}}{d_{i,j}}$).

We define $J_{i,j}^{r,q}$ as the value of $J_{i,j}^{t+1}$ between two elements i and j, the first with a change of type r and the second with a change of type q, for $r, q \in \{1, 2, 3, 4\}$, and $\hat{J}_{i,j}^{r,q}$ would be the difference between previous Jaccard value and the new one according to these changes. Moreover we let $z^{r,q}$ be an indicator for whether the Jaccard value was decreased or increased.

$$z^{r,q} = \begin{cases} 1 & \text{if } J_{i,j}^t > J_{i,j}^{t+1} \\ 0 & \text{o.w} \end{cases} \quad (7)$$

The change in coefficient $\hat{J}_{i,j}^{r,q}$ as a function of all possible changes r, q appears in Table 1, omitting cases where there is no change. These values are applied to the cost difference formula in (9).

In order to compute the cost change $\hat{C}_{i,j}^{r,q}$ according to $\hat{J}_{i,j}^{r,q}$, we need to take into consideration whether $J_{i,j}^t \geq s(i, j)$ and whether $C_{i,j}^t \geq |\hat{J}_{i,j}^{r,q}|$, thus we define the following indicators for every $i, j \in [n]$ and $r, q \in \{1, 2, 3, 4\}$:

$$g_{i,j} = \begin{cases} 1 & \text{if } J_{i,j}^t \geq s(i,j) \\ 0 & \text{o.w} \end{cases} \qquad k_{i,j}^{r,q} = \begin{cases} 1 & \text{if } C_{i,j}^t \geq |\hat{J}_{i,j}^{r,q}| \\ 0 & \text{o.w} \end{cases} \quad (8)$$

Claim. The cost difference $\hat{C}_{i,j}^{r,q}$ for changes of type r and q in elements v_i and v_j, respectively under the use of Jaccard similarity is:

$$\hat{C}_{i,j}^{r,q} = (g_{i,j} \cdot z^{r,q} + (1-g_{i,j})\cdot(1-z^{r,q}))\cdot\left(|\hat{J}_{i,j}^{r,q}|\cdot k_{i,j}^{r,q} + (2C_{i,j}^t - |\hat{J}_{i,j}^{r,q}|)\cdot(1-k_{i,j}^{r,q})\right)$$
$$+((1-g_{i,j})\cdot z^{r,q} + g_{i,j}\cdot(1-z^{r,q}))\cdot\left(-|\hat{J}_{i,j}^{r,q}|\right) \quad (9)$$

Table 1. Jaccard difference $\hat{J}_{i,j}^{r,q}$ according to type of changes. On time t the Jaccard coefficient is $J_{i,j}^t = \frac{n_{i,j}}{d_{i,j}}$, and on time $t+1$ it has the value $J_{i,j}^{r,q}$.

r	q	$J_{i,j}^{r,q}$	$\hat{J}_{i,j}^{r,q}$		$z^{r,q}$
1	1	$\frac{n_{i,j}+1}{d_{i,j}+1}$	$\begin{cases} \frac{n_{i,j}-d_{i,j}}{d_{i,j}\cdot(d_{i,j}+1)} & \text{if } d_{i,j} > 0 \\ -1 & \text{if } d_{i,j} = 0 \end{cases}$		0
1	2	$\frac{n_{i,j}}{d_{i,j}+1}$	$\begin{cases} \frac{n_{i,j}}{d_{i,j}\cdot(d_{i,j}+1)} & \text{if } d_{i,j} > 0 \\ 1 & \text{if } d_{i,j} = 0 \end{cases}$		1
1	4	$\frac{n_{i,j}+1}{d_{i,j}}$	$\frac{-1}{d_{i,j}} \; (d_{i,j} > 0)$		0
2	3	$\frac{n_{i,j}}{d_{i,j}-1}$	$\begin{cases} \frac{-n_{i,j}}{d_{i,j}\cdot(d_{i,j}-1)} & \text{if } d_{i,j} > 1 \\ -1 & \text{if } d_{i,j} = 1 \end{cases}$		0
3	3	$\frac{n_{i,j}-1}{d_{i,j}-1}$	$\begin{cases} \frac{d_{i,j}-n_{i,j}}{d_{i,j}\cdot(d_{i,j}-1)} & \text{if } d_{i,j} > 1 \\ 1 & \text{if } d_{i,j} = 1 \end{cases}$		1
3	4	$\frac{n_{i,j}-1}{d_{i,j}}$	$\frac{1}{d_{i,j}} \; (d_{i,j} > 0)$		1

Proof. – If $g_{i,j} = 1$ it means that at iteration t the Jaccard between the two elements exceeded their similarity, therefore: $C_{i,j}^t = J_{i,j}^t - s(i,j)$ and $C_{i,j}^{t+1} = |J_{i,j}^t - \hat{J}_{i,j}^{r,q} - s(i,j)| = |C_{i,j}^t - \hat{J}_{i,j}^{r,q}|$.

- If $z^{r,q} = 1$ then $\hat{J}_{i,j}^{r,q} = |\hat{J}_{i,j}^{r,q}|$,
  * If $k_{i,j}^{r,q} = 1$ then $C_{i,j}^t \geq |\hat{J}_{i,j}^{r,q}| = \hat{J}_{i,j}^{r,q}$, so: $\hat{C}_{i,j}^{r,q} = C_{i,j}^t - (C_{i,j}^t - \hat{J}_{i,j}^{r,q}) = \hat{J}_{i,j}^{r,q} = |\hat{J}_{i,j}^{r,q}|$.
  * If $k_{i,j}^{r,q} = 0$ then $C_{i,j}^t < |\hat{J}_{i,j}^{r,q}| = \hat{J}_{i,j}^{r,q}$, so: $\hat{C}_{i,j}^{r,q} = C_{i,j}^t - (\hat{J}_{i,j}^{r,q} - C_{i,j}^t) = 2C_{i,j}^t - \hat{J}_{i,j}^{r,q} = 2C_{i,j}^t - |\hat{J}_{i,j}^{r,q}|$.
- If $z^{r,q} = 0$ then $\hat{J}_{i,j}^{r,q} = -|\hat{J}_{i,j}^{r,q}|$, and we get that the cost at time $t+1$ is $C_{i,j}^{t+1} = |C_{i,j}^t + |\hat{J}_{i,j}^{r,q}|| = C_{i,j}^t + |\hat{J}_{i,j}^{r,q}|$, which means that $\hat{C}_{i,j}^{r,q} = -|\hat{J}_{i,j}^{r,q}|$.

– Conversely, if $g_{i,j} = 0$ then $C_{i,j}^t = s(i,j) - J_{i,j}^t$ and $C_{i,j}^{t+1} = |J_{i,j}^t - \hat{J}_{i,j}^{r,q} - s(i,j)| = |-C_{i,j}^t - \hat{J}_{i,j}^{r,q}|$.

- If $z^{r,q} = 1$ then $\hat{J}_{i,j}^{r,q} = |\hat{J}_{i,j}^{r,q}|$, and we get that the cost at time $t+1$ is $C_{i,j}^{t+1} = |-C_{i,j}^t - |\hat{J}_{i,j}^{r,q}|| = C_{i,j}^t + |\hat{J}_{i,j}^{r,q}|$, which means that $\hat{C}_{i,j}^{r,q} = -|\hat{J}_{i,j}^{r,q}|$.
- If $z^{r,q} = 0$ then $\hat{J}_{i,j}^{r,q} = -|\hat{J}_{i,j}^{r,q}|$, and we get that the cost at time $t+1$ is $C_{i,j}^{t+1} = ||\hat{J}_{i,j}^{r,q}| - C_{i,j}^t|$,
  * If $k_{i,j}^{r,q} = 1$ then $C_{i,j}^t \geq |\hat{J}_{i,j}^{r,q}|$, so: $\hat{C}_{i,j}^{r,q} = C_{i,j}^t - (C_{i,j}^t - |\hat{J}_{i,j}^{r,q}|) = |\hat{J}_{i,j}^{r,q}|$
  * If $k_{i,j}^{r,q} = 0$ then $C_{i,j}^t < |\hat{J}_{i,j}^{r,q}|$, so: $\hat{C}_{i,j}^{r,q} = C_{i,j}^t - (|\hat{J}_{i,j}^{r,q}| - C_{i,j}^t) = 2C_{i,j}^t - |\hat{J}_{i,j}^{r,q}|$

Combining together all these claims we can get the mentioned formula for $\hat{C}_{i,j}^{r,q}$.   □

Set-Intersection Label Fix. Similarly to the Jaccard case, we develop a formulation for the set-intersection case. If two elements v_i and v_j share at least one common label then their cost is $1 - s(v_i, v_j)$, and $s(v_i, v_j)$ otherwise. Thus the cost change for every pair of elements should be $\pm(1 - 2s(v_i, v_j))$ or zero, while the only interesting cases are when elements which share a single label do not share it anymore or when elements which do not have any common label now share the label l, which means when $|\mathcal{L}^t(v_i) \cap \mathcal{L}^t(v_j)|$ changes from one to zero or the other way around. To capture the first case we define the indicators $b_{i,j}$ for every $i, j \in [n]$ as follows:

$$b_{i,j} = \begin{cases} 1 & \text{if } |\mathcal{L}^t(v_i) \cap \mathcal{L}^t(v_j)| = 1 \\ 0 & \text{o.w} \end{cases} \tag{10}$$

Claim. The cost difference $\hat{C}_{i,j}^{r,q}$ for changes of type r and q in elements v_i and v_j respectively under the use of set-intersection similarity is:

$$\hat{C}_{i,j}^{r,q} = \begin{cases} (1 - I_{i,j}^t) \cdot (2\,s(v_i, v_j) - 1) & \text{if } r = 1, q \in \{1, 4\} \\ b_{i,j} \cdot (1 - 2\,s(v_i, v_j)) & \text{if } r = 3, q \in \{3, 4\} \\ 0 & \text{o.w} \end{cases} \tag{11}$$

Proof. As mentioned above we have only two major cases in which the cost by a single pair of elements may change:

- Two elements do not share any label at time t (means $I_{i,j}^t = 0$), but at time $t + 1$ start to share the label l only, either by both being newly assigned to it ($r = q = 1$) or when one of them joined it at time $t + 1$ ($r = 1$) and the other one was already assigned to it ($q = 4$). In this case their mutual cost changes from $s(v_i, v_j)$ to $1 - s(v_i, v_j)$ and so the difference is $2s(v_i, v_j) - 1$.
- Two elements are both assigned to label l at time t ($q \in \{3, 4\}$) and do not share any other label (means $b_{i,j} = 1$), but one of them is omitted from label l at time $t + 1$ ($r = 3$). In this case their mutual cost changes from $1 - s(v_i, v_j)$ to $s(v_i, v_j)$ and so the difference is $1 - 2s(v_i, v_j)$.

$\square$

In order to support the bound p on the number of clusters an element can be assigned to, we should let $x_i = 0$ if $e_i = 0$ and $|\mathcal{L}^t(v_i)| = p$, which means that no change should be performed regarding label l if this element is not currently part of this label and is already assigned to other p labels. Moreover we may validate if needed that every element is assigned to at least one label by letting $x_i = 0$ if $e_i = 1$ and $|\mathcal{L}^t(v_i)| = 1$.

We initialize the labeling using a randomized assignment which satisfies the condition of maximum p labels for each node (and $p \geq 1$ if required). We also test a variant of our algorithm which uses Bonchi's solution for initialization.

3.4 Performance Evaluation

The algorithms are foremost evaluated based on the total cost achieved, normalized by the size of the similarity matrix (number of elements pairs). In addition we use measures from [9] that account for overlaps between the computed clusters. The first measures are Sensitivity (Sen) and Positive predictive value (PPV) whose geometric average is the accuracy (Acc) measure. In order to define Sen and PPV we let $T_{i,j}$ be the number of elements which are present both in a true cluster N_i (out of the n clusters induced by the true labeling mapping $\mathcal{L}^*$) and in a suggested cluster M_j (out of m clusters induced by the computed labeling mapping $\mathcal{L}$). Then Sen and PPV are defined as follows:

$$Sen = \frac{\sum_{i=1}^{n} \max_{j=1}^{m} T_{i,j}}{\sum_{i=1}^{n} N_i} \quad PPV = \frac{\sum_{j=1}^{m} \max_{i=1}^{n} T_{i,j}}{\sum_{j=1}^{m} \sum_{i=1}^{n} T_{i,j}} \tag{12}$$

Sen reflects the weighted average coverage of the predicted labels by their best-matching true labels, and PPV reflects the weighted average reliability for the true labels to predict that an element belongs to their best-matching predicted labels [4]. The accuracy is balancing these two measures [17] and therefore explicitly penalizes predicted labels that do not match any of the true ones [13]. It is important to note that the value of PPV is relatively low in overlapping labels settings, and so would not be a good enough measure alone for checking the resulting clustering quality. Therefore we use also the maximum matching ratio (MMR) and $Fraction$ measures which together may overcome this difficulty.

$$MMR = \frac{\sum_{i=1}^{n} \max_{j=1}^{m} O(N_i, M_j)}{n} \tag{13}$$

$$Fraction = \frac{|\{i | i \in [n], \exists j \in [m], O(N_i, M_j) \geq \omega\}|}{n} \tag{14}$$

We may think of the MMR as finding the maximum weighted matching in a bipartite graph between the labels of $\mathcal{L}$ and the labels of $\mathcal{L}^*$ according the the overlap-score defined between two labels as $O(N_i, M_j) = \frac{|N_i \cap M_j|^2}{|N_i| \cdot |M_j|}$. As for the $Fraction$, it allows us to count the number of true labels which are highly overlaps with the predicted labels, given an overlap threshold ω which we set to 0.25 [17]. The overall performance is measured by summing the three measures of MMR, Acc and $Fraction$, which will be referred as the *composite score*.

4   Results

We examine the following algorithms:

- *ROW-BON* - vertex-by-vertex method with Bonchi's local steps.
- *ROW-FPT* - vertex-by-vertex method with naive FPT local steps.
- *FULL-ILP* - ILP for the entire *OCC* problem (either *ECLIP* or *ISEC-ILP*)
- *COL-ILP* - our proposed algorithm using a label-by-label method with Bonchi or random initialization.

For the assessment of our methods we use two different multi-label datasets from MULAN [16] which are most commonly used for training multi-labeling classifiers: EMOTIONS ($n = 593, k = 6, p = 3$) [15], and YEAST ($n = 2417, k = 14, p = 11$) [7]. In order to create the similarity matrices for a dataset we used Jaccard coefficient and set-intersection indicator between the given true labels.

All four algorithms where implemented in Python3.9 and run on a 3.2 GHz CPU with their relevant parameters k and p over 100 iterations. We apply all algorithms to subsets of different sizes from 10 to 100. The subsets were randomly chosen so as to preserve the relative sizes of the different clusters - we first select a cluster l with probability p_l which is the fraction of nodes assigned to the cluster from the total number of node assignments to clusters, and then choose uniformly an element l, repeating those steps until the desired number of elements is obtained. We calculated the mean value of every measure across all 100 iterations, omitting some of the results which ran for more than 1000 s.

Figure 1 shows that our new methods balance between the fast heuristic of *ROW-BON* to the slow fully exact method of *ECLIP* or *ISEC-ILP*. Interestingly, using Bonchi's labeling as an initialization for *COL-ILP* leads to a decrease in runtime. On the larger YEAST dataset, ROW-FPT and ECLIP are infeasible already for small subsets.

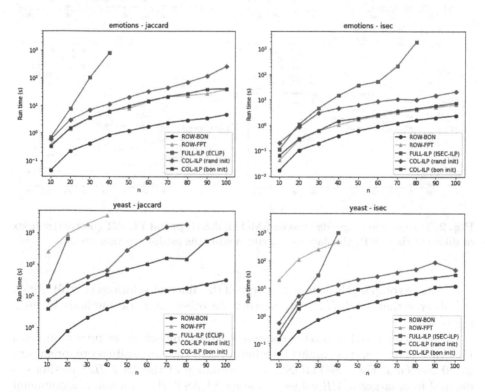

Fig. 1. The mean running time on EMOTIONS (top) and YEAST (bottom) subsets of different sizes. **left**: the Jaccard variant, **right**: the set-intersection variant.

Next, we assess the quality of the different solutions by computing the average edge loss which is defined as the total cost divided by n^2. As evident from Fig. 2, the exact methods of *ECLIP* and *ISEC-ILP* achieve zero cost, yet are too expensive to compute in part of the range. In contrast, *COL-ILP* is feasible across the range and outperforms the other heuristic searches in the vast majority of the cases, with the Bonchi initialization yielding smaller loss compared to the random one.

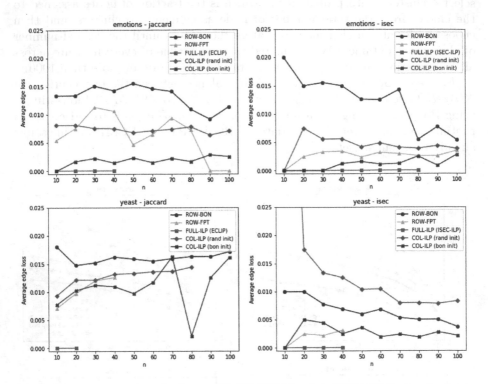

Fig. 2. The mean average edge loss on EMOTIONS (top) and YEAST (bottom) subsets of different sizes. **left**: the Jaccard variant, **right**: the set-intersection variant.

Last, we compare the composite score of the different solutions (Fig. 3). Again we observe that *COL-ILP* outperforms the other heuristic methods in most applications.

To get more insights into the composite score comparison, we present in Fig. 4 how the detailed parts combine together to yield the composite score for subsets of 20 elements. For EMOTIONS, the advantage of the *COL-ILP* methods is derived from larger *MMR* values, while for YEAST, the *Fraction* is a dominant factor.

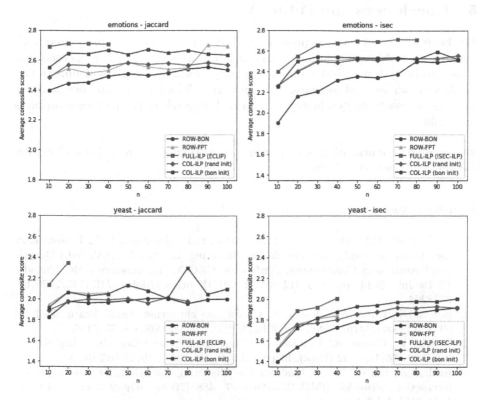

Fig. 3. The mean composite score on EMOTIONS (top) and YEAST (bottom) subsets of different sizes. **left**: the Jaccard variant, **right**: the set-intersection variant.

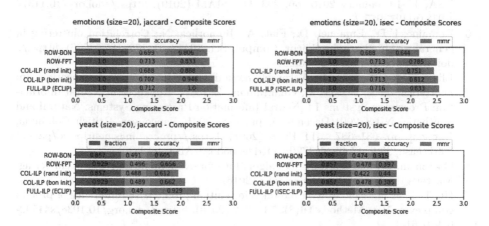

Fig. 4. The mean composite scores on EMOTIONS (top) and YEAST (bottom) with 20 elements. **left**: the Jaccard variant, **right**: the set-intersection variant.

5   Conclusions and Future Work

We have presented novel approaches for OCC that combine greedy iterations with exact ILP-based solutions for each iteration. Our *COL-ILP* method was shown to be both practical and effective, outperforming other algorithms on two datasets in an array of measures. Future work will be to generalize our methods to larger datasets and evaluate them on real data where the no true solution is available.

Acknowledgements. RS was supported by a research grant from the Israel Science Foundation (grant no. 715/18).

References

1. de Andrade, C.E., Resende, M.G.C., Karloff, H.J., Miyazawa, F.K.: Evolutionary algorithms for overlapping correlation clustering. In: Arnold, D.V. (ed.) Genetic and Evolutionary Computation Conference, GECCO '14, Vancouver, BC, Canada, 12–16 July 2014, pp. 405–412. ACM (2014). https://doi.org/10.1145/2576768.2598284
2. Bansal, N., Blum, A., Chawla, S.: Correlation clustering. Mach. Learn. **56**(1–3), 89–113 (2004). https://doi.org/10.1023/B:MACH.0000033116.57574.95
3. Bonchi, F., Gionis, A., Ukkonen, A.: Overlapping correlation clustering. Knowl. Inf. Syst. **35**(1), 1–32 (2013). https://doi.org/10.1007/s10115-012-0522-9
4. Brohée, S., van Helden, J.: Evaluation of clustering algorithms for protein-protein interaction networks. BMC Bioinform. **7**, 488 (2006). https://doi.org/10.1186/1471-2105-7-488
5. Chierichetti, F., Kumar, R., Pandey, S., Vassilvitskii, S.: Finding the Jaccard median. In: Charikar, M. (ed.) Proceedings of the Twenty-First Annual ACM-SIAM Symposium on Discrete Algorithms, SODA 2010, Austin, Texas, USA, 17–19 January 2010, pp. 293–311. SIAM (2010). https://doi.org/10.1137/1.9781611973075.25
6. Demaine, E.D., Emanuel, D., Fiat, A., Immorlica, N.: Correlation clustering in general weighted graphs. Theor. Comput. Sci. **361**(2–3), 172–187 (2006). https://doi.org/10.1016/j.tcs.2006.05.008
7. Elisseeff, A., Weston, J.: A kernel method for multi-labelled classification. In: Dietterich, T.G., Becker, S., Ghahramani, Z. (eds.) Advances in Neural Information Processing Systems 14 [Neural Information Processing Systems: Natural and Synthetic, NIPS 2001(December), pp. 3–8, 2001. Vancouver, British Columbia, Canada], pp. 681–687. MIT Press (2001). https://proceedings.neurips.cc/paper/2001/hash/39dcaf7a053dc372fbc391d4e6b5d693-Abstract.html
8. Gartzman, D., Sharan, R.: Exact algorithms for overlapping correlation clustering. Master's thesis, Tel Aviv University (2016)
9. Hu, L.Z., et al.: Epic: software toolkit for elution profile-based inference of protein complexes. Nat. Methods **16**(8), 737–742 (2019). https://doi.org/10.1038/s41592-019-0461-4
10. Johansson, F.D., Chattoraj, A., Bhattacharyya, C., Dubhashi, D.P.: Weighted theta functions and embeddings with applications to max-cut, clustering and summarization. In: Cortes, C., Lawrence, N.D., Lee, D.D., Sugiyama, M., Garnett,

R. (eds.) Advances in Neural Information Processing Systems 28: Annual Conference on Neural Information Processing Systems 2015(December), pp. 7–12, 2015. Montreal, Quebec, Canada, pp. 1018–1026 (2015). https://proceedings.neurips.cc/paper/2015/hash/4c27cea8526af8cfee3be5e183ac9605-Abstract.html

11. Macqueen, J.: Some methods for classification and analysis of multivariate observations. In: In 5-th Berkeley Symposium on Mathematical Statistics and Probability, pp. 281–297 (1967)

12. Miettinen, P.: On the positive-negative partial set cover problem. Inf. Process. Lett. **108**(4), 219–221 (2008). https://doi.org/10.1016/j.ipl.2008.05.007

13. Nepusz, T., Yu, H., Paccanaro, A.: Detecting overlapping protein complexes in protein-protein interaction networks. Nat. Methods **9**(5), 471–472 (2012). https://doi.org/10.1038/nmeth.1938

14. Shamir, R., Sharan, R., Tsur, D.: Cluster graph modification problems. In: Kucera, L. (ed.) Graph-Theoretic Concepts in Computer Science, 28th International Workshop, WG 2002, Cesky Krumlov, Czech Republic, 13–15 June 2002, Revised Papers. Lecture Notes in Computer Science, vol. 2573, pp. 379–390. Springer (2002). https://doi.org/10.1007/3-540-36379-3_33

15. Trohidis, K., Tsoumakas, G., Kalliris, G., Vlahavas, I.P.: Multi-label classification of music into emotions. In: Bello, J.P., Chew, E., Turnbull, D. (eds.) ISMIR 2008, 9th International Conference on Music Information Retrieval, Drexel University, Philadelphia, PA, USA, 14–18 September 2008, pp. 325–330 (2008). http://ismir2008.ismir.net/papers/ISMIR2008_275.pdf

16. Tsoumakas, G., Katakis, I., Vlahavas, I.P.: Mining multi-label data. In: Maimon, O., Rokach, L. (eds.) Data Mining and Knowledge Discovery Handbook, pp. 667–685. Springer, Boston (2010). https://doi.org/10.1007/978-0-387-09823-4_34

17. Wang, R., Liu, G., Wang, C., Su, L., Sun, L.: Predicting overlapping protein complexes based on core-attachment and a local modularity structure. BMC Bioinform. **19**(1), 1–15 (2018). https://doi.org/10.1186/s12859-018-2309-9

Deep Learning Models for LC-MS Untargeted Metabolomics Data Analysis

Francesco Russo[1] , Filip Ottosson[1] , Justin J. J. van der Hooft[2] ,
and Madeleine Ernst[1(✉)]

[1] Section for Clinical Mass Spectrometry, Danish Center for Neonatal Screening, Department of
Congenital Disorders, Statens Serum Institut, Copenhagen, Denmark
maet@ssi.dk
[2] Bioinformatics Group, Wageningen University, Wageningen, The Netherlands

Abstract. Metabolomics, the measurement of all metabolites in a given system,
is a growing research field with great potential and manifold applications in precision medicine. However, the high dimensionality and complexity of metabolomics
data requires expert knowledge, the use of proper methodology, and is largely
based on manual interpretation. In this book chapter, we discuss recent published
approaches using deep learning to analyze untargeted metabolomics data. These
approaches were applied within diverse stages of metabolomics data analysis, e.g.
to improve preprocessing, feature identification, classification, and other tasks.
We focus our attention on deep learning methods applied to liquid chromatography mass spectrometry (LC-MS), but these models can be extended or adjusted
to other applications. We highlight current deep learning-based computational
workflows that are paving the way toward high(er)-throughput use of untargeted
metabolomics, making it effective for clinical, environmental and other types of
applications.

Keywords: Metabolomics · Deep learning · Mac

1 Introduction

In recent years, biological research has become increasingly rich in data leading to
several new subfields of biology denoted by the suffix "-omics". These research fields aim
at characterizing and quantifying entire categories of biologically relevant molecules,
defined by the suffix "-ome". As such, genomics is the field of research where the
entire set of genes (i.e. genome) is characterized or analogously proteomics aims at
characterizing and quantifying the entire set of proteins (i.e. proteome).

One of the relatively newcomers in the omic-era is metabolomics. Thousands of
small molecules (<1500 Da), i.e. metabolites, are measured in a metabolomics experiment in a given biological sample [1]. By measuring the metabolome, an overview
of diverse metabolic processes in matrices as varied as plants, environmental samples,
tissues, cell or bacterial cultures, plasma, serum, skin or feces are obtained [2–7]. As

© The Author(s), under exclusive license to Springer Nature Switzerland AG 2024
D. Cantone and A. Pulvirenti (Eds.): *From Computational Logic to Computational Biology*,
LNCS 14070, pp. 128–144, 2024.
https://doi.org/10.1007/978-3-031-55248-9_7

such, among all the omics, metabolomics is the closest to the phenotype of an organism and thus has enormous potential in the field of precision medicine. The metabolome is a read-out of the combination of genetic, environmental, microbial and dietary factors that all influence the metabolic state of an individual. The metabolic state of an individual is closely related to the overall health status and an improved understanding could aid in prediction, diagnosis, prognosis and elucidation of molecular mechanisms of diseases [8]. However, the complexity and high dimensionality of data retrieved from metabolomics experiments, as well as the lack of sufficient public databases and automated data analysis workflows has hampered the integration of metabolomics data into the clinic. Deep learning methods applied to metabolomics data could aid the effective integration, and application of metabolomics data within a clinical context. This task is currently still largely dependent on expert knowledge with considerable investment in time for manual analysis, annotation and interpretation of the data.

In this book chapter we review deep learning methods applied to liquid chromatography mass spectrometry (LC-MS) metabolomics data, and highlight computational workflows that are paving the way towards a high(er)-throughput use of untargeted metabolomics data, making it effective for clinical, environmental and other types of applications.

1.1 Metabolomics

Metabolomics approaches can be grouped into targeted and untargeted measurements. In targeted metabolomics experiments certain metabolites of interest are pre-defined, while the aim of untargeted metabolomics approaches is to measure the entirety of all small molecules present in a given sample [9]. Advantages of targeted metabolomics approaches include higher reproducibility and ease of high throughput during data analysis and interpretation. The approach is however unable to realize the ambitious aims of metabolomics research, to measure the entire set of metabolites in a given sample. To get closer to this goal, untargeted metabolomics experiments are performed. Analysis and interpretation of untargeted metabolomics data is more challenging but allows for more explorative research and de-novo hypothesis generation. The number of metabolites measurable within the human body is unknown, but estimated at several thousand or more. The Human Metabolome Database (HMDB), an online database that aims to provide documentation for all known metabolites, has over 100,000 metabolite entries [10].

1.2 Liquid Chromatography Mass Spectrometry

Mass spectrometry (MS) is the analytical work horse of metabolomics measurements and remarkable developments in instrumentation technology have led to a rapid increase in the application of MS-based metabolomics platforms in clinical and research laboratories over the past decade [11]. High-resolution mass spectrometers allow for the simultaneous chemical structural elucidation and quantification of hundreds to thousands of small molecules in a sample with high sensitivity, specificity, throughput, and low sample consumption in a cost effective manner [11, 12]. Chemical characterization is further aided by integrating liquid chromatography with mass spectrometry [13]

(LC-MS), enabling quantitative and qualitative analysis of an increasing number of metabolites. Consequently, LC-MS techniques are widely used in clinical research for biomarker discovery [14], elucidation of molecular mechanisms of diseases [15] or monitoring disease therapy [16].

MS-based methods provide two levels of information, the mass-to-charge ratio (m/z) and abundance, which is summarized in a mass spectrum. When utilizing high resolution MS, the precision of the measured m/z is typically high enough for it to be mapped to a specific molecular formula, in particular when specific filters on the occurrence of elements are used, which in turn can be assigned to candidate molecules. The abundance is the measure of the number of times a specific m/z hits the detector and can be regarded as the relative concentration. However, measuring the exact mass (m/z) is often not enough to properly identify a metabolite, since numerous isomeric compounds could give rise to the same m/z [17]. To resolve the identity, it is usually necessary to perform MS/MS experiments, where a specific m/z is isolated, fragmented and subsequently subjected to an additional MS analysis. Fragmentation of the molecule of interest is useful because the fragmentation pattern is dependent on the molecular structure. The resulting fragmentation spectrum shows the m/z of the molecule fragments, which can be matched against databases of experimentally determined fragmentation patterns for different metabolites [18]. LC results in one additional dimension of data, the chromatographic retention time (RT). Mass spectra are continuously generated throughout the chromatographic separation of the samples, typically acquiring one or several mass spectra each second. Since the LC separates the compounds in the sample based on physico-chemical properties, each metabolite has a characteristic RT for the specific chromatographic setup it was generated on. Therefore, RT can be an additional variable that can be utilized when annotating untargeted metabolomics data [19–21].

Targeted metabolomics experiments typically measure up to a few hundred metabolites with known m/z and fragmentation patterns but are unable to detect metabolites that were not selected as targets. For untargeted metabolomics, high-resolution mass spectra are generated and are typically able to measure thousands of mass spectral features (i.e. metabolites before the identification) [13]. Since untargeted metabolomics is not hypothesis driven, much effort has to be put into detecting, aligning, and annotating each measured feature, noting its m/z and RT, relative abundance in a feature quantification table, and finally linking each feature to its fragmentation spectrum.

1.3 LC-MS Metabolomics Data Analysis and Interpretation

LC-MS approaches are powerful methods but the data that they generate are highly dimensional and extremely complex hindering several stages of metabolomics data treatment, including data preprocessing (the transformation of the multiple chemical signal dimensions into an easy-to-use format for subsequent statistical analysis) [22], analysis, chemical structural annotation and interpretation. Non-preprocessed LC-MS data files contain several thousands of MS spectra one after another and each spectrum is characterized by a sequence number which increases with the RT. Currently, several pipelines exist for automated LC-MS data preprocessing [23–27], peak detection and alignment across samples remains however challenging and often produces false positives. In addition, available modular framework algorithms are not scalable, but feasible for a few hundred samples, making them unsuitable for the analysis and processing of population-size clinical metabolomics cohorts. Another bottleneck is the chemical structural annotation of unknowns in untargeted metabolomics data. Compared to other omics sciences, chemical structural identification of metabolomics data is more challenging, as the estimated chemical space of small molecules is vast. Over 10 [60] possible carbon-based small molecules (<500 Da) are thought to exist [28], as compared to 20 unique amino acids, or four nucleotides, which form the building blocks of all proteins and DNA measured in proteomics and genomics experiments respectively. On average only 2–5% of the data collected in a typical LC-MS metabolomics experiment can be matched to known molecules [29] through private and public spectral libraries and the ever-increasing accumulation of unidentified metabolites reported in clinical metabolomics studies is a major bottleneck [30]. In more recent years, computational metabolomics workflows have been proposed based on substructure discovery, chemical compound class annotation, and mass spectral networking to make an inroad into the vast amount of yet unknown metabolite features in untargeted metabolomics profiles [31–38].

1.4 Machine Learning Applied to Untargeted Metabolomics

In the last few years, machine learning (ML)-based technologies entered every aspect of society, helping image and speech recognition, recommendation systems and many more tools that we use in our daily life. A typical form of ML is supervised learning, where the machine is trained with a large amount of data having labeled information and it gives as output a vector of scores for each category. Supervised learning includes models such as Random Forest [39] and Support Vector Machine (SVM) [40]. After training, the performance is evaluated on a different dataset, never seen during training (which is called test set) to verify the generalization ability of the ML model. In contrast, unsupervised learning is a form of ML where an algorithm learns patterns from unlabeled data. Typical examples of those methods are k-means clustering, Principal Component Analysis and Hierarchical Clustering [41].

Several ML-based approaches have been proposed for improving untargeted metabolomics data analysis (see Table 1). Taking inspiration from other fields of computer science and statistics, some supervised and unsupervised ML methods have been applied to substructure discovery and annotation (e.g., MS2LDA substructure discovery [38], MAGMa - ClassyFire - MS2LDA integration & MotifDB [42]), for improving spectral similarity scores (e.g., Spec2Vec [43]), and for large-scale mass spectral clustering and networking (e.g., falcon [44] and Molecular Networking [35]).

1.5 Deep Learning Applied to Untargeted Metabolomics

ML-based methods usually require careful feature engineering and a high level of knowledge of the specific field, in order to apply proper transformations to the raw data [46]. This step is crucial for most of the classical ML approaches and is needed for giving the proper inputs to ML methods. Therefore, alternative ML models have been proposed based on representation learning which are able to use data in the raw form and discover meaningful patterns. Deep learning (DL) models are a type of representation

Table 1. Recent ML methods applied to untargeted metabolomics.

Task	Name	Model	Availability	Year	Ref.
Spectral similarity score	Spec2Vec	Natural language processing algorithm (inspired by Word2Vec)	https://github.com/iomega/spec2vec	2021	[43]
Metabolite identification	MS2LDA	Topic modeling (latent Dirichlet allocation)	https://ms2lda.org/	2016	[38]
Metabolite identification	MAGMa - ClassyFire - MS2LDA integration	Unsupervised and supervised ML model - Multilayer Neural Network	https://github.com/sdrogers/lda https://github.com/iomega/motif_annotation https://github.com/sdrogers/nnpredict https://github.com/sdrogers/ms2ldaviz	2019	[42]
Metabolite identification	SIRIUS 4	Support vector machine (SVM)	https://bio.informatik.uni-jena.de/sirius/	2019	[45]
Spectrum clustering	Falcon	Tandem mass spectrum clustering using fast nearest neighbor searching	https://github.com/bittremieux/falcon	2021	[44]
Multiple tasks including metabolite identification and spectrum clustering	GNPS	Web-based MS platform including ML approaches for metabolite identification and clustering	https://gnps.ucsd.edu/	2016	[33, 35]

learning methods that are gaining more visibility in recent years. To improve some of the mentioned issues related to preprocessing and, overall, metabolomics data analyses, DL models have been proposed as a valid alternative for peak detection [47], identification of molecular structures [48] and, among other tasks, for batch effect removal [49].

In the last few years, DL has seen an impressive increase of applications in many scientific fields and we are observing a tremendous impact on society. In particular,

DL has been applied for improving classical ML-based approaches which are related to image and speech recognition and natural language processing. The computational biology field has shown a great interest in DL methods as well, particularly for discovering patterns in the data that could discriminate disease status and stratify patients according to meaningful features. However, for many years the low volume of data available and the challenging interpretation of DL models have limited the applications in computational biology and medicine. The increased availability of public datasets in the mass spectrometry field has opened new opportunities. However, in many cases it is still not enough for building proper DL models. In this context, we foresee more applications of transfer learning in the coming years, which is the capability to reuse knowledge obtained from other learning tasks for new and often unrelated tasks. A successful example is the use of the large ImageNet database (https://www.image-net.org/) containing 14,197,122 images, which has been instrumental for transfer learning applied to image recognition and advancing the DL field.

In the following, we present an overview of recent computational methods that implement DL models for different steps of metabolomics data analysis. These approaches were applied for improving preprocessing, feature identification and other tasks. Particularly, we focus our attention on DL methods applied to LC-MS.

2 Overview of DL Methods

Here, we present an overview of some of the recently published papers on DL models applied to the metabolomics field, in particular LC-MS methods (see Table 2). Overall, only few approaches have been published in the last few years, indicating that this is a novel and unexplored aspect of metabolomics research.

These methods include different steps of untargeted metabolomics data analysis such as batch effect correction, peak detection, spectra similarity and prediction of metabolite classes (Fig. 1). In the following sections, we will present each individual method in detail showing the innovative aspects of these approaches compared to traditional methods.

3 DL Methods for Pre- and Postprocessing Metabolomics Data

3.1 Peakonly: A DL Model for Detecting and Integrating Peaks

One of the main preprocessing tasks in untargeted metabolomics is peak detection and integration of LC-MS data, which is currently prone to several false positive signals. To overcome this challenging issue, Melnikov and co-authors developed peakonly [47], an algorithm which is able to detect and exclude low-intensity noisy peaks starting from raw data. Peakonly is based on a convolutional neural network (CNN) which classifies regions of interest (ROIs). ROIs were detected using a modified version of the *centWave* algorithm [54]. The algorithm classifies three classes: 1) noise (ROIs do not contain peaks), 2) ROI contains one or more peaks, 3) ROI might contain a peak but a particular attention is required and in this case the decision has to be taken by an expert. Furthermore, the peak integration step was considered as a segmentation problem using an additional CNN, which allowed the authors to define whether a point in a ROI belongs to a peak.

Table 2. DL methods published recently for improving different LC-MS based metabolomics data analyses.

Task	Name	Model	Availability	Year	Ref
Similarity measure to compare tandem mass spectra	DeepMASS	Customized deep neural network	https://github.com/hcji/DeepMASS	2019	[50]
Batch effect removal	NormAE	Deep Adversarial Learning Model	https://github.com/luyiyun/NormAE	2020	[49]
Peak detection	Peakonly	Convolutional neural network	https://github.com/arseha/peakonly	2020	[47]
Peak detection	NeatMS	Convolutional neural network	https://github.com/bihealth/NeatMS	2022	[51]
Prediction of compound classes	CANOPUS	Customized deep neural network	https://bio.inform atik.uni-jena.de/sof tware/canopus/	2020	[48]
Characterization and expansion of reference libraries for small molecule identification	DarkChem	Variational autoencoder (VAE)	https://github.com/pnnl/darkchem	2020	[52]
Similarity measure to compare tandem mass spectra	MS2DeepScore	Siamese neural network architecture	https://github.com/matchms/ms2dee pscore	2021	[53]
Prediction of liquid chromatographic retention times (RTs)	GNN-RT	Graph neural network	https://github.com/Qiong-Yang/GNN-RT	2021	[20]

This method showed high precision for solving the hard task of detecting and defining a peak area, discovering true positive peaks using in-house and publicly available datasets. The approach achieved a precision of 97% which is a very high value considering that the precision of existing algorithms without additional noise filtering ranges from 0.5 to 0.8 [47, 55]. The approach has been developed specifically for LC-MS but it is potentially applicable to other MS systems, even though some adjustments might be needed, especially the training of the CNNs.

The method was written in pytorch and it is available at https://github.com/arseha/peakonly.

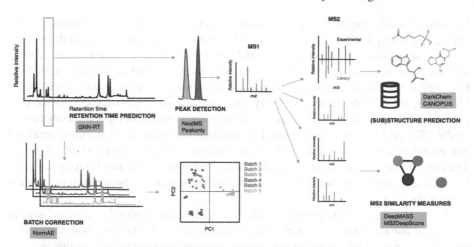

Fig. 1. Overview of recent deep learning methods applied to untargeted LC-MS MS1 and MS2 (MS/MS) metabolomics data analysis. Deep learning methods have been applied within four different subareas of untargeted metabolomics data analysis, including retention time prediction (GNN-RT), peak detection (NeatMS, Peakonly), (sub)structure prediction (DarkChem, CANOPUS) and tandem mass spectral similarity measures (DeepMASS, MS2DeepScore).

3.2 NeatMS: A DL-Based Post Processing Tool to Remove Low Quality Peaks

NeatMS is a tool to perform peak curation for LC-MS based metabolomics [51]. Where Peakonly removes low quality peaks from the raw data, NeatMS uses an alternative approach by filtering out poor peaks from data already preprocessed by conventional algorithms. NeatMS is based on a convoluted neural network architecture that was trained on datasets with a wide range of peak shapes. The algorithm comes with a pre-trained model but transfer learning functionality is also available. Peak filtering with NeatMS can be implemented after an existing metabolomics preprocessing workflow by categorizing detected peaks as high, acceptable or poor quality. The method managed to retain chemical standards that were spiked into a complex data matrix with high precision (94–97%). In a head-to-head comparison to Peakonly, the corresponding numbers were 79%. Also, NeatMS included a larger number of total peaks (acceptable quality or higher).

NeatMS is written in python and available at https://github.com/bihealth/NeatMS.

3.3 NormAE: The Normalized Autoencoder for Removing Batch Effects

Untargeted metabolomics data are usually affected by strong batch effects [56, 57], which is defined as systematic technical differences between samples due to the measurement of metabolites in several batches or analytical plates. To be considered as a technical difference, the variation observed between samples can not be explained by any biological difference. Batch effects limit downstream analyses and the interpretability of the data since they mask the real biological effect. This is a fundamental aspect that has to be taken into account when performing an untargeted metabolomics study, since this type of metabolomics gives global information about the metabolome without knowing absolute concentrations. Therefore, it is crucial to reduce batch effects in order to discover strong biological processes and biomarkers to be translated into the clinic.

Several methods aiming to remove batch effects have been proposed. Many of these methods take inspiration from other omics data and are based on linear models [58–61], which are not always based on the correct assumptions for metabolomics data. Moreover, some of the methods usually take into account inter-batch effects but they ignore intra-batch effects, which are commonly observed in untargeted metabolomics studies. To overcome these limitations, Rong and collaborators developed the normalized autoencoder (NormAE) [49], which is inspired by adversarial learning and autoencoders. The authors demonstrated that NormAE was able to remove non-linear batch effects from LC-MS untargeted metabolomics datasets without overfitting, better than the current state of the art methods.

NormAE takes as input preprocessed untargeted metabolomics data files and applies non-linear autoencoders and adversarial generative networks (GANs), with the goal to encode the original intensities of the peaks with latent representations. In this process, only the actual biological information is retained and during the decoding step the data are reconstructed without inter-batch effects. NormAE achieves the goal of removing the inter-batch effects by implementing an adversarial training step which aims to optimize the autoencoder by classifying the labels (i.e. batches) based on the latent representation. Furthermore, metabolomics-specific intra-batch effects, due for instance to injection order, are also taken into account and added to the model.

The authors evaluated the method on two datasets and based on the removal of the batch effect and the ability of the method to retain biological information, it outperformed the current state of the art approaches reducing the variance due to batch effect.

NormAE is publicly available at https://github.com/luyiyun/NormAE and it has been implemented in pytorch.

4   DL Methods for Metabolite Annotation

4.1   GNN-RT: Improving Metabolite Annotation by Predicting RT Using Graph Neural Networks

The chromatographic RT is strongly correlated to the molecular structure of metabolites and, therefore, it can be used for improving molecular identification and reducing false positives. However, in untargeted metabolomics RT is not commonly used for identifying small molecules because the current computational approaches are usually instrument-specific and most of the methods have been applied to small datasets. In order to improve RT prediction methods, the large METLIN small molecule retention time (SMRT) dataset [19] was released for boosting machine learning-based models and improving metabolite annotation.

GNN-RT is a method based on graph neural networks (GNNs), which have a high learning ability [20]. Recently, GNNs gained visibility in the metabolomics field because of the possibility of representing small molecules as graphs, which better describe the relation between atoms and bonds compared to fixed molecular descriptors and finger-prints. Therefore, methods based on graphs are more accurate and precise by taking into account the topological and chemical properties of small molecules [20].

The input of GNN-RT is a set of molecular graphs, which are based on the international chemical identifier (InChI) [62] and generated using the RDKit (https://www.rdkit.org/). Each graph can be described as vertices (i.e., atoms) and edges (i.e., chemical bonds between atoms). Additionally, the number of bonds was encoded as well. The molecular representation can be seen as a low dimensional vector which is learned by backpropagation. The learned representation is then used to predict RT and support metabolite annotation. Furthermore, an interesting and useful aspect of GNN-RT is its ability to perform transfer learning between different chromatographic systems. This is of particular importance because different systems will generate different RTs and makes it challenging to compare them. To overcome this issue, GNN-RT can be pre-trained with a large datasets such as SMRT, and used on a new dataset (even if generated using a different chromatographic system) by transfer learning, allowing to obtain high performance and avoiding the need of training on larger datasets.

Overall, GNN-RT achieved the highest performance based on different metrics including MAE = 39.87, MedAE = 25.24, and MRE = 4.9%, compared to multichannel-CNN (MC-CNN) [63], single channel-CNN (SC-CNN) [63], Bayesian ridge regression (BRR) [21] and Random Forest (RF) [21].

GNN-RT has been written in pytorch and is available at https://github.com/Qiong-Yang/GNN-RT.

4.2 DeepMASS: Structural Similarity Scoring of Unknown Metabolites Using Deep Neural Networks

As we mentioned in the previous paragraphs, the metabolomics field needs novel and advanced tools for metabolite annotation. MS/MS spectra generated in tandem mass spectrometry are a large source of knowledge, representing substructure information of metabolites. A classical approach for identifying metabolites consists of searching MS/MS spectra in a publicly available database, hoping to find a match. The databases containing MS/MS spectra are undoubtedly increasing, however they are still limited compared to the hypothetical large number of metabolites existing in nature.

In this context, DeepMASS [50] tries to solve the above limitations of current approaches using deep neural networks. The idea behind this model is based on biotransformations and the fact that reactant and product metabolites have similar substructures. Therefore, it is possible to search unknown metabolites in MS/MS spectra databases looking for transformational products of known metabolites.

The DeepMASS model trains and validates metabolite pairs characterized by high structure similarity (i.e., 'positive metabolite pair') retrieved from the KEGG database and random metabolite pairs (i.e., 'negative metabolite pair') generated randomly. Then, for each positive or negative metabolite pair their spectra are searched against MS/MS databases. Due to the limited number of experimental MS/MS spectra after matching, an additional tool (CFM-ID) was used for generating more spectra, increasing the total number of spectra needed for applying DeepMASS. Then, theoretical spectra pairs were collected in the same way as experimental spectra pairs and used for pretraining the deep neural network, while the experimental spectra pairs were used for fine-tuning the model. Using pretrained networks allows to apply DL models overcoming the issue of having a small number of experimental spectra pairs.

To validate the identification performance of DeepMASS, the authors performed a cross validation test based on 662 spectra. The percentage that the correct structure was found among the top 3, top 5, and top 10 hits was reported as 74.9%, 85.3%, and 92.0%, respectively, achieving remarkable performance compared to other methods (MetFrag [64] and CFM-ID [65]). Additionally, the authors expanded the search with the entire PubChem compound database achieving the highest percentage of identification compared to MetFrag and CFM-ID.

DeepMASS was implemented in Keras and Tensorflow backend and it is publicly available at https://github.com/hcji/DeepMASS.

4.3 MS2DeepScore: A Siamese Neural Network for Predicting the Structural Similarity of MS/MS Fragmentation Spectra

Following the recent development and improvements for predicting the structural similarity of MS/MS spectra using DL models, Huber and colleagues proposed a novel approach inspired by DeepMASS [50]. This new method, MS2DeepScore [53], uses a simpler architecture based on a Siamese neural network and it only relies on peak m/z positions and intensities instead of using mass and chemical formulas. An additional advantage of MS2DeepScore is that it creates mass spectral embeddings which can be the input for additional spectral clustering. The method has been evaluated on curated and cleaned spectra retrieved from the Global Natural Product Social Molecular Networking (GNPS) [35], reaching high performance.

The inputs of MS2DeepScore are pairs of MS/MS spectra. The model is based on a Siamese network which consists of two components, a base network creates the embeddings from the input spectra and another part of the network which is a cosine calculation of the embeddings. Additionally, MS2DeepScore applies Monte-Carlo Dropout ensembles to estimate the uncertainty of a prediction.

The advantage of using MS2DeepScore relies on the fact that it is trained to predict structural similarities, usually obtained by applying Tanimoto or Dice scores based on molecular fingerprints, directly from pairs of MS/MS spectra without the need to compute molecular fingerprints. In the introducing paper, the authors show that MS2DeepScore can predict, based on MS/MS spectral pairs, the Tanimoto scores (between 0.1 and 0.9) of the fragmented molecules with an RMSE between 0.13 and 0.2.

MS2DeepScore is available as a python library at https://github.com/matchms/ms2 deepscore.

4.4 CANOPUS: A Computational Tool for Systematic Compound Class Annotation

One of the most recent works on deep learning applied to metabolite annotation is CANOPUS [48]. CANOPUS is based on deep neural networks. It uses predicted fingerprints as input for assigning classes and ontology to metabolites. The authors used the probabilistic molecular fingerprint predicted by CSI:FingerID as well as the molecular formula computed by SIRIUS [45] as input. In particular, MS/MS spectra are the input of support vector machines (SVMs), which are trained with reference MS/MS spectra and used to predict a probabilistic fingerprint. Then, the probabilistic fingerprint is used as the input of a deep neural network trained on 4.1 million compounds without needing MS/MS spectra as input and gives predicted classes as output.

CANOPUS showed a very high prediction performance with an average accuracy of 99.7% in cross-validation, using reference data.

Advantages of CANOPUS include the ability to assign putative compound classes to every mass spectral feature in a LC-MS experiment, including molecules which have not been previously reported in any database. The algorithm is available as source code and software, making it also available for users with limited bioinformatics skills.

The source code of CANOPUS is available at https://github.com/boecker-lab/sirius-libs and the software implementation at https://bio.informatik.uni-jena.de/software/canopus/.

4.5 DarkChem: A Variational Autoencoder for Creating a Massive in Silico Library

One of the most important approaches for small metabolite annotation relies on the comparison of experimental features characterized by *m/z* and RT to libraries containing reference values, which help the identification of known molecules. However, these libraries are not able to identify unknown molecules and therefore most of the available commercial reference standards are not enough for untargeted metabolomics experiments. On the other hand, modern *in silico* approaches have the potential of building large libraries and boosting the identification process.

A recent approach, DarkChem [52], aims to create a massive *in silico* library using a variational autoencoder (VAE). One of the key qualities of DarkChem is the ability to include collision cross section (CCS) in the model, which is obtained from ion mobility spectrometry and measures the interaction between the ionized molecule and a buffer gas. CCS represents an additional dimension for small metabolite annotation, enabling the measuring of the mobility of the molecule in the mass spectrometer. Furthermore, DarkChem contains a 3-stage transfer learning method which allows it to learn important molecular structure representations from millions of molecules. Then, an optimization step improves the ability of the model to predict chemical properties. As we mentioned in the previous paragraphs, transfer learning is a useful approach in cases where the experimental datasets are too small and the risk of overfitting is too high.

The network architecture of DarkChem included an encoder consisting of canonical SMILES, a character embedding and convolutional layers. The latent representation was a fully connected dense layer and the decoder was convolutional layers and a linear layer with softmax activation for giving the outputs (i.e. canonical SMILES).

The model achieved a validation reconstruction accuracy of 98.9% for the experimental dataset and 99.0% for the in silico dataset, demonstrating the ability of transfer learning to improve performances in case of small experimental datasets.

DarkChem was written in python using Keras and Tensorflow backend and it is available at https://github.com/pnnl/darkchem.

5 Conclusions

In this book chapter, we presented the most recent works regarding DL and its applications to untargeted metabolomics. We introduced the main characteristics of the DL models and their strength for solving several bottlenecks in the metabolomics field. As we have shown, different deep neural network architectures share a common strategy based on transfer learning. The amount of publicly available datasets in this field is growing through platforms such as MassBank [66], GNPS [35] or MetaboLights [67] where raw, processed, or annotated mass spectrometry data are shared. However, DL methods require a large amount of data to perform and to avoid overfitting. Transfer learning is an effective approach for learning fundamental aspects of the data and for generalizing models that can be applied to different and often unrelated tasks.

We observe the development of DL models spanning several topics related to computational metabolomics, but undoubtedly the small metabolite annotation has seen enormous progress in recent years. With the ever growing amount of mass spectrometry data being collected and shared in public repositories, we and others [68] foresee the development of many more DL methods aiming to identify the unknown molecules and finally reconstruct complex metabolic pathways related to biological processes and diseases.

In this book chapter, we focused our attention on DL methods applied to LC-MS-based metabolomics. However, DL is having an impact on other techniques and research fields such as gas chromatography mass spectrometry (GC-MS) [69], nuclear magnetic resonance (NMR) spectroscopy [70, 71], Matrix Assisted Laser Desorption-Ionization Time-of-Flight mass spectrometry (MALDI-TOF) [72] and proteomics [73, 74]. We believe that these approaches can inspire each other to improve future methods and generate new ideas, for solving bottlenecks within and between the research fields.

Sharing metabolomics data will help increase the number of available datasets for training ML models and become better at performing classification tasks [75]. However, in the context of translational research several ethical considerations remain to be addressed, therefore data sharing remains challenging. This is an aspect of research which has been faced by genetic data [76, 77], which are considered personal information and therefore have to follow the data protection regulations [78]. However, the ability of metabolomics studies to identify research participants is largely unknown [79]. It has been proposed that a more extensive controlled data access might be needed for metabolomics data in order to be shared [79]. This requires that the availability of data will be limited to researchers that have been authorized by appropriate data access committees and a proper infrastructure will be needed.

References

1. Fiehn, O.: Metabolomics — the link between genotypes and phenotypes. In: Functional Genomics, pp. 155–171, Springer, Netherlands (2002). https://doi.org/10.1007/978-94-010-0448-0_11
2. Zierer, J., et al.: The fecal metabolome as a functional readout of the gut microbiome. Nat. Genet. **50**, 790–795 (2018)
3. Psychogios, N., et al.: The human serum metabolome. PLoS ONE **6**, e16957 (2011)

4. Dame, Z.T., et al.: The human saliva metabolome. Metabolomics **11**, 1864–1883 (2015)
5. Beltran, A., et al.: Assessment of compatibility between extraction methods for NMR- and LC/MS-based metabolomics. Anal. Chem. **84**, 5838–5844 (2012)
6. Dietmair, S., Timmins, N.E., Gray, P.P., Nielsen, L.K., Krömer, J.O.: Towards quantitative metabolomics of mammalian cells: development of a metabolite extraction protocol. Anal. Biochem. **404**, 155–164 (2010)
7. Elpa, D.P., Chiu, H.-Y., Wu, S.-P., Urban, P.L.: Skin Metabolomics. Trends Endocrinol Metab **32**, 66–75 (2021)
8. Beger, R.D., et al.: Metabolomics enables precision medicine: 'a white Paper. Community Perspect. Metabolomics **12**, 149 (2016)
9. Fiehn, O.: Metabolomics–the link between genotypes and phenotypes. Plant Mol. Biol. **48**, 155–171 (2002)
10. Wishart, D.S., et al.: HMDB 4.0: the human metabolome database for 2018. Nucleic Acids Res. **46**, D608–D617 (2018)
11. Jannetto, P.J., Fitzgerald, R.L.: Effective use of mass spectrometry in the clinical laboratory. Clin. Chem. **62**, 92–98 (2016)
12. Chace, D.H., Kalas, T.A., Naylor, E.W.: Use of tandem mass spectrometry for multianalyte screening of dried blood specimens from newborns. Clin. Chem. **49**, 1797–1817 (2003)
13. Wishart, D.S.: Metabolomics for investigating physiological and pathophysiological processes. Physiol. Rev. **99**, 1819–1875 (2019)
14. Liang, Q., Liu, H., Xie, L.-X., Li, X., Zhang, A.-H.: High-throughput metabolomics enables biomarker discovery in prostate cancer. RSC Adv. **7**, 2587–2593 (2017)
15. Johnson, C.H., Ivanisevic, J., Siuzdak, G.: Metabolomics: beyond biomarkers and towards mechanisms. Nat. Rev. Mol. Cell Biol. **17**, 451–459 (2016)
16. van der Hooft, J.J.J., Padmanabhan, S., Burgess, K.E.V., Barrett, M.P.: Urinary antihypertensive drug metabolite screening using molecular networking coupled to high-resolution mass spectrometry fragmentation. Metabolomics **12**, 125 (2016)
17. Sumner, L.W., et al.: Proposed minimum reporting standards for chemical analysis chemical analysis working Group (CAWG) metabolomics standards initiative (MSI). Metabolomics **3**, 211–221 (2007)
18. Xiao, J.F., Zhou, B., Ressom, H.W.: Metabolite identification and quantitation in LC-MS/MS-based metabolomics. Trends Analyt. Chem. **32**, 1–14 (2012)
19. Domingo-Almenara, X., et al.: The METLIN small molecule dataset for machine learning-based retention time prediction. Nat. Commun. **10**, 5811 (2019)
20. Yang, Q., Ji, H., Lu, H., Zhang, Z.: Prediction of liquid chromatographic retention time with graph neural networks to assist in small molecule identification. Anal. Chem. **93**, 2200–2206 (2021)
21. Bouwmeester, R., Martens, L., Degroeve, S.: Comprehensive and empirical evaluation of machine learning algorithms for small molecule LC retention time prediction. Anal. Chem. **91**, 3694–3703 (2019)
22. Katajamaa, M., Oresic, M.: Data processing for mass spectrometry-based metabolomics. J. Chromatogr. A **1158**, 318–328 (2007)
23. Pluskal, T., Castillo, S., Villar-Briones, A., Oresic, M.: MZmine 2: modular framework for processing, visualizing, and analyzing mass spectrometry-based molecular profile data. BMC Bioinf. **11**, 395 (2010)
24. Tsugawa, H., et al.: MS-DIAL: data-independent MS/MS deconvolution for comprehensive metabolome analysis. Nat. Methods **12**, 523–526 (2015)
25. Smith, C.A., Want, E.J., O'Maille, G., Abagyan, R., Siuzdak, G.: XCMS: processing mass spectrometry data for metabolite profiling using nonlinear peak alignment, matching, and identification. Anal. Chem. **78**, 779–787 (2006)

26. Röst, H.L., et al.: OpenMS: a flexible open-source software platform for mass spectrometry data analysis. Nat. Methods **13**, 741–748 (2016)
27. Protsyuk, I., et al.: 3D molecular cartography using LC-MS facilitated by optimus and 'ili software. Nat. Protoc. **13**, 134–154 (2018)
28. Bohacek, R.S., McMartin, C., Guida, W.C.: The art and practice of structure-based drug design: a molecular modeling perspective. Med. Res. Rev. **16**, 3–50 (1996)
29. da Silva, R.R., Dorrestein, P.C., Quinn, R.A.: Illuminating the dark matter in metabolomics. Proc. Natl. Acad. Sci. U.S.A. **112**, 12549–12550 (2015)
30. Wood, P.L.: Mass spectrometry strategies for clinical metabolomics and lipidomics in psychiatry, neurology, and neuro-oncology. Neuropsychopharmacology **39**, 24–33 (2014)
31. Beniddir, M.A., et al.: Advances in decomposing complex metabolite mixtures using substructure- and network-based computational metabolomics approaches. Nat. Prod. Rep. **38**, 1967–1993 (2021)
32. Ernst, M., et al.: MolNetEnhancer: enhanced molecular networks by integrating metabolome mining and annotation tools. Metabolites **9**, 144 (2019)
33. Nothias, L.-F., et al.: Feature-based molecular networking in the GNPS analysis environment. Nat. Methods **17**, 905–908 (2020)
34. Wang, M., et al.: Mass spectrometry searches using MASST. Nat. Biotechnol. **38**, 23–26 (2020)
35. Wang, M., et al.: Sharing and community curation of mass spectrometry data with global natural products social molecular networking. Nat. Biotechnol. **34**, 828–837 (2016)
36. Mohimani, H., et al.: Dereplication of microbial metabolites through database search of mass spectra. Nat. Commun. **9**, 4035 (2018)
37. Scheubert, K., et al.: Significance estimation for large scale metabolomics annotations by spectral matching. Nat. Commun. **8**, 1494 (2017)
38. van Der Hooft, J.J.J., Wandy, J., Barrett, M.P., Burgess, K.E.V., Rogers, S.: Topic modeling for untargeted substructure exploration in metabolomics. Proc. Natl. Acad. Sci. **113**, 13738–13743 (2016)
39. Breiman, L.: Random forests. Mach. Learn. **45**, 5–32 (2001)
40. Cortes, C., Vapnik, V.: Support-vector networks. Mach. Learn. **20**, 273–297 (1995)
41. Malik, A., Tuckfield, B.: Applied unsupervised learning with R: uncover hidden relationships and patterns with k-means clustering, hierarchical clustering, and PCA. Packt Publishing Ltd (2019)
42. Rogers, S., et al.: Deciphering complex metabolite mixtures by unsupervised and supervised substructure discovery and semi-automated annotation from MS/MS spectra. Faraday Discuss. **218**, 284–302 (2019)
43. Huber, F., et al.: Spec2Vec: Improved mass spectral similarity scoring through learning of structural relationships. PLoS Comput. Biol. **17**, e1008724 (2021)
44. Bittremieux, W., Laukens, K., Noble, W.S., Dorrestein, P.C.: Large-scale tandem mass spectrum clustering using fast nearest neighbor searching. Rapid Commun. Mass Spectrom. e9153 (2021)
45. Dührkop, K., et al.: SIRIUS 4: a rapid tool for turning tandem mass spectra into metabolite structure information. Nat. Methods **16**, 299–302 (2019)
46. LeCun, Y., Bengio, Y., Hinton, G.: Deep learning. Nature **521**, 436–444 (2015)
47. Melnikov, A.D., Tsentalovich, Y.P., Yanshole, V.V.: Deep learning for the precise peak detection in high-resolution LC–MS data. Anal. Chem. **92**, 588–592 (2020)
48. Dührkop, K., et al.: Systematic classification of unknown metabolites using high-resolution fragmentation mass spectra. Nat. Biotechnol. **39**, 462–471 (2020)
49. Rong, Z., et al.: NormAE: deep adversarial learning model to remove batch effects in liquid chromatography mass spectrometry-based metabolomics data. Anal. Chem. **92**, 5082–5090 (2020)

50. Ji, H., Xu, Y., Lu, H., Zhang, Z.: Deep MS/MS-aided structural-similarity scoring for unknown metabolite identification. Anal. Chem. **91**, 5629–5637 (2019)
51. Gloaguen, Y., Kirwan, J.A., Beule, D.: Deep learning-assisted peak curation for large-scale LC-MS metabolomics. Anal. Chem. **94**, 4930–4937 (2022)
52. Colby, S.M., Nuñez, J.R., Hodas, N.O., Corley, C.D., Renslow, R.R.: Deep learning to generate chemical property libraries and candidate molecules for small molecule identification in complex samples. Anal. Chem. **92**, 1720–1729 (2020)
53. Huber, F., van der Burg, S., van der Hooft, J.J.J., Ridder, L.: MS2DeepScore: a novel deep learning similarity measure to compare tandem mass spectra. J. Cheminform. **13**, 84 (2021)
54. Tautenhahn, R., Böttcher, C., Neumann, S.: Highly sensitive feature detection for high resolution LC/MS. BMC Bioinformatics **9**, 504 (2008)
55. Tengstrand, E., Lindberg, J., Åberg, K.M.: TracMass 2–a modular suite of tools for processing chromatography-full scan mass spectrometry data. Anal. Chem. **86**, 3435–3442 (2014)
56. Liu, Q., et al.: Addressing the batch effect issue for LC/MS metabolomics data in data preprocessing. Sci. Rep. **10**, 13856 (2020)
57. Wehrens, R., et al.: Improved batch correction in untargeted MS-based metabolomics. Metabolomics **12**, 88 (2016)
58. Johnson, W.E., Li, C., Rabinovic, A.: Adjusting batch effects in microarray expression data using empirical Bayes methods. Biostatistics **8**, 118–127 (2007)
59. Zhang, Y., Parmigiani, G., Johnson, W.E.: ComBat-seq: batch effect adjustment for RNA-seq count data. NAR Genom Bioinform **2**, lqaa078 (2020)
60. Leek, J. T. svaseq: removing batch effects and other unwanted noise from sequencing data. Nucleic Acids Res. **42**, e161–e161 (2014)
61. Pang, Z., Chong, J., Li, S., Xia, J.: MetaboAnalystR 3.0: toward an optimized workflow for global metabolomics. Metabolites **10**, 186 (2020)
62. Heller, S.R., McNaught, A., Pletnev, I., Stein, S., Tchekhovskoi, D.: InChI, the IUPAC international chemical identifier. J. Cheminform. **7**, 23 (2015)
63. Matyushin, D.D., Sholokhova, A.Y., Buryak, A.K.: A deep convolutional neural network for the estimation of gas chromatographic retention indices. J. Chromatogr. A **1607**, 460395 (2019)
64. Ruttkies, C., Schymanski, E.L., Wolf, S., Hollender, J., Neumann, S.: MetFrag relaunched: incorporating strategies beyond in silico fragmentation. J. Cheminform. **8**, 3 (2016)
65. Allen, F., Pon, A., Wilson, M., Greiner, R., Wishart, D.: CFM-ID: a web server for annotation, spectrum prediction and metabolite identification from tandem mass spectra. Nucleic Acids Res. **42**, W94–W99 (2014). https://doi.org/10.1093/nar/gku436
66. Horai, H., et al.: MassBank: a public repository for sharing mass spectral data for life sciences. J. Mass Spectrom. **45**, 703–714 (2010)
67. Haug, K., et al.: MetaboLights: a resource evolving in response to the needs of its scientific community. Nucleic Acids Res. **48**, D440–D444 (2020)
68. Liu, Y., De Vijlder, T., Bittremieux, W., Laukens, K., Heyndrickx, W.: Current and future deep learning algorithms for tandem mass spectrometry (MS/MS)-based small molecule structure elucidation. Rapid Commun. Mass Spectrom. e9120 (2021)
69. Li, M., Wang, X.R.: Peak alignment of gas chromatography–mass spectrometry data with deep learning. J. Chromatogr. A **1604**, 460476 (2019)
70. Qu, X., et al.: Accelerated nuclear magnetic resonance spectroscopy with deep learning. Angew. Chem. Int. Ed. Engl. **59**, 10297–10300 (2020)
71. Hansen, D.F.: Using deep neural networks to reconstruct non-uniformly sampled NMR spectra. J. Biomol. NMR **73**(10–11), 577–585 (2019). https://doi.org/10.1007/s10858-019-002 65-1
72. Normand, A.-C., et al.: Identification of a clonal population of aspergillus flavus by MALDI-TOF mass spectrometry using deep learning. Sci. Rep. **12**, 1575 (2022)

73. Meyer, J.G.: Deep learning neural network tools for proteomics. Cell Reports Methods **1**, 100003 (2021)
74. Mund, A., et al.: AI-driven deep visual proteomics defines cell identity and heterogeneity. bioRxiv 2021.01.25.427969 (2021). https://doi.org/10.1101/2021.01.25.427969
75. Jarmusch, S.A., van der Hooft, J.J.J., Dorrestein, P.C., Jarmusch, A.K.: Advancements in capturing and mining mass spectrometry data are transforming natural products research. Nat. Prod. Rep. **38**, 2066–2082 (2021)
76. Gymrek, M., McGuire, A.L., Golan, D., Halperin, E., Erlich, Y.: Identifying personal genomes by surname inference. Science **339**, 321–324 (2013)
77. Erlich, Y., Narayanan, A.: Routes for breaching and protecting genetic privacy. Nat. Rev. Genet. **15**, 409–421 (2014)
78. Shabani, M., Borry, P.: Rules for processing genetic data for research purposes in view of the new EU general data protection regulation. Eur. J. Hum. Genet. **26**, 149–156 (2018)
79. Keane, T.M., O'Donovan, C., Vizcaíno, J.A.: The growing need for controlled data access models in clinical proteomics and metabolomics. Nat. Commun. **12**, 5787 (2021)

The Search for Cancer Drivers

Alessandro Laganà[1,2,3](✉) iD

[1] Department of Oncological Sciences, Icahn School of Medicine at Mount Sinai, New York, NY 10029, USA
alessandro.lagana@mssm.edu
[2] Department of Genetics and Genomic Sciences, Icahn School of Medicine at Mount Sinai, New York, NY 10029, USA
[3] Tisch Cancer Institute, Icahn School of Medicine at Mount Sinai, New York, NY 10029, USA

Abstract. During the past decade, significant technological and computational advances have provided unprecedented opportunities for gaining a better understanding of cancer biology and translating this knowledge into meaningful and concrete clinical benefits. Research has made considerable progress in identifying genes and molecular changes that promote cancer initiation and progression. Large-scale genomic and transcriptomic studies involving thousands of patients have prompted the development of sophisticated computational approaches and tools that enable identifying key driver alterations and characterizing their impact on tumor development. This chapter provides an overview of the basic principles that describe the complexity of cancer, which are known as cancer hallmarks, as well as the most relevant computational techniques and tools that have been designed to investigate the genes and genomic changes that contribute to these hallmarks.

Keywords: Cancer hallmarks · oncogene · tumor suppressor · driver · mutation · copy number alteration · gene fusion · pathway · regulatory network · non-coding driver · multiomics

1 Oncogenes, Tumor Suppressors and the Hallmarks of Cancer

Cancer is a complex disease characterized by genetic and epigenetic aberrations that disrupt signaling and metabolic pathways, leading to uncontrolled cell growth and proliferation. Most individual tumors are made up by heterogeneous populations of somatic cells resulting from an evolutionary process initiated by one or more genetic changes acquired by a single cell and carried to its offspring. A close look at the genome of a cancer cell, may reveal hundreds, if not thousands, of single-nucleotide variations (SNV), or point mutations, which are changes of single bases in the genome sequence compared to the original germline genome. Other frequent somatic alterations observed in cancer cells are insertions and deletions of larger stretches of DNA (indels), copy number alterations (CNA) and chromosomal rearrangements, as well as changes in DNA methylation and chromatin structure which, in turn, impact gene expression [1]. Such

© The Author(s), under exclusive license to Springer Nature Switzerland AG 2024
D. Cantone and A. Pulvirenti (Eds.): *From Computational Logic to Computational Biology*,
LNCS 14070, pp. 145–171, 2024.
https://doi.org/10.1007/978-3-031-55248-9_8

genetic and epigenetic events can affect genes called proto-oncogenes, which encode for proteins that are essential for physiologic cell growth regulation, for example during organs development, and tumor suppressor genes, which slow down cell division, inhibit cell proliferation and repair DNA. An alteration in one of the two alleles of a proto-oncogene may be sufficient to cause its oncogenic activation, which permanently *turns it on*, enhancing its function and promoting uncontrolled cell growth, cell division and proliferation. However, both alleles of a tumor suppressor genes must be inactivated, for example through SNVs or deletion, to promote tumorigenesis [2].

The complexity of cancer can be conceptually decomposed into ten hallmarks, which are defined by Hanahan and Weinberg as distinct and complementary capabilities and characteristics enabling tumor growth, progression, and invasion. The ten cancer hallmarks proposed in two articles published in 2000 and 2011, comprise the overarching enabling characteristics of "genome instability and mutation" and "tumor-promoting inflammation", and the acquired functional capabilities for "sustaining proliferative signaling", "evading growth suppressors", "resisting cell death", "enabling replicative immortality", "inducing/accessing vasculature", "activating invasion and metastasis", "cellular metabolism" and "avoiding immune destruction" [3, 4]. While not all hallmarks contribute equally to disease pathogenesis in all cancer types, they represent features and capabilities that cancer cells must acquire to foster tumor growth and aggression.

The two enabling characteristics of "Genome Instability and Mutation" and "Tumor-promoting inflammation" explain for the most part the acquisition of all the other hallmark capabilities. A variety of events, such as copying mistakes during DNA replication, exposure to mutagenic chemicals, or viral infections, may result in mutations and other genomic alterations causing the inactivation of genes deputed to maintain genomic stability, detect and repair DNA damage, or stop cell cycle and trigger apoptosis [5]. Conversely, cells may acquire alterations in oncogenes promoting neoplastic progression [6]. Inflammatory processes can also create an environment conducive to the development of cancer, for example by releasing growth factors that sustain proliferative signaling and survival factors that inhibit cell death [7–9]. The capability of cancer cells to sustain proliferative signaling or evade growth suppression are perhaps their most fundamental traits. Growth factors bind to cell surface receptors, triggering the activation of intra-cellular signaling pathways which regulate cell cycle and lead to proliferation and increased survival. Cancer cells may also achieve proliferative capabilities through constitutive activation of components of signaling pathways, e.g. through mutations (e.g. KRAS in colon cancer or BRAF in melanoma), by secreting its own growth factors (autocrine signaling), or by increasing the number of receptors in the surface [10–13]. Cancer cells can additionally resist cell death by up-regulating anti-apoptotic genes (e.g. BCL2, MCL1) or down-regulating pro-apoptotic genes (e.g. BAX, BAK) [14].

One mechanism to control indefinite cell divisions is through replicative senescence, a process that limits the number of cell divisions by triggering cell cycle arrest in response to the progressive shortening of the protective structures located at chromosomes ends, called telomeres, with each replication [15]. The telomerase enables cancer cells to grow quickly and replicate indefinitely by keeping the telomeres long. Without telomerase, cancer cells would be deactivated, stop dividing and eventually undergo apoptosis.

Another hallmark of cancer cells is the capability to avoid senescence and achieve replicative immortality, which is typically acquired by chromosome telomeres lengthening by increased telomerase production or activation of the alternative lengthening of telomeres (ALT) pathway [16].

The cross-talk between cancer cells and the surrounding stromal and immune cells in the tumor microenvironment (TME) is involved in the acquisition of other hallmark capabilities, such as the induction of angiogenesis, the activation of the metastatic process and the evasion of immune destruction. Cancer cells can hijack physiological processes of embryonic morphogenesis to further grow and disseminate. For example, to sustain their growth, solid tumors can induce angiogenesis, the process by which new blood vessels are formed, ensuring the supply of essential nutrients and oxygen to their cells [17, 18]. Angiogenesis can also facilitate tumor dissemination, which is another hallmark. When cancer cells penetrate a blood or lymphatic vessel wall and enter circulation (intravasation), they can reach distant tissues and invade them (extravasation), forming small micrometastasis at first and then macroscopic tumors (colonization) [19, 20]. One mechanism through which cancer cells achieve this capability is the epithelial–mesenchymal transition (EMT), a process by which epithelial cells lose their adhesion properties and acquire migratory and invasive capacity [21]. Cancer cells can also evade immune destruction by activating or inhibiting signaling with the immune microenvironment. Innate and adaptive immune cells, in fact, can identify cancer cells and eradicate them in their early development stages. However, cancer cells may evade immune destruction by either secreting immunosuppressive factors or expressing immunosuppressive cell-surface ligands, such as PD-L1, therefore inhibiting immune effector cells or preventing the activation of cytotoxic T cells [22, 23]. Finally, cancer cells can reprogram their metabolism to support the increased energy request caused by their growth and proliferation and adapt to challenging microenvironments characterized by hypoxia and hypo-nutrient conditions [24].

Very recently, four additional prospective hallmarks and enabling characteristics were proposed [25]. The hallmark "unlocking phenotypic plasticity" refers to the ability of cancer cells to exploit the proliferative potential of stem and progenitor cells in several ways. For example, tumors originating from fully differentiated cells may de-differentiate back to a progenitor-like state, thus removing the proliferative brakes activated with the differentiated state. In other cases, tumors originating in progenitor cells may be capable of maintaining a partially differentiated, progenitor-like state. There is increasing evidence of such mechanisms, for example in melanoma and colon cancer [26–28]. The enabling characteristic of "Nonmutational epigenetic programming" explains the non-mutational epigenetic alterations that contribute to the acquisition of hallmark capabilities throughout tumorigenesis and malignant progression. For example, aberrant properties and factors of the microenvironment, such as hypoxia or cytokines secreted by stromal cells, can cause changes in the epigenome, like hypermethylation or the induction of specific chromatin modifiers [29, 30]. A third recently proposed enabling characteristic of tumors is "Polymorphic microbiomes". There is increasing evidence that the composition of the microbiota, the community of microorganisms such as bacteria and fungi residing in a specific ecologic niche like the colon, can have an impact on cancer phenotypes. Recent studies have shown that certain microorganisms can have protective or deleterious effects on cancer development, as well as affect response to therapy [31–33]. Finally, the hallmark of "Senescent cells" refers to the increasingly recognized fact that cellular senescence can, in certain contexts, promote malignant transformation and tumor progression, rather than arresting proliferation [34, 35]. Indeed, senescent cells can secrete high levels of inflammatory cytokines, immune modulators and growth factors that promote proliferative signaling, angiogenesis, evasion of apoptosis and immunity suppression (SASP, senescence-associated secretory phenotype).

Tumors are defined by a constellation of features involving both the cancer cells and the microenvironment that surround them. Multiple hallmarks are activated to different degrees at any given time in different clonal and subclonal population of cancerous cells. The question is: how do we separate the signal from the noise? What are the drivers of these processes?

2 The Investigation of Driver Mutations

We start our journey to explore the many types of cancer drivers, and the methods to identify them, with what surely may be considered the quintessential and most extensively studied drivers: single nucleotide variations (SNV).

Somatic SNVs, or point mutations, are changes of a single nucleotide in the DNA of a somatic cell, which can be switched with another nucleotide (e.g. C -> A), deleted or inserted. These changes can occur because of mistakes during DNA replication and defects in the DNA repair machinery, and facilitated by exposure to mutagenic substances, such as radioactive materials and carcinogenic chemicals. During mitosis, the process of cell division which produces two genetically identical cells, DNA is replicated so that each daughter cell can receive a copy of the DNA. Errors can occur during this process, which are usually detected and corrected by the DNA repair machinery, which ensures that the correct genetic material is passed on. However, when mistakes are not corrected, they become permanent mutations after cell division. There are two types of SNV: transitions and transversions. Transitions occurs when a base is replaced with another base of the same class, e.g. purine-to-purine (A -> G) or pyrimidine-to-pyrimidine (T -> C). Transversions consists in substituting a base of a class with one of the other class, e.g. purine-to-pyrimidine or vice versa (A -> C). Point mutations can have very different effects depending on the location where they occur. For example, intronic mutations are unlikely to have any consequence, while exonic mutations can have potential deleterious effects, particularly when they change a codon in the protein coding sequence, or when they affect a splice site. Functional mutations can also occur in non-coding areas of a gene, such as the promoter, which may affect gene expression. Deleterious mutations can either cause cell death, when incompatible with cell survival, or lead to the development of cancer, when they affect oncogenes and/or tumor suppressors. SNVs can be detected by targeted panel sequencing, which focuses on a specific set of genes of interest, for example relevant in a certain cancer type, and by whole-exome (WES) and whole-genome sequencing (WGS), which allow to identify most mutations in the exons or across the full genome, respectively. Many tools have been developed to analyze sequencing data and identify mutations in a tumor sample, which is most effective when the sample is compared with a normal (germline) sample from the same individual. Mutation callers are bioinformatics programs that implements various algorithms, including machine learning and statistical approaches, to detect SNVs and determine whether an SNV is real or a sequencing artifact [36, 37]. Genome browsers also allow visual inspection of sequencing data, where sequences can be visualized and compared, and mutations are highlighted [38]. Once SNVs are detected, their functional consequences can be assessed using variant interpretation tools, which calculate the likelihood of each mutation being deleterious or benign [39].

Not all cancer types carry the same number of mutations. A seminal paper on mutational signatures from 2013 analyzed close to 5 million mutations from over 7,000 cancer samples and estimated the prevalence of somatic mutations across human cancer types, which ranges from the 0.05 mutations per megabase (mut/Mb) observed in pilocytic astrocytoma to the over 10 mut/Mb observed in melanoma [40]. The study went on to develop sigProfiler, an algorithm to detect characteristic patterns of mutation types, termed mutational signatures, and compile a catalogue of such signatures in human cancers [41]. The original set of 21 signatures was later extended to include 57 additional signatures following a new analysis of over 20,000 cancer samples [42]. Each signature reflects the footprint of specific mutational process to the cancer genome, and the authors of the study hypothesized the possible etiology of most signatures using previous knowledge from the literature or performing association studies with epidemiological and biological features of the particular cancer types. For example, signature SBS1 (single-base substitution 1) was characterized by a prevalence of $C > T$ (or, equivalently, $G > A$) mutations and was attributed to a clock-like mechanism, as the number of mutations in both cancers and normal cells correlated with the age of the individual. Other signatures were attributed to exposure to carcinogenic agents, such as SBS4, which was associated to tobacco smoking.

Among all potentially deleterious SNVs, a (small) subset are true drivers, e.g. mutations that activate or enable one or more cancer hallmarks, thus conferring selective advantages to the mutant cell population and leading to its clonal expansion. Most other SNVs will be passengers and, although not functional, they can still be informative, as they capture global mutational patterns that reflect the underlying biology and have been reported to accurately classify human cancers [43]. It is also important to mention that what is proven a driver in a cancer type, may not be a driver in a different cancer type, due their specificity and functional relevance to a specific tissue.

Finding driver SNVs is a challenging task and many computational tools have been developed to address it. Several methods are based on assessing the difference between the observed frequency of an SNV, for example in a specific tumor type, and the baseline mutational rate of the genome at that particular locus expected by random chance, adjusting for replication timing and other relevant covariates. The rationale for this choice is that functional mutations are more likely to become stable and recur across cancers. These methods usually work at the cohort-level, as the large population size allows to assess the statistical significance of candidate drivers, allowing to detect true mutations and to distinguish them from background random mutations. **MutSigCV** and **MuSiC** are good examples of tools in this category [44, 45]. One of the challenges in this type of approach is that mutation rates across cancer types and across patients with the same cancer type are heterogeneous, which underlies different paths to oncogenesis, whether endogenous, e.g. defects in repair pathways, or exogenous, like exposure to carcinogenic factors. Therefore, it is essential to account for the heterogeneity within each tumor type. The study that led to the development of MutSigCV, revealed that mutation rate varies broadly across the genome and that it correlates with DNA replication time during the cell cycle and expression level, with late-replicating regions and genes with low expression levels having much higher mutation rates.

Another approach for the identification of driver SNVs is based on assessing the composition of mutations normalized by the total mutations in a gene. Different mutation patterns and metrics are employed, such as mutations clustering, functional impact bias, or the proportion of inactivating and recurrent missense mutations. For example, the 20/20 rule approach considers a gene as an oncogene if >20% of its mutations are at recurrent positions and are missense, while >20% of the mutations in a tumor suppressor gene should be inactivating [46]. The algorithm implemented in **OncodriveCLUST**, instead, considers the presence of large mutation clusters in a gene as a marker of positive selection, and utilizes silent mutations as a background reference, assuming they are not under selective pressure [47]. The tool **OncodriveFM** selects candidate driver genes by assessing the bias toward the accumulation of mutations with high functional impact, e.g. missense mutations in critical protein domains, across a cohort of tumor samples [48]. Therefore, genes with a high ratio of deleterious mutations, even in a small subset of samples, will have a higher likelihood of being drivers than genes carrying low-impact mutations in a larger sample population. A more recent extension of OncodriveFM, called **OncodriveFML**, implements a local simulation of the mutational process in both coding and non-coding regions, therefore enabling the discovery of non-coding driver mutations [49]. Other methods implementing similar ratiometric and functional impact approaches include **ActiveDriver**, **dNdScv**, **DriverPower**, **e-Driver**, **SomInaClust** and **TUSON** [50–54].

More recently, the tool **MutPanning** was developed following the observation that passenger mutations occurred in characteristic nucleotide contexts, while driver mutations were enriched in functionally relevant positions not necessarily located in particular contexts [55]. Therefore, it was possible to identify driver mutations based on the deviation from the characteristic context of passenger mutations. MutPanning implements a probabilistic approach which combines this feature with other signals implemented in earlier methods, such as functional relevance and mutation frequency. Through this novel approach, the authors identified 460 driver genes in ~12,000 samples across 28 tumor types.

Machine learning approaches have also been successfully employed to predict driver mutations. **DeepDriver** is a deep learning-based method that implements a convolutional neural network (CNN) which learns information on mutation data and gene expression similarity simultaneously [56]. The CNN is trained on a mutation feature matrix constructed based on the topological structure of a gene similarity network. More specifically, the k-nearest neighbors (kNN) algorithm is used to build an undirected network based on gene expression, then the mutation feature matrix is generated by arranging the feature vectors of each gene and its k nearest neighbors in a matrix, and includes features like the fraction of missense mutations, the ratio of missense to silent mutations or the normalized mutation entropy.

The tool **DRIVE** implements a machine learning approach based on features at both gene and mutation levels integrated with characteristics derived from pathways and protein-protein interactions (PPI) to distinguish between driver and passenger mutations [57]. Mutation-level features were included based on the observation that passenger mutations were frequently observed in cancer driver genes and were based on functional impact as assessed by Variant Effect Predictor (VEP) [58]. Gene-level features included the typical ratiometric features as well as additional features chosen, again, based on the potential functional impact of the mutations, such as the maximum number of PPIs involving the gene under consideration, the biological pathways and processes where these PPIs were enriched, and the types of post-transcriptional modification that the protein was subjected to. The authors evaluated the performance of several machine learning algorithms trained on data from the Genomics Evidence Neoplasia Information Exchange (GENIE) project and tested on The Cancer Genome Atlas (TCGA) data, revealing that Random Forest was the most robust and accurate method [59, 60].

Finally, a recent study applied an *in silico* saturation mutagenesis approach combined with machine learning to accurately determine novel drivers and their characteristics [61]. The intuition behind this approach, which resulted in the tool **BoostDM** is that the observation of a specific mutation in thousands of tumors can be considered like a series of natural experiments that test its oncogenic potential across individuals and tissues. The authors performed an *in silico* saturation mutagenesis experiment, that is the computational evaluation of all possible changes in a gene, through a systematic machine learning approach to identify the mutations that are likely to drive tumorigenesis. The models, gradient-boosted decision trees, were trained on a set of 282 gene-tissue pairs with an excess of observed mutations. For each gene, synthetic mutations generated following the pattern of trinucleotide changes observed across samples of the specific tumor type were used as a negative set. The results of this study showed that mutational clusters around functionally important regions were highly predictive of tumorigenesis in many gene-tissue pairs. However, many genes exhibited different driver regions in different tumor types, possibly reflecting different mechanisms of pathogenesis. Moreover, the probability of occurrence of a potential driver mutation across tissues was influenced by mutational signatures, which are linked to specific active processes like aging or exposure to carcinogenic factors.

2.1 Consensus Approaches for Driver Discovery

A study from Conway et al. published in 2020, employed a consensus approach to determine active mutational processes and molecular drivers in melanoma [62]. The authors analyzed data from over 1,000 Whole-Exome Seq melanoma samples with matched germline controls. Melanoma has a high mutational rate, and distinguishing drivers from passengers is particularly challenging. The authors applied three tools, **MutSigCV2**, **MutPanning** and **OncodriveFML** to identify potential driver SNVs, then combined the P values from each method, filtered the mutations based on a false discovery rate (FDR) cutoff of 1%, and estimated the transcriptional activity of these genes in bulk and single-cell RNA sequencing data, in order to reduce false positives as much as possible. Of 178 candidate mutations, only 8% were predicted by all three tools. Moreover, only 38% of genes that were previously reported as drivers in other studies with large patient cohorts, were classified as drivers in this analysis. This highlights the intrinsic complexity in driver prediction, even when tools implementing variations of the same approach are considered. The authors classified patients into four subtypes based on the presence of mutations in NF1, NRAS/KRAS, BRAF and TWT, then evaluated mutational burden and secondary driver mutations in each subtype. The NF1 subtype had the highest mutational burden, while TWT had the lowest. Specific secondary mutations within each subtype had significant prognostic implications. For example, mutations in genes in the PBAF complex (Polybromo-associated BAF complex) responsible for chromatin remodeling, were significantly enriched in the NRAS subtype, where they were associated with much better response to immunotherapy than in other subtypes.

Another study performed a comprehensive assessment of 18 driver prediction methods on more than 3,000 tumor samples from 15 cancer types, which showed moderate overall predictive power and resulted in the generation of a consensus tool and database called **ConsensusDriver**, which will be described in Sect. 10.

3 Exploiting Pathway and Gene Set Information to Power the Discovery of Driver Mutations

While recurrent mutations observed in a significant portion of cancer patients, such as BRAF V600E in melanoma or EGFR L858R in lung cancer, are easier to detect based on statistical and ratiometric considerations, the identification of rare mutations pose a great challenge because of the small population of carriers and consequent lack of statistical support in data analysis. As explained in Sect. 1, tumors may acquire several hallmarks to achieve their growth and proliferation goals, and do so by several mechanisms, including selecting for oncogenic SNVs. However, their ultimate goal is to hijack or disrupt a whole mechanism, e.g. disabling apoptosis, and such goals can be achieved in multiple ways, sometimes by selecting an advantageous mutant in a central gene, other times by selecting different mutants participating in the same pathways and cellular processes. In the latter case, there may not be strong evidence of any mutation being particularly enriched in patients, but significant accumulation of oncogenic mutations can be observed at the pathway level.

Several methods and tools have been developed to evaluate candidate driver SNVs in the context of pathways and functional gene sets. The tool **DrGaP** integrates several mutational metrics, including differences in background mutation rates, multiple mutations in the same gene and variation in mutation types, with pathway information into statistical models based on a Poisson process [63]. Specifically, mutated genes in each pathway are modeled as if they were in a unique gene representing the variations in the pathway. Simulations and validation on public cancer data showed high accuracy and sensitivity of this approach. The tools **PathScan**, **PathScore** and **SLAPenrich** identify pathways that are consistently mutated across a sample population [64–66].

SLAPenrich further classifies pathways into then 10 cancer hallmarks, assesses the predominance of mutations in pathways associated to the same canonical hallmark and provides a score that can be interpreted as a probability that each pathway is significantly mutated in each sample of the cohort. The tool **Dendrix** introduces two combinatorial properties, coverage and exclusivity, to distinguish driver vs passenger groups of genes [67]. Coverage ensures that most patients have at least one mutation in the gene set, while exclusivity ensures that nearly all patients have one mutation at most in the gene set. This particular feature of mutual exclusivity is based on the observation that cancer cells often achieve disruption of a pathway through a minimal, non-redundant set of alternative variants and alterations. Dendrix implements a heuristic greedy algorithm to find gene sets that optimize these measures, that is driver gene sets or pathways that are mutated in many patients, but whose mutations are mutually exclusive. A further optimization of Dendrix, called **Multi-Dendrix**, was later proposed where the algorithm was improved by formulating the problem as an ILP (Integer Linear Program) and generalize it to simultaneously identify multiple driver pathways [68]. The concept of mutual exclusivity is also implemented in the tool **MEMo**, which aims to identify groups of genes that are significantly altered in patients, likely to participate in the same pathways, and whose alterations are mutually exclusive [69]. Variations and further improvements on this strategy were implemented in **CoMEt**, which calculated an exact statistical test for mutual exclusivity that was more sensitive to combinations of low frequency alterations and allowed to analyze alterations [70]; **WeSME**, a method for fast genome-wide calculation of statistical significance of mutual exclusivity controlling for mutation frequencies in patients and mutation rates of genes [71]; **TiMEx**, which models the mutational process as a Poisson process, where mutations in a group of genes appear over time [72]; **QuaDMutEx**, which uses a quadratic penalty for sets with excessive number of mutations, allowing higher sensitivity to detect low-frequency driver genes [73].

4 Identification of Driver Copy Number Alterations

Copy Number Alterations (CNA) are another recurrent category of driver lesion in cancer, consisting in the gain or loss in copies of regions of DNA [74]. Events that involve a whole chromosome or one of its arms are termed broad CNAs, while those restricted to small areas are termed focal CNAs. While the terms *gain* and *loss* are commonly used to events involving single-copy alterations, i.e. the acquisition of an extra copy of a chrososome/region or the loss of one of the two copies, the terms *amplification* and *deletion* refer to the acquisition of more than one copy and the loss of both copies of a chromosome or DNA region. Like for SNVs, not all CNAs are drivers. Carcinogenesis, in fact, often involves large genomic rearrangements acquired through a single catastrophic event, which is known as chromothripsis, which result in different kinds of aberrations, including CAN [75]. However, some of these events will not have any functional impact and constitute passenger alterations.

   To distinguish driver from passenger CNAs, different approaches have been proposed. The most popular tool is called **GISTIC** and implements a statistical approach to identify DNA regions that are more frequently altered than expected by random chance, similarly to frequency-based tools for the discovery of driver SNVs [76]. Amplification and homozygous deletions are considered less likely to represent random aberrations and are, therefore, given a greater weight. **GISTIC 2.0,** which was introduced a few years later, implemented several methodological improvements that increased sensitivity and specificity of driver detection, and included the capability to separate broad arm-level and focal CNAs based explicitly on length [77]. More recently, a different approach was proposed by the tool **RUBIC,** which detects recurrent copy number breaks rather than recurrent alterations [78]. A break is defined as the transition in copy number from neutral to gain/loss (positive break) and vice versa (negative break). The idea behind this approach is that driver alterations are more likely to be found in regions enclosed between recurrent positive and negative breaks. Moreover, focusing on breaks rather than regions, eliminated the need to determine the correct size of each region, which was one of the main challenges encountered by previous region-based tools. The authors demonstrated that RUBIC was able to call more focal recurrent regions and identify more known cancer genes than other tools, including GISTIC.

A different approach for driver CNA identification is based on the assumption that driver copy number changes will result in significant changes in the expression of the affected genes [79]. **CONEXIC** was one of the first algorithms integrating CNA and gene expression and was based on the assumptions that driver CNAs should occur in tumors more often than expected by chance, that these CNAs may be correlated with the expression of a group (module) of genes, and that the expression of the genes in the module is affected by changes in the expression of the driver [80]. First, the algorithm identifies significant CNAs using a variation of GISTIC. Then, a Single Modulator Step creates a model that associates each target gene with the single driver gene that explains its variation best, which results in modules of genes whose expression is associated with the same candidate drivers. Next, the Network Learning Step implements an iterative approach based on a Bayesian score and the Expectation Maximization (EM) algorithm to learn a regulation program for each module (a regression tree) and then reassign each gene to the module that best model its behavior. The authors demonstrated the advantages of this integrated approach over earlier methods based on the analysis of CNA alone, and successfully identified known drivers and novel dependencies of melanoma. A study on oral squamous cell carcinoma determined that relative copy number, i.e. copy number normalized to tumor ploidy, was more significantly correlated to gene expression than absolute copy number, i.e. normalized to a ploidy of 2, suggesting that tumors may adjust their overall gene expression to compensate for ploidy [81]. The tool **Oncodrive-CIS** measures the *in cis* effect of CNAs by computing a gene-wise impact score, which considers the expression deviation of each sample with CNAs as compared to normal samples and tumor diploid samples and, therefore, reflects the bias towards gene expression misregulation due to CNAs [82]. Another tool, **DriverNet**, identifies transcriptional patterns disrupted by genomic aberrations, whether SNV or CNA, using a bipartite graph and a greedy optimization approach which selects the smallest number of potential driver aberrations explaining the highest number of gene expression changes [83]. The authors showed that a high-level amplification of *EGFR* in a glioblastoma multiforme sample was associated to significant changes in the expression of genes known to interact with *EGFR*. A more recent study proposed a pathway-based pipeline called **ProcessDriver**, which combines different tools and algorithms to identify driver CNAs by calculating their impact on dysregulated pathways [84]. ProcessDriver first applies GISTIC to detect significant CNAs, then identifies *cis* and *trans* genes affected by these CNAs, clusters them at the pathway level and employs a Multi-Task LASSO regression to find *cis* genes which predict changes of *trans* genes in the enriched pathways. In a more recent study, the authors identified key CNAs in lower-grade glioma patients by random walk with restart (RWR), a guilt-by-association approach [85]. First, they identified CNAs associated with significant changes in the expression of cis genes. Then, they built a weighted protein-protein interaction (PPI) network by annotating each edge in the PPI with the correlation between the corresponding gene pairs. Candidate drivers in the PPI network were nominated through RWR based on their stable transfer probabilities and the enrichment for cancer hallmark gene sets of the neighboring genes. Finally, Partial Least Squares Regression Analysis was carried out to identify driver CNAs that cooperatively contributed to the disruption of a hallmark gene set.

5 Identification of Driver Gene Fusions

Chromosomal rearrangements such as translocations, deletions and inversions can result in the creation of hybrid genes which arise from the fusion of two separate genes formed as a product of chromosomal rearrangements. Gene fusions can be functional and encode chimeric proteins with oncogenic potential [86]. For example, *BCR-ABL*, a gene fusion found in most patients with chronic myelogenous leukemia (CML), is created from the translocation between the long arms of chromosomes 9 and 22 t(9; 22), known as the Philadelphia chromosome, which merges the 5′ part of the BCR gene, normally located on chromosome 22, with the 3′ part of the ABL1 gene, located on chromosome 9 [87]. This fusion is translated into a hybrid protein with constitutive kinase and oncogenic activity. The chimeric transcripts originating from gene fusions can be detected by either mapping the reads to the genome capturing discordant read pairs and chimeric alignments or by performing *de novo* RNA-seq assembly of the transcripts and then identifying the chimeric alignments.

Many tools have been developed to detect gene fusions from RNA-seq data, including **ChimeraScan**, **TopHat-Fusion**, **STAR-Fusion**, **Fusion-Catcher** and **Arriba** [88–93]. However, the question remains as to which gene fusions are actual drivers and which are non-functional passengers. Most tools for fusion detection implement some version of scoring function based on different metrics, such as the number of supporting junction reads or the consistency of the breakpoint position, which is then use to rank the fusion candidates and filter out possible artifacts. Candidate high-confidence fusions are then annotated using curated databases, to determine the genes involved and prioritize candidates which were previously identified in other studies and have proven clinical implications. However, only a relatively small number of gene fusions have been fully characterized as oncogenic drivers, therefore further analysis is necessary to identify novel fusion drivers. The tool **Oncofuse** implements a Naïve Bayes network classifier to predict the oncogenic potential of novel fusions [94]. The classifier was trained on a set of 24 features extracted from genomic hallmarks of oncogenic fusions, including gene expression and functional profile (e.g. kinase activity). For each fusion, the tool provides detailed scores for each feature and a probability of it being a driver. A more recent tool called **driverFuse** integrates gene fusions identified by RNA-seq analysis with orthogonal data types, such as CNV and SV, to distinguish driver from passenger fusions [95]. Gene fusions that map onto CNVs and SVs are flagged as drivers and further annotated with genomic features such as the functional domains involved in the chimera.

6 Driver Identification Through Regulatory Network Analysis

The analysis of regulatory networks provides a different approach to the identification of cancer drivers. A regulatory network consists of a set of interactions between gene transcripts or proteins, which can be inferred through gene or protein expression, and can be as comprehensive as to include the full cell interactome or limited to specific regulatory programs and pathways. Key drivers (KD), also known as master regulators (MR), are genes that most significantly modulate the transcriptional programs governing a specific phenotype and can be inferred through sophisticated analysis of regulatory networks [96].

A node in a regulatory network usually represents a gene or a protein, although hybrid networks including both types of molecules and extended to further incorporate other molecules such as metabolites, have been generated and employed in many studies. The edges connecting the nodes can represent different types of interactions, the most common being co-expression. A gene co-expression network is formed by calculating the correlation between the expressions of all pairs of gene transcripts (e.g. from a microarray or an RNA-seq experiment) and connecting genes that are significantly correlated with undirected edges [97]. These networks can then be transformed to approximate a scale-free topology characterized by a power law distribution of node degrees and edge strengths, which confers upon them important properties and allows the use of different algorithms for their analysis. One such property of scale-free networks is the presence of a few large *hubs* of connectivity, which are high-degree nodes with more interactions than regular nodes [98]. The typical analysis of a gene co-expression network aims to identify sets of highly correlated genes, which are usually termed modules and are likely to represent functional units sharing common regulatory mechanisms. Co-expression modules that significantly correlate with specific phenotypic traits (e.g. disease stage or survival) or genetic alterations (e.g. SNV), represent candidate transcriptional programs with pathogenic properties, and the hub genes within these modules are more likely to be key drivers of such programs [99, 100].

However, co-expression networks are based on correlations and cannot truly model the causal relationships between genes and proteins, which is essential to determine KDs. The next step is then to add directionality to these networks, so that the direction of an edge indicates the causal flow between the two connected genes. One approach to infer directionality is through genetic, biochemical, or temporal perturbations. The tool **RIMBANet** (Reconstructing Integrative Molecular Bayesian Networks) learns probabilistic Bayesian networks that allow to distinguish upstream regulators from downstream effects using genetic data as a source of perturbation [101, 102]. More specifically, eQTLs (Expression quantitative trait loci), DNA variants that affect gene expression, are used as structural priors, such that a gene whose expression is associated with an eQTL will be considered a parent node, given that a genetic variation cannot be the effect of a gene expression change [103]. Then, the tool **KDA** (Key Driver Analysis) can be used to identify the nodes in the network that are most likely to regulate a set of target nodes G, representing for example a specific transcriptional program, by maximizing an enrichment statistic of the expanded network neighborhood obtained by moving k-steps away from each node in G [104]. This approach has been widely applied to determine KD genes in many diseases, including cancer [99, 105–109].

The tool **iMaxDriver** defines driver genes as the nodes in a transcriptional regulatory network (TRN) with the greatest influence on other nodes [110]. The TRN is built based on RegNetwork, a repository of transcription factor and microRNA-mediated gene regulations, and TRRUST, a manually curated database of transcriptional regulatory networks [111–113]. Nodes in the network represent genes and are annotated with their expression. Then, the Influence Maximization (IM) algorithm is applied to rank genes based on their coverage, i.e. the number of genes that they activate.

MARINA (Master Regulator Inference Algorithm) infers the activity of transcription factors (TFs) controlling the transition between two phenotypes and the maintenance of the latter phenotype from the global transcriptional activation of their regulons, i.e. is the sets of their activated and repressed targets [114]. MARINA takes as input a set of interactions, e.g. from a regulatory network, and a set of differentially expressed (DE) genes representing the changes observed in the transition from one phenotype (e.g. normal) to another (e.g. cancer). For each TF, the algorithm first identifies its activated targets (positive regulon) and its repressed targets (negative regulon). Then, it computes the TF activity by gene set enrichment analysis (GSEA) of the DE genes in its regulon, and its biological relevance by the overall contribution of the regulon genes to the specific pathways of interest [115]. False positives are minimized by comparing the enrichment of TFs with overlapping targets and discarding the TFs whose enrichment is significantly reduced when the shared targets are removed (termed shadow regulators, SR). The final MR candidates are either non-SRs or SRs that are synergistic with a non-SR, i.e. TFs whose common targets are more enriched than their individual regulons. MARINA was later incorporated in a pipeline to determine causal genetic drivers called **DIGGIT** (driver-gene inference by genetical-genomics and information theory) [116]. In DIGGIT, a regulatory network is first computed using the tool ARACNe, which is based on gene co-regulation inferred by mutual information, using a gene expression cancer signature [117]. Then, MARINA is applied to the network to infer candidate MRs. Next, functional CNAs are selected based on either mutual information between copy number and expression or DE between cancer and control samples and the tool MINDy is applied to interrogate pathways upstream of the MRs and identify candidate CNA modulators of MR activity by conditional mutual information analysis [118]. The candidate CNA modulators are also analyzed to identify aQTL (activity QTL), i.e. loci predicting MR activity. Finally, the CNA modulators inferred by MINDy and aQTL analysis are integrated by Fisher's method to obtain a finalized list of candidate CNA modulators.

7 Multi-omics Approaches for the Identification of Cancer Drivers

The approaches to determine cancer drivers presented so far have been designed to identify a specific type of driver, such as SNVs or CNA, in some cases with the support of gene expression data as a readout of functional impact. However, many cancers are driven by multiple types of aberrations that act synergistically to develop and maintain the malignant phenotype. Therefore, when multiple omics data is available, it is worth considering integrative approaches designed to identify and characterize groups of driver alterations that may better explain the observed downstream effects on the transcriptome, the proteome and the overall behavior of cancer cells.

ModulOmics is a method for the identification of cancer driver pathways, defined as modules, which integrates mutual exclusivity of SNVs and CNAs, proximity in a protein-protein interaction network, and transcriptional co-regulation and co-expression into a single probabilistic model [119]. It implements an optimization procedure that combines integer linear programming to identify good initial solutions and a stochastic search to refine them. The ModulOmics probabilistic score of a set of genes G and a collection of models, e.g. mutual exclusivity of SNVs or gene co-expression, is computed as the mean of the individual probabilistic scores for each model, which represent how strongly the genes are functionally connected. For example, the mutual exclusivity model uses the tool TiMEx to calculate the mutual exclusivity intensity of the gene set G, while the co-regulation model is defined as the fraction of genes in G that are co-regulated by at least one common active transcription factor. The authors have shown that driver modules identified by ModulOmics in public cancer datasets, such as TCGA, are enriched with known drivers, are functionally coherent, reliably separate cancer from normal tissue, recapitulate known mechanisms and suggest novel hypotheses.

AMARETTO is another multi-omics method for the identification of cancer drivers which first models the effects of genomic and epigenomic data on gene expression and then generates a module network to connect cancer drivers with their downstream targets [120, 121]. The AMARETTO algorithm identifies candidate drivers by implementing a linear model to capture the relationship between significant CNAs or hyper/hypomethylated genes, detected by GISTIC and MethylMix, respectively, and disease-specific gene expression. Then, it models the trans-effects of the candidate drivers by generating co-expression modules associated with specific regulatory programs, which are defined as sparse linear combinations of the candidate drivers predicting the module's mean expression. The algorithm runs iteratively, learning the regulatory programs and reassigning genes to modules until less than 1% of the genes are reassigned to a new module. Application of AMARETTO to TCGA data demonstrated its ability to identify novel candidates in addition to known drivers, and to detect modules discriminating good and poor outcome.

The tool **Moonlight** integrates multi-omics data with information from literature and pathway databases and is based on the concept of *protein moonlighting*, a phenomenon by which a protein can perform multiple functions [122]. The Moonlight algorithm analyzes gene expression data from tumor and normal samples to identify differentially expressed genes, then it maps these genes onto a gene regulatory network and performs functional enrichment analysis to identify altered oncogenic processes enriched in the network. Next, it runs an upstream regulatory analysis to identify the candidate genes responsible for the activation or inhibition of the altered processes, defined as oncogenic mediators, by comparing the differentially expressed genes with the information extracted from a manually curated library of biological processes associated with cancer, including proliferation and apoptosis. Then, it uses additional layers of data, such as DNA methylation, CNA, SNV and clinical data, to prioritize the candidate oncogenic mediators through a pattern recognition analysis. The authors applied Moonlight to TCGA data and showed that the genes identified as oncogenic mediators were associated with specific tumor subtypes, had potential dual roles, were prioritized in accessible regions, exhibited differences in mutations and copy number changes, and were associated with survival outcomes.

Multi-omics MasterRegulator Analysis (**MOMA**) is a more recent tool which integrates gene expression and genomic alterations, such as SNVs, CNAs and gene fusions, to identify master regulator proteins and modules representing the key effectors responsible for implementing and maintaining the transcriptional identity of cancer cells as a function of their genomic alterations [123]. The MOMA pipeline identifies candidate MR proteins by Fisher's integration of p values (MOMA score) from four separate analyses: (1) analysis of gene expression data to infer protein activity profiles using the tool VIPER (Virtual Proteomics by Enriched Regulon Analysis) [124]; (2) analysis of functional genetic alterations in their upstream pathways inferred by the tool DIGGIT (Driver-Gene Inference by Genetical-Genomic Information Theory) (DIGGIT) [116]; (3) analysis of literature-based evidence which supports direct protein-protein interactions between MRs and proteins harboring genetic alterations using the PrePPI (Predicting Protein-Protein Interactions) algorithm [125]. Then the PAM algorithm (partitioning around medoids) is used to identify tumor subtypes representing distinct transcriptional entities regulated by the same MRs [126]. In this step, a protein activity profile similarity between samples is calculated using the MOMA scores to weigh the contribution of each candidate MR. Lastly, a genomic saturation analysis is performed to identify the top candidate MRs that are most likely to control the subtype transcriptional identity. The application of MOMA to TCGA data identified 24 pan-cancer MR modules, each regulating key cancer hallmarks and with prognostic impact. Moreover, more than half of the somatic alterations detected in each individual sample, were predicted to induce aberrant MR activity.

8 Identification of Non-coding Drivers

As discussed in the previous sections, most efforts to identify cancer drivers have been focused, with good reason, on coding regions of the genome, that is areas where protein-coding genes are located. While large structural drivers, like broad CNAs, encompass coding and non-coding areas, the driver effects are mostly determined by their impact on the expression of the affected genes and by their prognostic value. Within genes, SNV drivers are mostly identified in the coding regions, as they are associated with protein-level changes, and thus more likely to be functionally relevant. However, while the majority of SNV and indel drivers are probably harbored in coding areas, only less than 2% of the human genome encodes for proteins, and there is growing evidence supporting the presence of non-coding drivers [127–130]. Like for coding alterations, functional non-coding variants can be found in both germline and somatic DNA. However, most germline variants are not as disruptive as the large and complex events commonly observed in cancer cells, as they are expressed in all cells of the body and, therefore, must be compatible with life [130].

Non-coding variants with functional potential involve regulatory regions, e.g. promoters, enhancers and silencers, and non-coding RNA (ncRNA) genes [131]. The first non-coding cancer driver mutations were reported in 2013 and involve the promoter region of TERT (telomerase reverse transcriptase), the gene coding for the catalytic subunit of telomerase [132, 133]. Two distinct mutations occurring in 71% of the melanomas analyzed in the study, created *de novo* consensus binding motifs for ETS transcription factors, which increased transcriptional activity of the TERT promoter by two- to four-fold. Another type of regulatory non-coding driver is generated by genomic rearrangements involving gene regulatory regions. The translocations commonly observed in multiple myeloma, a cancer of bone marrow plasma cells, juxtapose regulatory regions of immunoglobulin heavy chain genes (IgH) on chromosome 14 with oncogenes such as MMSET (translocation t(4;14)), MAF (translocation t(14;16)) and CCND1 (translocation t(11;14)), leading to their overexpression [134, 135]. Other potential non-coding driver events may involve intronic regions and affect splicing, affect regulatory ncRNAs, or generate and disrupt binding sites for regulatory ncRNAs, e.g. microRNAs, in non-coding gene regions like the 3′UTR [136–138].

Several tools and databases are available to prioritize non-coding variants with potential functional effects, such as ANNOVAR, VEP and Onco-cis [58, 139, 140]. A few algorithms have been specifically developed to identify non-coding drivers in cancer, including **ncdDetect** and **ncDriver** [141, 142]. The tool ncdDetect analyzes the mutations observed in a candidate genomic region of interest by comparing them with a sample- and position-specific background mutation rate estimated based on different genomic annotations correlating with mutation rate, such as the flanking bases of the mutated nucleotides, the tissue-specific gene expression level, and replication timing. The authors provide downloadable model estimates obtained from a pan-cancer set of 505 whole genomes, from which position- and sample-specific probabilities can be predicted. A scoring scheme is then applied to further evaluate the functional impact of a candidate cancer driver element. The tool provides three different scoring schemes. The first evaluates the mutation burden and is defined by the average number of mutations in the candidate region. The second one evaluates the goodness of fit of the observed mutations to the null model, where mutations unlikely to be observed have a higher score, which is defined as a log-likelihood. The third scheme evaluates the functional impact of the mutations and uses a position-specific score of evolutionary conservation, phyloP, as a proxy for functional impact [143]. Alternatively, ncdDetect allows users to define their own scoring schemes.

ncDriver is another tool specifically designed to identify non-coding drivers in regions with significantly elevated conservation and cancer specificity. To develop ncDriver, the authors used a set of >3 million SNVs and >200 K INDELs from 507 whole-genome samples from 10 different cancer types. A mutational recurrence test was performed to evaluate the significance of the total number of mutations observed in each non-coding element, including long ncRNAs, short ncRNAs and promoters, considering their lengths and the background mutation rate for the specific type of element based on a binomial distribution. Then, three separate tests were performed to evaluate the conservation of the mutations within each element and the specificity to cancer. In particular, a sampling procedure based on the mean phyloP scores calculated across the observed mutations was used to assess local conservation, i.e. by considering random samples with the same number of mutations and distribution of phyloP scores as the specific element analyzed, and global conservation, i.e. by considering to the observed distribution across all elements of the same given type. Then, Fisher's method was used to combine the three individual p-values to obtain an overall significance measure.

9 Patient-Specific Drivers and Precision Medicine

All the methods and tools described so far have been designed to discover driver genes and alterations in large sample dataset, which often provide sufficient statistical power to differentiate between drivers and passengers. However, cohort-based methods fail to detect rare driver alterations. Cancer cells may disrupt hallmark pathways in different ways, even within the same cancer type, and patients often have no alterations in known cancer genes. In other cases, it may be necessary to prioritize different co-occurring cancer drivers for therapeutic decisions. These considerations are at the core of precision oncology, a novel translational research area whose goal is to develop and apply methods to devise treatment strategies that are tailored to the individual tumors through in-depth characterization of patient's specific features [144]. Therefore, several tools have been developed over the past few years to support this endeavor and allow the discovery and prioritization of drivers in individual tumor samples.

The objective of **DawnRank** is to rank mutated genes in a patient according to their potential to be drivers, through an iterative network approach based on the PageRank algorithm, which was originally developed by Google's founders to rank web pages based on their importance [145]. The algorithm requires a gene interaction directed network, the somatic SNVs observed in the patient, and the DE genes between cancer and control samples. Driver genes are then prioritized based on their connectivity to downstream DE genes, under the assumption that alterations in highly connected genes are more likely to be impactful. The authors provide a gene interaction network based on the network used in the tool MEMo for mutual exclusivity analysis, and curated data from pathway databases such as Reactome, PID and KEGG. Similarly, the tool **SCS** (Single-sample Controller Strategy) integrates mutations (SNV and CNA) and gene expression data to identify a minimal number of mutations (drivers) with the highest impact on the DE genes through a gene interaction network [146]. The tool employs a network control theory algorithm which models the mutations as controllers and gene expression as network states. Another tool, **PRODIGY** (Personalized Ranking of DrIver Genes analYsis), ranks mutations in an individual patient based on their impact on dysregulated pathways, which is calculated separately for each pathway using a variant of the prize-collecting Steiner tree (PCST) model, where node prizes reflect the extent of differential expression of DE genes in the pathway, and node penalties reflect the degrees of the other nodes [147]. The influence scores for each gene in each pathway are then aggregated and ranked.

Other methods are based on machine learning approaches. The tool **sysSVM**, and its upgrade **sysSVM2**, implement a one-class support vector machine (SVM) trained on a set of canonical cancer drivers to learn their gene-level and molecular features [148, 149]. The model is then used to predict driver alterations in individual patient profiles that best match the learned features. The tool driveR implements a multi-task learning classification model which uses different features of drivers such as CNA and SNV scores, pathway membership and a feature generated by a coding impact metapredictor model which prioritizes coding variants according to their pathogenicity [150].

One limitation of tools that detect and rank driver genes as separate events is that in many cases cancer is driven by a set of co-occurring alterations, which are all synergistic and necessary to achieve the oncogenic transformation. For example, activating mutations in NRAS and KRAS often co-occur with inactivation of CDKN2A and CDKN2B in several cancers, including lung and melanoma, which promotes cellular proliferation, cancer cell differentiation and metastasic dissemination [151]. The tool **CRSO** (Cancer Rule Set Optimization) was designed to specifically identify modules of cooperating alterations that are necessary and sufficient to drive cancer in individual patients [152]. CRSO first identifies candidate driver SNVs and CNAs in a cohort of patients using the tools dNdScv and GISTIC2, respectively, and represents them as a categorical matrix, where each entry indicates the specific type of event, e.g. hotspot mutation or strong amplification. Next, a penalty matrix is calculated based on the observed alterations, where each entry is the negative logarithm of the probability of the corresponding event of being a passenger. A rule is then defined as a minimally sufficient combination of two or more co-occurring alterations that drive the tumors which harbor it. The CRSO algorithm employs a stochastic procedure which evaluates and optimizes an objective scoring function, based on statistical penalty, over the set of all possible rules.

10 Databases of Cancer Drivers

Several resources are available online that collect and catalogue data on mutations and driver alterations in cancer. **COSMIC** (Catalogue of Somatic Mutations in Cancer) is the most comprehensive resource of somatic mutations in human cancer [153]. It comprises several projects, each dedicated to the presentation of different aspects of mutation data. The COSMIC core database is a manually curated collection of somatic mutations discovered in human cancers, which can also be explored through an interactive view of 3D structures (COSMIC-3D) [154]. The collection includes SNVs, structural rearrangements, gene fusions, CNAs, non-coding variants as well as gene expression and methylation data from the TCGA (The Cancer Genome Atlas) project. The COSMIC Cancer Gene Census (CGC) database is a collection of driver genes whose mutations are causally implicated in cancer [155]. The data is organized in two different tiers: high confidence driver genes (tier I) and recently characterized genes with a less defined role in cancer (tier II). The COSMIC actionability database aims at enabling precision oncology by further connecting somatic mutations with potential therapeutic options. Furthermore, COSMIC includes a collection of mutation profiles of >1,000 cell lines used in cancer research. The data can be browsed online and downloaded as flat text file.

Additional resources are mainly focused on collecting proven and predicted cancer drivers. **IntOGen** (Integrative OncoGenomics) is a compendium of mutational cancer drivers obtained through the analysis of public domain data collected from over 28,000 samples of 66 cancer types [156]. The IntOGen analytical pipeline consists of a pre-filtering and quality control step, where low quality variants and duplicated samples are discarded, and the application of six different tools for driver prediction: dNdScv, CBaSE, OncodriveCLUSTL, HotMAPS3D, smRegions, OncodriveFML and MutPanning [49, 51, 55, 157–159]. The predictions are then combined through a weighted vote in which each method is assigned a weight based on its perceived credibility. The collection can be browsed online and downloaded for offline processing.

Other collections of cancer drivers include **CancerMine** and **OncoVar**. CancerMine is a database of literature-mined cancer drivers, oncogenes, and tumor suppressors [160]. The CancerMine authors extracted titles, abstracts, and full-text articles available from PubMed and PubMed Central, then identified 1,500 sentences that contained a gene name, a cancer type and keywords suggesting a possible role of the gene in cancer. These sentences were manually reviewed to annotate drivers, oncogenes, and tumor suppressors. A Kindred relation classifier was trained to learn the characteristics of these sentences through a logistic regression of word frequencies and semantic features. The classifier was then applied to a larger pool of >38,000 sentences from >28,000 papers and determined >4,000 genes linked to 425 cancer types as drivers, oncogenes, and tumor suppressors. Like COSMIC and IntOGen, CancerMine offers an online interface to browse and query the data as well as text files for free download.

OncoVar implements a driver prioritization approach which employs several prediction algorithms and known driver events [161]. Somatic missense SNVs identified from the TCGA and ICGC (International Cancer Genome Consortium) projects were analyzed by the machine learning method AI-Driver, which combines predictions from MutSigCV and MutPanning [162]. Mutations with an AI-Driver score $>= 0.95$ and observed in at least two patients were considered driver mutations. Additionally, a five-tiered consensus score based on an improved Borda counting approach was used to rank the genes, where each gene was assigned a score equal to the sum of its weight across 16 driver sets, including OncoKB, COSMIC CGC, and IntOGen [163]. The adjusted p values from AI-Driver and the five-tiered consensus classifiers were then combined using the Fisher's Test to obtain a unique final p value. OncoVar is available both through an online interface and as text files.

11 Conclusions

Enormous progress has been made in the past decade towards the definition, identification, and characterization of cancer drivers. The term "driver" has been applied both generally to describe genes that are often altered in cancer, and more narrowly to describe genes or alterations with a proven causative effect in the (dys)regulation of cancer hallmarks. This latter class is harder to determine and investigate, requiring more sophisticated modeling and experimental validation. Hundreds of computational resources, methods, tools, and databases have been developed to process, model, analyze, interpret, and catalogue the massive amount of data produced every year by cancer sequencing projects. Many times, important discoveries are made by using *ad hoc* approaches that implement combinations of the methodologies and tools described in this chapter. The results of such endeavors are then translated into new prognostic and therapeutic tools:

novel clinically meaningful classifications that allow more precise patient stratification and the implementation of precision medicine strategies, as well as novel targets whose investigation elucidates complex molecular mechanisms and enables the development of new therapeutic agents.

References

1. Tabassum, D.P., Polyak, K.: Tumorigenesis: it takes a village. Nat. Rev. Cancer **15**, 473–483 (2015)
2. Kontomanolis, E.N., et al.: Role of oncogenes and tumor-suppressor genes in carcinogenesis: a review. Anticancer Res. **40**, 6009–6015 (2020)
3. Hanahan, D., Weinberg, R.A.: The hallmarks of cancer. In: Oxford Textbook of Oncology, pp. 3–10. Oxford University Press (2016)
4. Hanahan, D., Weinberg, R.A.: Hallmarks of cancer: the next generation. Cell **144**, 646–674 (2011)
5. Sherr, C.J.: Principles of tumor suppression. Cell **116**, 235–246 (2004)
6. Shortt, J., Johnstone, R.W.: Oncogenes in cell survival and cell death. Cold Spring Harb. Perspect. Biol. **4**, a009829–a009829 (2012)
7. Jang, J.-H., Kim, D.-H., Surh, Y.-J.: Dynamic roles of inflammasomes in inflammatory tumor microenvironment. NPJ Precis. Oncol. **5**, 18 (2021)
8. Greten, F.R., Grivennikov, S.I.: Inflammation and cancer: triggers, mechanisms, and consequences. Immunity **51**, 27–41 (2019)
9. Zhao, H., et al.: Inflammation and tumor progression: signaling pathways and targeted intervention. Signal Transduct. Target. Ther. **6**, 263 (2021)
10. Zhu, G., Pei, L., Xia, H., Tang, Q., Bi, F.: Role of oncogenic KRAS in the prognosis, diagnosis and treatment of colorectal cancer. Mol. Cancer **20**, 143 (2021)
11. Alqathama, A.: BRAF in malignant melanoma progression and metastasis: potentials and challenges. Am. J. Cancer Res. **10**, 1103–1114 (2020)
12. Ungefroren, H.: Autocrine TGF-β in cancer: review of the literature and caveats in experimental analysis. Int. J. Mol. Sci. **22**, 977 (2021)
13. Zeromski, J.: Significance of tumor-cell receptors in human cancer. Arch. Immunol. Ther. Exp. (Warsz.) **50**, 105–110 (2002)
14. Wong, R.S.Y.: Apoptosis in cancer: from pathogenesis to treatment. J. Exp. Clin. Cancer Res. **30**, 87 (2011)
15. Kumari, R., Jat, P.: Mechanisms of cellular senescence: cell cycle arrest and senescence associated secretory phenotype. Front. Cell Dev. Biol. **9**, 645593 (2021)
16. Recagni, M., Bidzinska, J., Zaffaroni, N., Folini, M.: The role of alternative lengthening of telomeres mechanism in cancer: translational and therapeutic implications. Cancers (Basel) **12**, 949 (2020)
17. De Palma, M., Biziato, D., Petrova, T.V.: Microenvironmental regulation of tumour angiogenesis. Nat. Rev. Cancer **17**, 457–474 (2017)
18. Lugano, R., Ramachandran, M., Dimberg, A.: Tumor angiogenesis: causes, consequences, challenges and opportunities. Cell. Mol. Life Sci. **77**, 1745–1770 (2020)
19. Klein, C.A.: Tumour cell dissemination and growth of metastasis. Nat. Rev. Cancer **10**, 156 (2010)
20. Fares, J., Fares, M.Y., Khachfe, H.H., Salhab, H.A., Fares, Y.: Molecular principles of metastasis: a hallmark of cancer revisited. Signal Transduct. Target. Ther. **5**, 28 (2020)
21. Brabletz, T., Kalluri, R., Nieto, M.A., Weinberg, R.A.: EMT in cancer. Nat. Rev. Cancer **18**, 128–134 (2018)

22. Munn, D.H., Bronte, V.: Immune suppressive mechanisms in the tumor microenvironment. Curr. Opin. Immunol. **39**, 1–6 (2016)

23. Shimizu, K., Iyoda, T., Okada, M., Yamasaki, S., Fujii, S.-I.: Immune suppression and reversal of the suppressive tumor microenvironment. Int. Immunol. **30**, 445–455 (2018)

24. Faubert, B., Solmonson, A., DeBerardinis, R.J.: Metabolic reprogramming and cancer progression. Science **368**, eaaw5473 (2020)

25. Hanahan, D.: Hallmarks of cancer: new dimensions. Cancer Discov. **12**, 31–46 (2022)

26. Perekatt, A.O., et al.: SMAD4 suppresses WNT-driven dedifferentiation and oncogenesis in the differentiated gut epithelium. Cancer Res. **78**, 4878–4890 (2018)

27. Köhler, C., et al.: Mouse cutaneous melanoma induced by mutant BRaf arises from expansion and dedifferentiation of mature pigmented melanocytes. Cell Stem Cell **21**, 679-693.e6 (2017)

28. Shah, M., et al.: A role for ATF2 in regulating MITF and melanoma development. PLoS Genet. **6**, e1001258 (2010)

29. Thienpont, B., Van Dyck, L., Lambrechts, D.: Tumors smother their epigenome. Mol. Cell. Oncol. **3**, e1240549 (2016)

30. Skrypek, N., Goossens, S., De Smedt, E., Vandamme, N., Berx, G.: Epithelial-to-mesenchymal transition: epigenetic reprogramming driving cellular plasticity. Trends Genet. **33**, 943–959 (2017)

31. Dzutsev, A., et al.: Microbes and cancer. Annu. Rev. Immunol. **35**, 199–228 (2017)

32. Helmink, B.A., Khan, M.A.W., Hermann, A., Gopalakrishnan, V., Wargo, J.A.: The microbiome, cancer, and cancer therapy. Nat. Med. **25**, 377–388 (2019)

33. Sears, C.L., Garrett, W.S.: Microbes, microbiota, and colon cancer. Cell Host Microbe **15**, 317–328 (2014)

34. Kowald, A., Passos, J.F., Kirkwood, T.B.L.: On the evolution of cellular senescence. Aging Cell **19**, e13270 (2020)

35. Wang, B., Kohli, J., Demaria, M.: Senescent cells in cancer therapy: friends or foes? Trends Cancer **6**, 838–857 (2020)

36. Chang, T.-C., Xu, K., Cheng, Z., Wu, G.: Somatic and germline variant calling from next-generation sequencing data. Adv. Exp. Med. Biol. **1361**, 37–54 (2022)

37. Xu, C.: A review of somatic single nucleotide variant calling algorithms for next-generation sequencing data. Comput. Struct. Biotechnol. J. **16**, 15–24 (2018)

38. Robinson, J.T., et al.: Integrative genomics viewer. Nat. Biotechnol. **29**, 24–26 (2011)

39. Privitera, G.F., Alaimo, S., Ferro, A., Pulvirenti, A.: Computational resources for the interpretation of variations in cancer. Adv. Exp. Med. Biol. **1361**, 177–198 (2022)

40. Alexandrov, L.B., et al.: Signatures of mutational processes in human cancer. Nature **500**, 415–421 (2013)

41. Bergstrom, E.N., et al.: SigProfilerMatrixGenerator: a tool for visualizing and exploring patterns of small mutational events. BMC Genom. **20**, 685 (2019)

42. Alexandrov, L.B., et al.: The repertoire of mutational signatures in human cancer. Nature **578**, 94–101 (2020)

43. Salvadores, M., Mas-Ponte, D., Supek, F.: Passenger mutations accurately classify human tumors. PLoS Comput. Biol. **15**, e1006953 (2019)

44. Rao, Y., Ahmed, N., Pritchard, J., O'Brien, E.: MutSigCVsyn: identification of thirty synonymous cancer drivers. bioRxiv (2022). https://doi.org/10.1101/2022.01.16.476507

45. Dees, N.D., et al.: MuSiC: identifying mutational significance in cancer genomes. Genome Res. **22**, 1589–1598 (2012)

46. Vogelstein, B., et al.: Cancer genome landscapes. Science **339**, 1546–1558 (2013)

47. Tamborero, D., Gonzalez-Perez, A., Lopez-Bigas, N.: OncodriveCLUST: exploiting the positional clustering of somatic mutations to identify cancer genes. Bioinformatics **29**, 2238–2244 (2013)

48. Gonzalez-Perez, A., Lopez-Bigas, N.: Functional impact bias reveals cancer drivers. Nucl. Acids Res. **40**, e169 (2012)
49. Mularoni, L., Sabarinathan, R., Deu-Pons, J., Gonzalez-Perez, A., López-Bigas, N.: OncodriveFML: a general framework to identify coding and non-coding regions with cancer driver mutations. Genome Biol. **17**, 1–13 (2016)
50. Reimand, J., Bader, G.D.: Systematic analysis of somatic mutations in phosphorylation signaling predicts novel cancer drivers. Mol. Syst. Biol. **9**, 637 (2013)
51. Martincorena, I., et al.: Universal patterns of selection in cancer and somatic tissues. Cell **173**, 1823 (2018)
52. Porta-Pardo, E., Godzik, A.: E-Driver: a novel method to identify protein regions driving cancer. Bioinformatics **30**, 3109–3114 (2014)
53. Van den Eynden, J., Fierro, A.C., Verbeke, L.P.C., Marchal, K.: SomInaClust: detection of cancer genes based on somatic mutation patterns of inactivation and clustering. BMC Bioinform. **16**, 125 (2015)
54. Davoli, T., et al.: Cumulative haploinsufficiency and triplosensitivity drive aneuploidy patterns and shape the cancer genome. Cell **155**, 948–962 (2013)
55. Dietlein, F., et al.: Identification of cancer driver genes based on nucleotide context. Nat. Genet. **52**, 208–218 (2020)
56. Luo, P., Ding, Y., Lei, X., Wu, F.-X.: DeepDriver: predicting cancer driver genes based on somatic mutations using deep convolutional neural networks. Front. Genet. **10**, 13 (2019)
57. Dragomir, I., et al.: Identifying cancer drivers using DRIVE: a feature-based machine learning model for a pan-cancer assessment of somatic missense mutations. Cancers (Basel) **13**, 2779 (2021)
58. McLaren, W., et al.: The ensemble variant effect predictor. Genome Biol. **17**, 1–14 (2016)
59. Micheel, C.M., et al.: American Association for Cancer Research Project Genomics Evidence Neoplasia Information Exchange: from inception to first data release and beyond-lessons learned and member institutions' perspectives. JCO Clin. Cancer Inform. **2**, 1–14 (2018)
60. Cancer Genome Atlas Research Network et al.: The Cancer Genome Atlas Pan-Cancer analysis project. Nat. Genet. **45**, 1113–1120 (2013)
61. Muiños, F., Martínez-Jiménez, F., Pich, O., Gonzalez-Perez, A., Lopez-Bigas, N.: In silico saturation mutagenesis of cancer genes. Nature **596**, 428–432 (2021)
62. Conway, J.R., et al.: Integrated molecular drivers coordinate biological and clinical states in melanoma. Nat. Genet. **52**, 1373–1383 (2020)
63. Hua, X., et al.: DrGaP: a powerful tool for identifying driver genes and pathways in cancer sequencing studies. Am. J. Hum. Genet. **93**, 439–451 (2013)
64. Wendl, M.C., et al.: PathScan: a tool for discerning mutational significance in groups of putative cancer genes. Bioinformatics **27**, 1595–1602 (2011)
65. Gaffney, S.G., Townsend, J.P.: PathScore: a web tool for identifying altered pathways in cancer data. Bioinformatics **32**, 3688–3690 (2016)
66. Iorio, F., et al.: Pathway-based dissection of the genomic heterogeneity of cancer hallmarks' acquisition with SLAPenrich. Sci. Rep. **8**, 6713 (2018)
67. Vandin, F., Upfal, E., De Raphael, B.J.: Novo discovery of mutated driver pathways in cancer. Genome Res. **22**, 375–385 (2012)
68. Leiserson, M.D.M., Blokh, D., Sharan, R., Raphael, B.J.: Simultaneous identification of multiple driver pathways in cancer. PLoS Comput. Biol. **9**, e1003054 (2013)
69. Ciriello, G., Cerami, E., Sander, C., Schultz, N.: Mutual exclusivity analysis identifies oncogenic network modules. Genome Res. **22**, 398–406 (2012)
70. Leiserson, M.D.M., Wu, H.-T., Vandin, F., Raphael, B.J.: CoMEt: a statistical approach to identify combinations of mutually exclusive alterations in cancer. Genome Biol. **16**, 160 (2015)

71. Kim, Y.-A., Madan, S., Przytycka, T.M.: WeSME: uncovering mutual exclusivity of cancer drivers and beyond. Bioinformatics **33**, 814–821 (2016). btw242

72. Constantinescu, S., Szczurek, E., Mohammadi, P., Rahnenführer, J., Beerenwinkel, N.: TiMEx: a waiting time model for mutually exclusive cancer alterations. Bioinformatics **32**, 968–975 (2016)

73. Bokhari, Y., Arodz, T.: QuaDMutEx: quadratic driver mutation explorer. BMC Bioinform. **18**, 458 (2017)

74. Nabavi, S., Zare, F.: Identification of copy number alterations from next-generation sequencing data. Adv. Exp. Med. Biol. **1361**, 55–74 (2022)

75. Leibowitz, M.L., Zhang, C.-Z., Pellman, D.: Chromothripsis: a new mechanism for rapid karyotype evolution. Annu. Rev. Genet. **49**, 183–211 (2015)

76. Beroukhim, R., et al.: Assessing the significance of chromosomal aberrations in cancer: methodology and application to glioma. Proc. Natl. Acad. Sci. U. S. A. **104**, 20007–20012 (2007)

77. Mermel, C.H., et al.: GISTIC2.0 facilitates sensitive and confident localization of the targets of focal somatic copy-number alteration in human cancers. Genome Biol. **12**, R41 (2011)

78. van Dyk, E., Hoogstraat, M., ten Hoeve, J., Reinders, M.J.T., Wessels, L.F.A.: RUBIC identifies driver genes by detecting recurrent DNA copy number breaks. Nat. Commun. **7**, 12159 (2016)

79. Fan, B., et al.: Integration of DNA copy number alterations and transcriptional expression analysis in human gastric cancer. PLoS ONE **7**, e29824 (2012)

80. Akavia, U.D., et al.: An integrated approach to uncover drivers of cancer. Cell **143**, 1005–1017 (2010)

81. Pickering, C.R., et al.: Integrative genomic characterization of oral squamous cell carcinoma identifies frequent somatic drivers. Cancer Discov. **3**, 770–781 (2013)

82. Tamborero, D., Lopez-Bigas, N., Gonzalez-Perez, A.: Oncodrive-CIS: a method to reveal likely driver genes based on the impact of their copy number changes on expression. PLoS ONE **8**, e55489 (2013)

83. Bashashati, A., et al.: DriverNet: uncovering the impact of somatic driver mutations on transcriptional networks in cancer. Genome Biol. **13**, R124 (2012)

84. Baur, B., Bozdag, S.: ProcessDriver: a computational pipeline to identify copy number drivers and associated disrupted biological processes in cancer. Genomics **109**, 233–240 (2017)

85. Zhou, Y., et al.: Identifying key somatic copy number alterations driving dysregulation of cancer hallmarks in lower-grade glioma. Front. Genet. **12**, 654736 (2021)

86. Wu, H., Li, X., Li, H.: Gene fusions and chimeric RNAs, and their implications in cancer. Genes Dis. **6**, 385–390 (2019)

87. Kang, Z.-J., et al.: The Philadelphia chromosome in leukemogenesis. Chin. J. Cancer **35**, 48 (2016)

88. Hedges, D.J.: RNA-seq fusion detection in clinical oncology. Adv. Exp. Med. Biol. **1361**, 163–175 (2022)

89. Iyer, M.K., Chinnaiyan, A.M., Maher, C.A.: ChimeraScan: a tool for identifying chimeric transcription in sequencing data. Bioinformatics **27**, 2903–2904 (2011)

90. Kim, D., Salzberg, S.L.: TopHat-Fusion: an algorithm for discovery of novel fusion transcripts. Genome Biol. **12**, R72 (2011)

91. Haas, B.J., et al.: STAR-fusion: fast and accurate fusion transcript detection from RNA-Seq. bioRxiv (2017). https://doi.org/10.1101/120295

92. Nicorici, D., et al.: FusionCatcher - a tool for finding somatic fusion genes in paired-end RNA-sequencing data. bioRxiv (2014). https://doi.org/10.1101/011650

93. Uhrig, S., et al.: Accurate and efficient detection of gene fusions from RNA sequencing data. Genome Res. **31**, 448–460 (2021)

94. Shugay, M., Ortiz de Mendíbil, I., Vizmanos, J.L., Novo, F.J.: Oncofuse: a computational framework for the prediction of the oncogenic potential of gene fusions. Bioinformatics **29**, 2539–2546 (2013)

95. Roy, S., Gupta, D.: DriverFuse: an R package for analysis of next-generation sequencing datasets to identify cancer driver fusion genes. PLoS ONE **17**, e0262686 (2022)

96. Karlebach, G., Shamir, R.: Modelling and analysis of gene regulatory networks. Nat. Rev. Mol. Cell Biol. **9**, 770–780 (2008)

97. Langfelder, P., Horvath, S.: WGCNA: an R package for weighted correlation network analysis. BMC Bioinform. **9**, 559 (2008)

98. Langfelder, P., Mischel, P.S., Horvath, S.: When is hub gene selection better than standard meta-analysis? PLoS ONE **8**, e61505 (2013)

99. Zhang, B., et al.: Integrated systems approach identifies genetic nodes and networks in late-onset Alzheimer's disease. Cell **153**, 707–720 (2013)

100. Laganà, A., et al.: Integrative network analysis identifies novel drivers of pathogenesis and progression in newly diagnosed multiple myeloma. Leukemia **32**, 120–130 (2018)

101. Zhu, J., et al.: Stitching together multiple data dimensions reveals interacting metabolomic and transcriptomic networks that modulate cell regulation. PLoS Biol. **10**, e1001301 (2012)

102. Zhu, J., et al.: Increasing the power to detect causal associations among genes and between genes and complex traits by combining genotypic and gene expression data in segregating populations. PLoS Comput. Biol. **preprint**, e69 (2005)

103. Cohain, A., et al.: Exploring the reproducibility of probabilistic causal molecular network models. Pac. Symp. Biocomput. **22**, 120–131 (2017)

104. Bin Zhang, J.Z.: Identification of key causal regulators in gene networks. In: Proceedings of the World Congress on Engineering 2013, vol. II (2013)

105. Beckmann, N.D., et al.: Multiscale causal networks identify VGF as a key regulator of Alzheimer's disease. Nat. Commun. **11**, 3942 (2020)

106. Peters, L.A., et al.: A functional genomics predictive network model identifies regulators of inflammatory bowel disease. Nat. Genet. **49**, 1437–1449 (2017)

107. Liu, Y., et al.: A network analysis of multiple myeloma related gene signatures. Cancers (Basel) **11**, 1452 (2019)

108. Watson, C.T., et al.: Integrative transcriptomic analysis reveals key drivers of acute peanut allergic reactions. Nat. Commun. **8**, 1943 (2017)

109. Gong, Y., et al.: Constructing Bayesian networks by integrating gene expression and copy number data identifies NLGN4Y as a novel regulator of prostate cancer progression. Oncotarget **7**, 68688–68707 (2016)

110. Rahimi, M., Teimourpour, B., Marashi, S.-A.: Cancer driver gene discovery in transcriptional regulatory networks using influence maximization approach. Comput. Biol. Med. **114**, 103362 (2019)

111. Liu, Z.-P., Wu, C., Miao, H., Wu, H.: RegNetwork: an integrated database of transcriptional and post-transcriptional regulatory networks in human and mouse. Database (Oxford) **2015**, bav095 (2015)

112. Han, H., et al.: TRRUST: a reference database of human transcriptional regulatory interactions. Sci. Rep. **5**, 11432 (2015)

113. Han, H., et al.: TRRUST v2: an expanded reference database of human and mouse transcriptional regulatory interactions. Nucleic Acids Res. **46**, D380–D386 (2018)

114. Lefebvre, C., et al.: A human B-cell interactome identifies MYB and FOXM1 as master regulators of proliferation in germinal centers. Mol. Syst. Biol. **6**, 377 (2010)

115. Subramanian, A., et al.: Gene set enrichment analysis: a knowledge-based approach for interpreting genome-wide expression profiles. Proc. Natl. Acad. Sci. U. S. A. **102**, 15545–15550 (2005)

116. Chen, J.C., et al.: Identification of causal genetic drivers of human disease through systems-level analysis of regulatory networks. Cell **159**, 402–414 (2014)
117. Margolin, A.A., et al.: ARACNE: an algorithm for the reconstruction of gene regulatory networks in a mammalian cellular context. BMC Bioinform. **7 Suppl 1**, S7 (2006)
118. Wang, K., et al.: Genome-wide identification of post-translational modulators of transcription factor activity in human B cells. Nat. Biotechnol. **27**, 829–839 (2009)
119. Silverbush, D., et al.: ModulOmics: integrating multi-omics data to identify cancer driver modules. *bioRxiv* (2018). https://doi.org/10.1101/288399
120. Gevaert, O., Villalobos, V., Sikic, B.I., Plevritis, S.K.: Identification of ovarian cancer driver genes by using module network integration of multi-omics data. Interface Focus **3**, 20130013 (2013)
121. Champion, M., et al.: Module analysis captures pancancer genetically and epigenetically deregulated cancer driver genes for smoking and antiviral response. EBioMedicine **27**, 156–166 (2018)
122. Colaprico, A., et al.: Interpreting pathways to discover cancer driver genes with Moonlight. Nat. Commun. **11**, 69 (2020)
123. Paull, E.O., et al.: A modular master regulator landscape controls cancer transcriptional identity. Cell **184**, 334-351.e20 (2021)
124. Alvarez, M.J., et al.: Functional characterization of somatic mutations in cancer using network-based inference of protein activity. Nat. Genet. **48**, 838–847 (2016)
125. Zhang, Q.C., et al.: Structure-based prediction of protein-protein interactions on a genome-wide scale. Nature **490**, 556–560 (2012)
126. Park, H.-S., Jun, C.-H.: A simple and fast algorithm for K-medoids clustering. Expert Syst. Appl. **36**, 3336–3341 (2009)
127. Weinhold, N., Jacobsen, A., Schultz, N., Sander, C., Lee, W.: Genome-wide analysis of noncoding regulatory mutations in cancer. Nat. Genet. **46**, 1160–1165 (2014)
128. Khurana, E., et al.: Integrative annotation of variants from 1092 humans: application to cancer genomics. Science **342**, 1235587 (2013)
129. Fredriksson, N.J., Ny, L., Nilsson, J.A., Larsson, E.: Systematic analysis of noncoding somatic mutations and gene expression alterations across 14 tumor types. Nat. Genet. **46**, 1258–1263 (2014)
130. Khurana, E., et al.: Role of non-coding sequence variants in cancer. Nat. Rev. Genet. **17**, 93–108 (2016)
131. Melton, C., Reuter, J.A., Spacek, D.V., Snyder, M.: Recurrent somatic mutations in regulatory regions of human cancer genomes. Nat. Genet. **47**, 710–716 (2015)
132. Horn, S., et al.: TERT promoter mutations in familial and sporadic melanoma. Science **339**, 959–961 (2013)
133. Huang, F.W., et al.: Highly recurrent TERT promoter mutations in human melanoma. Science **339**, 957–959 (2013)
134. Morgan, G.J., Walker, B.A., Davies, F.E.: The genetic architecture of multiple myeloma. Nat. Rev. Cancer **12**, 335–348 (2012)
135. Bergsagel, P.L., Kuehl, W.M.: Chromosome translocations in multiple myeloma. Oncogene **20**, 5611–5622 (2001)
136. Cao, S., et al.: Discovery of driver non-coding splice-site-creating mutations in cancer. Nat. Commun. **11**, 5573 (2020)
137. Jayasinghe, R.G., et al.: Systematic analysis of splice-site-creating mutations in cancer. Cell Rep. **23**, 270-281.e3 (2018)
138. Urbanek-Trzeciak, M.O., et al.: Pan-cancer analysis of somatic mutations in miRNA genes. EBioMedicine **61**, 103051 (2020)
139. Yang, H., Wang, K.: Genomic variant annotation and prioritization with ANNOVAR and wANNOVAR. Nat. Protoc. **10**, 1556–1566 (2015)

140. Perera, D., et al.: OncoCis: annotation of cis-regulatory mutations in cancer. Genome Biol. **15**, 485 (2014)

141. Juul, M., et al.: NcdDetect2: improved models of the site-specific mutation rate in cancer and driver detection with robust significance evaluation. Bioinformatics **35**, 189–199 (2019)

142. Hornshøj, H., et al.: Pan-cancer screen for mutations in non-coding elements with conservation and cancer specificity reveals correlations with expression and survival. NPJ Genom. Med. **3**, 1 (2018)

143. Pollard, K.S., Hubisz, M.J., Rosenbloom, K.R., Siepel, A.: Detection of nonneutral substitution rates on mammalian phylogenies. Genome Res. **20**, 110–121 (2010)

144. Hodson, R.: Precision oncology. Nature **585**, S1 (2020)

145. Hou, J.P., Ma, J.: DawnRank: discovering personalized driver genes in cancer. Genome Med. **6**, 56 (2014)

146. Guo, W.-F., et al.: Discovering personalized driver mutation profiles of single samples in cancer by network control strategy. Bioinformatics **34**, 1893–1903 (2018)

147. Dinstag, G., Shamir, R.: PRODIGY: personalized prioritization of driver genes. Bioinformatics **36**, 1831–1839 (2020)

148. Mourikis, T.P., et al.: Patient-specific cancer genes contribute to recurrently perturbed pathways and establish therapeutic vulnerabilities in esophageal adenocarcinoma. Nat. Commun. **10**, 3101 (2019)

149. Nulsen, J., Misetic, H., Yau, C., Ciccarelli, F.D.: Pan-cancer detection of driver genes at the single-patient resolution. Genome Med. **13**, 12 (2021)

150. Ülgen, E., Sezerman, O.U.: DriveR: a novel method for prioritizing cancer driver genes using somatic genomics data. BMC Bioinform. **22**, 263 (2021)

151. Schuster, K., et al.: Nullifying the CDKN2AB locus promotes mutant K-ras lung tumorigenesis. Mol. Cancer Res. **12**, 912–923 (2014)

152. Klein, M.I., et al.: Identifying modules of cooperating cancer drivers. Mol. Syst. Biol. **17**, e9810 (2021)

153. Tate, J.G., et al.: COSMIC: the catalogue of somatic mutations in cancer. Nucleic Acids Res. **47**, D941–D947 (2019)

154. Jubb, H.C., Saini, H.K., Verdonk, M.L., Forbes, S.A.: COSMIC-3D provides structural perspectives on cancer genetics for drug discovery. Nat. Genet. **50**, 1200–1202 (2018)

155. Sondka, Z., et al.: The COSMIC Cancer Gene Census: describing genetic dysfunction across all human cancers. Nat. Rev. Cancer **18**, 696–705 (2018)

156. Martínez-Jiménez, F., et al.: A compendium of mutational cancer driver genes. Nat. Rev. Cancer **20**, 555–572 (2020)

157. Weghorn, D., Sunyaev, S.: Bayesian inference of negative and positive selection in human cancers. Nat. Genet. **49**, 1785–1788 (2017)

158. Arnedo-Pac, C., Mularoni, L., Muiños, F., Gonzalez-Perez, A., Lopez-Bigas, N.: OncodriveCLUSTL: a sequence-based clustering method to identify cancer drivers. Bioinformatics **35**, 5396 (2019)

159. Martínez-Jiménez, F., Muiños, F., López-Arribillaga, E., Lopez-Bigas, N., Gonzalez-Perez, A.: Systematic analysis of alterations in the ubiquitin proteolysis system reveals its contribution to driver mutations in cancer. Nat. Cancer **1**, 122–135 (2020)

160. Lever, J., Zhao, E.Y., Grewal, J., Jones, M.R., Jones, S.J.M.: CancerMine: a literature-mined resource for drivers, oncogenes and tumor suppressors in cancer. Nat. Methods **16**, 505–507 (2019)

161. Wang, T., et al.: OncoVar: an integrated database and analysis platform for oncogenic driver variants in cancers. Nucleic Acids Res. **49**, D1289–D1301 (2021)

162. Wang, H., et al.: AI-Driver: an ensemble method for identifying driver mutations in personal cancer genomes. NAR Genom. Bioinform. **2**, lqaa084 (2020)
163. Chakravarty, D., et al.: OncoKB: a precision oncology knowledge base. JCO Precis. Oncol. **1**, 1–16 (2017)

Inferring a Gene Regulatory Network from Gene Expression Data. An Overview of Best Methods and a Reverse Engineering Approach

Vincenzo Cutello$^{(\boxtimes)}$, Mario Pavone , and Francesco Zito

Department of Mathematics and Computer Science, University of Catania,
v.le Andrea Doria 6, 95125 Catania, Italy
cutello@unict.it, mpavone@dmi.unict.it, francesco.zito@phd.unict.it

Abstract. Gene Regulatory Networks (GRNs) are widely used to understand processes in cellular organisms. The spread of viruses and the development of new unknown diseases require the employment of algorithmic tools to support research in this direction. Several methods have been developed to infer a GRN from gene expression data observed in the field, each with its own features. In this article, we provide an overview of the most popular methods in this field to highlight their advantages and weaknesses. In addition, a reverse engineering framework is presented in order to facilitate the inference process and provide researchers with an artificial environment capable of replicating gene expression from genes by simulating their behavior in the real world.

Keywords: Gene Regulatory Network · Reverse Engineering ·
Machine Learning · Gene Correlation · Ordinary Differential
Equations · Boolean Network

1 Introduction

In the last few years, recent innovations in the field of information processing have played an important role in various application areas. With the proliferation of "open data", researchers are finding increasingly accurate ways to process raw data to obtain useful information to share with the scientific community. Valuable information that can then be analyzed by experts in that field to gain new knowledge. The development of inference methods for Gene Regulatory Networks lies in this context. The goal of researchers is to make advances in the regulatory mechanisms of cellular organisms, which are largely unknown today [16]. Gene regulatory networks provide a visual description of these phenomena and allow researchers to extract information and interpret it in a biological sense, thus gaining knowledge that is extremely useful in identifying genes involved in the life processes of cellular organisms [2].

© The Author(s), under exclusive license to Springer Nature Switzerland AG 2024
D. Cantone and A. Pulvirenti (Eds.): *From Computational Logic to Computational Biology*,
LNCS 14070, pp. 172–185, 2024.
https://doi.org/10.1007/978-3-031-55248-9_9

In recent years, the computer science community has employed state-of-the-art methods to help biologists better understand these mechanisms. Several papers have been recently published analyzing the proposed methods for inferring a gene regulatory network, such as [7,18,30]. However, in addition to the methods analyzed in these works, we present here a new different point of view based on the idea of Reverse Engineering. In our previous works [31,32], we presented a novel method for inferring a gene regulatory network, essentially based on the principle of creating an artificial environment capable of perfectly replicating the behavior of the interaction between genes and simulating their gene expression levels over time. The goal is to provide researchers a framework to artificially simulate real experiments and observe the system's response to external perturbations.

This article is organized as follows. In the first section, we introduce a gene regulatory network and define a common structure for inferring a regulatory network. In the second section, we provide an overview of the methods commonly used in this field and analyze the advantages and weaknesses of each method. In the third section, we analyze the validation criteria used to validate an inference method. Finally, in the last section, we compare the results obtained with our approach with those obtained with other methods presented in the literature and reported in [30].

2   Background

Regulatory networks are widely used in biology to help researchers understand the mechanisms of genetic regulation [13]. They reveal the interactions between genes, proteins, mRNAs, or cellular processes and provide important information about the development of diseases [9]. This section provides the reader a better understanding of what a gene regulatory network is. The focus is not so much on the biological aspect, but rather on the development of inference methods using computerized tools.

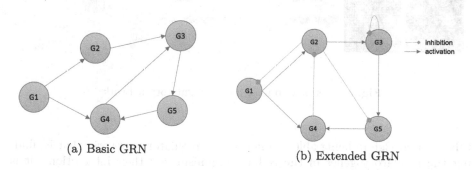

(a) Basic GRN   (b) Extended GRN

Fig. 1. Two examples of GRN. Figure 1a shows a simple GRN where the interactions between genes are clearly visible. Figure 1b instead shows an example of an extended GRN that includes gene regulation information.

2.1   Gene Regulatory Network

A gene regulatory network (GRN), modeled as a directed graph, is a regulatory network concerned with identifying gene interactions. The nodes of this network are genes, and the directed arcs indicate the interaction between them [25]. Moreover, an arc between two nodes, i.e., genes, mainly provides information about the regulatory process. Namely, if there is an arc from gene G_i to gene G_j, then G_i is referred to as regulatory gene or simply as regulator [10]. This means that a change in the gene expression of G_i will cause a significant change in the expression of gene G_j, according to the principle of cause-effect. Figure 1a shows an example of a gene regulatory network.

A gene regulatory network can combine more detailed information about regulation (Fig. 1b). Actually, a regulatory gene controls the expression of genes associated with it in a positive or negative way. When the expression level of the regulator reaches a threshold, another gene can be activated or inhibited based on that level [27]. This results in a change in the expression level of the regulated gene: if the gene expression decreases, the gene is inhibited, otherwise it is activated. In addition, genes that are self-regulating are common in a gene regulatory network. In this particular, but not uncommon, case, a gene has an inhibitory/activating effect that affects its expression level itself [8] (see gene G3 in Fig. 1b).

2.2   Inferring Gene Regulatory Networks

The process of inferring a gene regulatory network for a cellular organism can be divided into two phases, which for simplicity we can refer to as observing and inferring. This process is illustrated in Fig. 2 and is explained in this subsection.

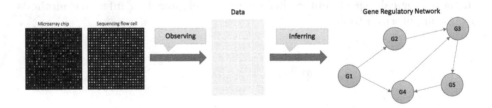

Fig. 2. Overview to infer a gene regulatory network

Observing. Before being able to infer a gene regulatory network, that is, finding the regulatory genes of the cellular organism and their interactions, it is necessary to know how the gene expression levels of a group of genes evolve over time. Some strategies to obtain these information include the use of micro-array technology [15] or more recently next generation sequencing technology [3]. A detailed description of these technologies is beyond the scope of this research.

However, their goal is to determine the presence and gene expression level in a biological sample at a given time point [7]. At the end of this phase, a time series dataset can be created containing observations at different time points. Multiple experiments can be performed on the same biological sample to improve the precision of the measurement and have more data available in the subsequent phase [19]. As an alternative to time-series data, a gene regulatory network can also be inferred using steady-state data. In this case, gene expression of genes is measured at successive time points after the system has been perturbed [26]. In this article however, we wish to clarify that all results were obtained with time-series datasets and not with steady-state datasets.

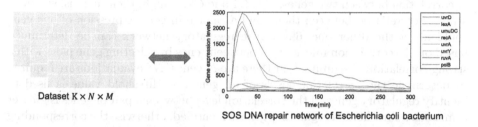

Fig. 3. An example of a time series dataset typically used to infer a gene regulatory network. On the left, a graphical representation of a gene expression dataset. On the right, a time plot related to an experiment from the dataset.

Inferring. A time series dataset obtained from the previous phases is used to build a gene regulatory network. Such a dataset can be represented as a three-dimensional matrix $D \in \mathbb{R}^{K \times N \times M}$, where K is the number of experiments, N is the number of genes considered and observed in K experiments, and M is the number of observations of each gene, which is the same for all K experiments performed (Fig. 3). A gene expression dataset is used to trace back the actual relationships between genes that have caused that gene expression [16]. Basically, all methods for this type of problem use these data to identify the regulatory genes and figure out the relationships between them. More details on the common methodologies used for this purpose are described in the next section.

3   Methodologies

This section focuses on analyzing the evolution of methods commonly used to infer a genetic regulatory network from observations of gene expressions. Although it seems to be only a biological problem, the process to infer a gene regulatory network, knowing only how gene expression levels change over time, borrows techniques from computer science and, more precisely, data science [23]. Since their introduction, a variety of methods have been presented to infer a gene regulatory network. An interesting aspect is that initially, information theory methods were often used for this task, allowing gene interactions to be found out using statistical and correlation models. Only in recent years, the rapid spread of artificial intelligence has enabled a change in the methods used to infer a gene regulatory network. However, before presenting the machine learning methods and our point of view, different categories of methods are analyzed in terms of their performance in solving this task, highlighting some critical aspects.

3.1   Information Theory Methods

These methods use statistical information, such as variance or correlation, to make assumptions about the degree of correlation between genes. If the degree of correlation between two genes, e.g., $G1$ and $G2$, is high, then there is obviously a direct correlation between them, and a change in gene expression of one gene could affect the other one [6]. A gene regulatory network can be determined based on the correlation information between genes by selecting gene pairs with a strong correlation. Assuming that there are N genes, N^2 evaluations are required to determine the correlation levels of all genes. A threshold value is used to identify regulatory genes. If the correlation level between a pair of genes is greater than such a threshold, a regulatory gene is identified, otherwise the corresponding arc is discarded.

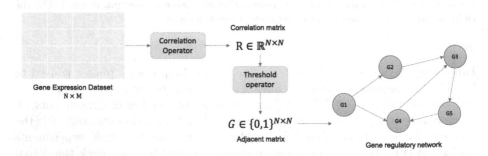

Fig. 4. A general diagram to infer a gene regulatory network with information theory methods

Figure 4 shows a possible representation of how information theory methods work. The correlation operator can implement any correlation method, such as Pearson, Euclidean distances, mutual information, and so on. Hence, a correlation matrix $R \in \mathbb{R}^{N \times N}$ showing the relationships between pairs of genes [17] is obtained. The threshold operator is then used to construct a gene regulatory network based on that correlation matrix.

Although these methods are more or less able to correctly predict some regulatory genes and have high precision, their accuracy is low due to several factors. First, correlation methods are known to work very well with a large amount of data that may not always be available. Second, a threshold must be set in advance that is not determined by the algorithm. And finally, the number of false positives is quite high, so their accuracy is somewhat lower than that of other methods, as it can be seen from the results of [30].

3.2 Boolean Networks

In the boolean network models, all biological components are described by binary states and their interactions by boolean functions [22]. Each gene is represented as a node in a directed-graph (i.e. boolean network) such than it can assume only one of two possible states (expressed or not expressed). The arcs of a boolean network represents the regulatory interactions. Boolean functions instead define the regulations of genes/nodes. More precisely, boolean functions are used to update the state of all nodes at a specified time, given the state of the nodes at the previous time. The process of state updating can be synchronous (all boolean functions are executed simultaneously), asynchronous (only one boolean function is executed per step), and probabilistic (boolean functions are executed in one step based on a probability) [14]. Boolean networks are easy to implement, but first require discretization of the data and also data cannot contain noise, which is difficult to achieve with real-world data [7]. For this reason, it is better to describe the entire model with differential equations that use continuous variables instead of discretizing the gene expression data for Boolean networks.

3.3 Differential Equations

Ordinary Differential Equation (ODE) model defines the dynamics of gene expression over time for each gene under consideration. The gene expression of a gene can be modeled as a differential equation that takes into account the relationships with its regulators. Knowledge of an ordinary differential equation model makes it possible to mathematically describe the evolution of gene expression over time and indirectly determine the regulatory genes for each gene, and in this way design a gene regulatory network (Fig. 5). According to [4] and assuming that there are N genes in the system, the ordinary equation models can be written as follows:

$$\frac{dX_j(t)}{dt} = f_j\left(X_1(t), X_2(t), \ldots, X_I(t) \mid \theta\right) \quad \forall j = 1, \ldots, N \qquad (1)$$

where $f_j(\cdot)$ is a parametric function whose structure is known, I represents the number of components of the ODE model, and $X_j(\cdot)$ is the gene expression level related to jth gene at time t. The final ODE model is determined based on parameters θ, which can be estimated by various methods, e.g., metaheuristics or log-likelihood function. An example of this application can be found in [1], where the *CMA-ES algorithm* is used to find the parameters of a S-System model used to model gene expression.

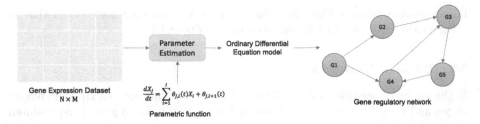

Fig. 5. A general diagram to infer a gene regulatory network with ordinary differential equation model

There are in literature several research paper to improve the accuracy of standard ODE. One of them is [29], where a grammar-guided genetic programming was used to build a complex-valued ordinary differential equation model (CVODE). Although the use of ordinary differential equations to estimate regulatory genes is an improvement over methods based only on correlation of genes, it has some weaknesses. The first, which requires special attention, is the size of the parameter space, which increases exponentially with the increase of genes in the model. As a result, the search for the optimal parameters can take a long time. In addition, due to the size of the parameter space, a large amount of data is required to infer a GRN with acceptable accuracy. However, for a small network with few parameters, the use of ODE could be a suitable solution.

3.4 Machine Learning

Other methods that are gaining ground in this area are those based on machine learning techniques. As is well known, machine learning is nowadays employed in a wide range of applications, such as finance, computer vision, forecasting, and others. The reason for its spread lies in its ability to recognize patterns from data better than traditional statistical methods, as discussed in Sect. 3.1. Therefore, several proposals can be found in literature to infer a gene regulatory network using machine learning techniques.

Overall, these methods can be divided into three categories. In the first category, the problem of determining a gene regulatory network from expression data is considered as a classification or regression problem. In this case, a machine learning model directly forecasts the relationship between genes, i.e., the regulators are predicted. An example of method that can be considered in of this category is dynGENIE3 [11], which is an improvement of GENIE3 [12]. In the second category, instead, a gene regulatory network is modeled as a neural network. According to [5], a neural network can basically be defined as follows:

$$\frac{de_i}{dt} = \frac{1}{\tau_i} \left(g \left(\sum_{j=1}^{N} w_{ij} e_j + \beta_i \right) - \lambda_i e_i \right) \tag{2}$$

The neurons in a neural network represent the genes and their connections define the relationships between the genes. Instead, the weights w_{ij} of the connections indicate the concentration of interaction between the ith gene and the jth gene. The sign of the weights reveals the nature of the relationship: a positive value means that j is repressing i and vice versa [5]. In [20], an Extreme Learning Machine (ELM) technique was used to model a GRN through a deep neural network. The last category considered in this section is the use of the Support Vector Machine (SVM) to infer a gene regulatory network. As reported in [28], the main idea is to map the samples to the high-dimensional feature space (kernel space) to convert non-separable data into separable data. Such a method is used to classify the interaction between gene +1 positive and -1 negative. For this type of application, it is more appropriate to use steady-state data instead of time series.

3.5   Reverse Engineering

In a sense, the problem of predicting a complex network containing important information about the interactions between genes as well as the gene regulators, i.e., a gene regulatory network, can be described as an inverse problem better known as reverse engineering. Basically, an inverse problem analyzes the results/effects that arise from a classical problem (also called a forward problem) to determine the inputs/causes that brought to those effects.

Starting from a collection of observations of one or more variables of interest over time, denoted by the symbol d, an inverse problem is to find a model m which describes the real system better.

$$d = G(m) \tag{3}$$

where $G(\cdot)$ is an operator describing the explicit relation between the observed data d and the model parameters [24]. The basic idea is to use gene expression time series to discover the internal relationships between variables/genes that can be represented, for example, by nonlinear functions. To achieve this results, such a methodology can be divided into two distinguish steps (Fig. 6). In the initial step, machine learning is used to determine a model that describes the behavior of gene expression given time series data. The goal is to create a model that can simulate gene expression in a way that approximates the results obtained with micro-array technology. For more details, see our previous works [31, 32], which present an artificial environment capable of forecasting gene expression of a group of genes over time. In the second step, such a model is used to infer gene regulatory network by determining the gene regulators and estimating the nature of their interactions, i.e., activation/inhibition. To obtain these results, our artificial environment is to perturbed at a specific time point by manually changing the expression level of target genes. By recording how the gene expression of each gene in the artificial environment changes at the next time points, the nature of the interaction between genes can be determined and it is possible to extract a gene regulatory network [31].

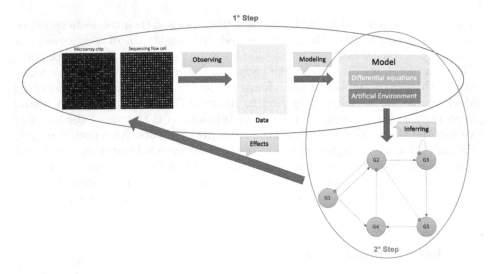

Fig. 6. Reverse engineering concept for inferring a gene regulatory network.

4   Validation

A method for predicting a genetic regulatory network should consider several requirements before being used in a real application. First, it would be useful to validate such a method using an artificial dataset rather than a real dataset in the first instance. The reason for this is that a real dataset usually contains noise caused by various factors such as the surrounding environment or the accuracy of the measurement instruments. These factors can negatively or positively influence our judgment about the goodness and effectiveness of our method. Therefore, it is not recommended to first validate a method to infer a GRN directly on a real dataset.

Unlike a real dataset, an artificial dataset is created ad hoc for this purpose. With software such as GeneNetWeaver [21], it is possible to generate an in silico gene expression dataset in both ways: time-series or steady-state. However, a peculiarity of this type of software is that an artificial dataset is modeled according to in silico gene network and properties of the genes such as the rate of change of mRNA concentration and the rate of change of protein concentration. These values are extrapolated from a real biological network.

By comparing the target network with the predicted network, it is possible to measure the accuracy and precision of our methodology. The validation process is as important as the process to infer a gene regulatory network itself. Indeed, for an interaction between a gene and its predicted regulator to be considered valid, biological feedback of these relationships is required, i.e., perturbation of one or both elements triggers a change in gene expression. [7].

Performance metrics are often used to validate a GRN method. The basic idea is to compare the predicted gene regulatory network with the target network

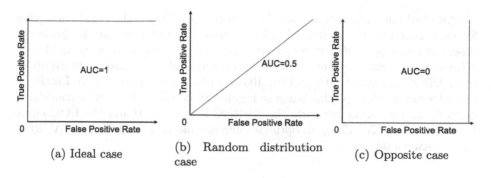

Fig. 7. Example of Receiver Operating Characteristic

obtained by our method in terms of true positives, true negatives, false positives, and false negatives. However, in the field of gene regulation, the typical elements of a confusion matrix also have biological significance.

- True Positive (TP) denotes the number of regulatory mechanisms correctly predicted.
- True Negative (TN) represents the number of arcs that are not present in both the predicted GRN and the target GRN.
- False Positive (FP) denotes the number of regulatory mechanisms predicted that are incorrect.
- False Negative (FN) denotes the number of regulatory mechanisms not detected.

These information obtained from the comparison can be used to compute classical metrics typically used in machine learning classification tasks, such as accuracy (Eq. 4), precision (Eq. 5), sensitivity (Eq. 6), and specificity (Eq. 7). Their definition is reported below.

$$Accuracy = \frac{TP + TN}{TP + TN + FP + FN} \tag{4}$$

$$Precision = \frac{TP}{TP + FP} \tag{5}$$

$$Sensitivity = \frac{TP}{TP + FN} \tag{6}$$

$$Specificity = \frac{TN}{TN + FP} \tag{7}$$

The most commonly index to compare different methods using the same gene expression dataset is the ROC (Receiver Operating Characteristic) curve and more specifically the AUC (Area Under The Curve). Therefore, AUROC (Area Under Receiver Operating Characteristic) is the most commonly used metric in this field, as it combines all of the above metrics into a single real value between 0 and 1.

An ideal inference method is able to correctly estimate the regulatory genes for each gene and the estimated GRN is equal to the target one. If this is the case, the resulting ROC curve corresponds to Fig. 7a and AUROC is equal to 1. If instead the regulatory genes are estimated according to a random distribution, the AUROC is equal to 0.5 and the ROC curve is similar to Fig. 7b. Finally, if the inference method produces a gene regulatory network that can be considered opposite to the target network, the AUROC is equal to 0 and the ROC curve is similar to Fig. 7c. Thus, an optimal inference method must have an AUROC value between 0.5 and 1.

5   Results

As mentioned in the introduction, the purpose of this article is to compare our method based on artificial environment, which we briefly described in Sect. 3.5, with other methods in order to evaluate its performance and highlighted its quality. Basically, to achieve this result, we have extended the results reported in [30] with those of the reverse engineering method we presented in [31, 32]. To compare our method with the others, we used two of the four benchmarks used in [30]. More specifically, the gene expression datasets used in our experiment are *In Silico Size 10* and *In Silico Size 100*, both of which are part of DREAM4 and available in the authors' repository.

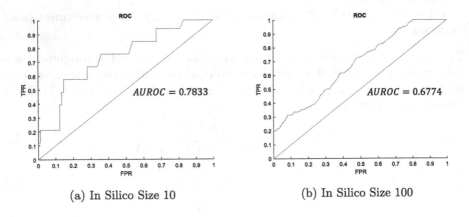

(a) In Silico Size 10   (b) In Silico Size 100

Fig. 8. ROC Curve

Figure 9 shows the ROC curve for both datasets considered. Figure 8 instead shows a comparison with the methods listed in [30], with our approach marked with a dark color and named with *Agent Based Model*.   As it can be seen, the results of our reverse engineering method are consistent with the results of other methods that have already been published in the literature.

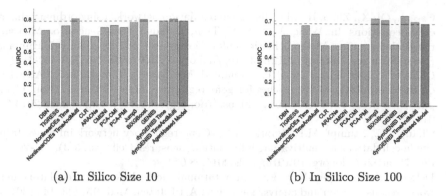

(a) In Silico Size 10   (b) In Silico Size 100

Fig. 9. Comparison with the state-of-the-art

6 Conclusions

As described in previous sections of this article, we have introduced a different perspective for inferring a gene regulatory network: In contrast to traditional methods that primarily aim to infer GRN based on data, our approach provides researchers with a framework in which it is possible to simulate different scenarios that are difficult to perform in practice due to external factors. Our proposed framework to infer a GRN, as well as the methods, allows experiments to be conducted. Gene expression of genes can be altered at a particular time point in order to observe the response in terms of gene expression of all genes at next time points. By measuring these variations, it is possible to quantify the intensity of the variation and determine under what terms a change in the expression level of a regulator involves a change in the regulated genes.

Acknowledgements. This research is supported by the project Future Artificial Intelligence Research (FAIR) – PNRR MUR Cod. PE0000013 - CUP: E63C220019 40006.

References

1. Agostini, D., et al.: Effective calibration of artificial gene regulatory networks. In: ECAL 2011: The 11th European Conference on Artificial Life, p. 11. Artificial Life Conference Proceedings, 08 2011. https://doi.org/10.7551/978-0-262-29714-1-ch011
2. Brazhnik, P., de la Fuente, A., Mendes, P.: Gene networks: how to put the function in genomics. Trends Biotechnol. **20**(11), 467–472 (2002). https://doi.org/10.1016/S0167-7799(02)02053-X
3. Buermans, H., den Dunnen, J.: Next generation sequencing technology: advances and applications. Biochimica et Biophysica Acta (BBA) - Mol. Basis Dis. **1842**(10), 1932–1941 (2014). https://doi.org/10.1016/j.bbadis.2014.06.015

4. Cao, J., Qi, X., Zhao, H.: Modeling gene regulation networks using ordinary differential equations. In: Wang, J., Tan, A., Tian, T. (eds.) Next Generation Microarray Bioinformatics. Methods in Molecular Biology, vol. 802, pp. 185–197. Humana Press+Springer, Totowa (2012). https://doi.org/10.1007/978-1-61779-400-1_12

5. Chai, L.E., Loh, S.K., Low, S.T., Mohamad, M.S., Deris, S., Zakaria, Z.: A review on the computational approaches for gene regulatory network construction. Comput. Biol. Med. **48**, 55–65 (2014). https://doi.org/10.1016/j.compbiomed.2014.02.011

6. Chan, T.E., Stumpf, M.P., Babtie, A.C.: Gene regulatory network inference from single-cell data using multivariate information measures. Cell Syst. **5**(3), 251-267.e3 (2017). https://doi.org/10.1016/j.cels.2017.08.014

7. Delgado, F.M., Gómez-Vela, F.: Computational methods for gene regulatory networks reconstruction and analysis: a review. Artif. Intell. Med. **95**, 133–145 (2019). https://doi.org/10.1016/j.artmed.2018.10.006

8. Gebert, J., Radde, N., Weber, G.W.: Modeling gene regulatory networks with piecewise linear differential equations. Eur. J. Oper. Res. **181**(3), 1148–1165 (2007). https://doi.org/10.1016/j.ejor.2005.11.044

9. Glubb, D.M., Innocenti, F.: Mechanisms of genetic regulation in gene expression: examples from drug metabolizing enzymes and transporters. WIREs Syst. Biol. Med. **3**(3), 299–313 (2011). https://doi.org/10.1002/wsbm.125

10. Hecker, M., Lambeck, S., Toepfer, S., van Someren, E., Guthke, R.: Gene regulatory network inference: data integration in dynamic models-a review. Biosystems **96**(1), 86–103 (2009). https://doi.org/10.1016/j.biosystems.2008.12.004

11. Huynh-Thu, V.A., Geurts, P.: dyngenie3: dynamical genie3 for the inference of gene networks from time series expression data. Sci. Rep. **8**(1), 3384 (2018). https://doi.org/10.1038/s41598-018-21715-0

12. Huynh-Thu, V.A., Irrthum, A., Wehenkel, L., Geurts, P.: Inferring regulatory networks from expression data using tree-based methods. PLoS ONE **5**(9), e12776 (2010)

13. Karlebach, G., Shamir, R.: Modelling and analysis of gene regulatory networks. Nat. Rev. Mol. Cell Biol. **9**(10), 770–780 (2008). https://doi.org/10.1038/nrm2503

14. Lähdesmäki, H., Shmulevich, I., Yli-Harja, O.: On learning gene regulatory networks under the Boolean network model. Mach. Learn. **52**(1), 147–167 (2003). https://doi.org/10.1023/A:1023905711304

15. Li, Y., et al.: Comparative study of discretization methods of microarray data for inferring transcriptional regulatory networks. BMC Bioinform. **11**(1), 520 (2010). https://doi.org/10.1186/1471-2105-11-520

16. Mercatelli, D., Scalambra, L., Triboli, L., Ray, F., Giorgi, F.M.: Gene regulatory network inference resources: a practical overview. Biochimica et Biophysica Acta (BBA) - Gene Regul. Mech. **1863**(6), 194430 (2020). https://doi.org/10.1016/j.bbagrm.2019.194430

17. Mohamed Salleh, F.H., Arif, S.M., Zainudin, S., Firdaus-Raih, M.: Reconstructing gene regulatory networks from knock-out data using gaussian noise model and Pearson correlation coefficient. Comput. Biol. Chem. **59**, 3–14 (2015). https://doi.org/10.1016/j.compbiolchem.2015.04.012

18. Nguyen, H., Tran, D., Tran, B., Pehlivan, B., Nguyen, T.: A comprehensive survey of regulatory network inference methods using single cell RNA sequencing data. Brief. Bioinform. **22**(3) (2020). https://doi.org/10.1093/bib/bbaa190

19. Ronen, M., Rosenberg, R., Shraiman, B.I., Alon, U.: Assigning numbers to the arrows: parameterizing a gene regulation network by using accurate expression kinetics. Proc. Natl. Acad. Sci. U.S.A. **99**(16), 10555–10560 (2002). https://doi.org/10.1073/pnas.152046799

20. Rubiolo, M., Milone, D.H., Stegmayer, G.: Extreme learning machines for reverse engineering of gene regulatory networks from expression time series. Bioinformatics **34**(7), 1253–1260 (2017). https://doi.org/10.1093/bioinformatics/btx730

21. Schaffter, T., Marbach, D., Floreano, D.: GeneNetWeaver: in silico benchmark generation and performance profiling of network inference methods. Bioinformatics **27**(16), 2263–2270 (2011)

22. Schwab, J.D., Kühlwein, S.D., Ikonomi, N., Kühl, M., Kestler, H.A.: Concepts in Boolean network modeling: what do they all mean? Comput. Struct. Biotechnol. J. **18**, 571–582 (2020). https://doi.org/10.1016/j.csbj.2020.03.001

23. Thomas, S.A., Jin, Y.: Reconstructing biological gene regulatory networks: where optimization meets big data. Evol. Intel. **7**(1), 29–47 (2014). https://doi.org/10.1007/s12065-013-0098-7

24. Tveito, A., Langtangen, H.P., Nielsen, B.F., Cai, X.: Parameter estimation and inverse problems. In: Elements of Scientific Computing. Texts in Computational Science and Engineering, vol. 7, pp. 411–421. Springer, Berlin, Heidelberg (2010). https://doi.org/10.1007/978-3-642-11299-7_9

25. Vijesh, N., Chakrabarti, S.K., Sreekumar, J.: Modeling of gene regulatory networks: a review. J. Biomed. Sci. Eng. **06**(02), 9 (2013)

26. Wang, Y.K., Hurley, D.G., Schnell, S., Print, C.G., Crampin, E.J.: Integration of steady-state and temporal gene expression data for the inference of gene regulatory networks. PLOS ONE **8**(8), 1–11 (2013). https://doi.org/10.1371/journal.pone.0072103

27. Wang, Y.R., Huang, H.: Review on statistical methods for gene network reconstruction using expression data. J. Theoret. Biol. **362**, 53–61 (2014). https://doi.org/10.1016/j.jtbi.2014.03.040

28. Yang, B., Bao, W., Chen, B., Song, D.: Single-cell-GRN: gene regulatory network identification based on supervised learning method and single-cell RNA-seq data. BioData Mining **15**(1), 13 (2022). https://doi.org/10.1186/s13040-022-00297-8

29. Yang, B., et al.: Reverse engineering gene regulatory network based on complex-valued ordinary differential equation model. BMC Bioinform. **22**(3), 448 (2021). https://doi.org/10.1186/s12859-021-04367-2

30. Zhao, M., He, W., Tang, J., Zou, Q., Guo, F.: A comprehensive overview and critical evaluation of gene regulatory network inference technologies. Brief. Bioinform. **22**(5), bbab009 (2021). https://doi.org/10.1093/bib/bbab009

31. Zito, F., Cutello, V., Pavone, M.: A novel reverse engineering approach for gene regulatory networks. In: Cherifi, H., Mantegna, R.N., Rocha, L.M., Cherifi, C., Micciché, S. (eds.) Complex Networks and Their Applications XI, pp. 310–321. Springer International Publishing, Cham (2023). https://doi.org/10.1007/978-3-031-21127-0_26

32. Zito, F., Cutello, V., Pavone, M.: Optimizing multi-variable time series forecasting using metaheuristics. In: Di Gaspero, L., Festa, P., Nakib, A., Pavone, M. (eds.) Metaheuristics, pp. 103–117. Springer International Publishing, Cham (2023). https://doi.org/10.1007/978-3-031-26504-4_8

Efficient Random Strategies for Taming Complex Socio-economic Systems

Alessio E. Biondo[1], Alessandro Pluchino[2,3],
and Andrea Rapisarda[2,3,4]

[1] Department of Economics and Business, University of Catania, Catania, Italy
[2] Department of Physics and Astronomy "Ettore Majorana", University of Catania, Catania, Italy
andrea.rapisarda@unict.it
[3] Sezione INFN of Catania, Catania, Italy
[4] Complexity Science Hub, Vienna, Austria

Abstract. In this paper we review some recent results about the efficiency of random strategies for complex systems. Inspired by physical systems, we discuss several successful examples of the beneficial role of randomness in the realm of socio-economic systems.

We dedicate this paper to Alfredo Ferro to celebrate his distinguished career.

Keywords: random strategies · complex systems · socio-economic systems · agent-based models

1 Introduction

The use of random numbers and random strategies in science goes back to the Manhattan Project and to the Monte Carlo algorithms developed within that project by Ulam, Teller and Metropolis for the construction of the first atomic bomb in Los Alamos [1]. Since then, random numbers have had a large and very successful use in physics and in many other scientific disciplines [2]. In addition, in 1980, studying climate series and the periodic oscillations of glaciation periods, the group of Giorgio Parisi in Rome [3] discovered the phenomenon of stochastic resonance, another wonderful example of the beneficial effects that a random noise can have. Since then, after these studies, such a phenomenon has had a lot of very important applications in different scientific fields, see ref [4] for a comprehensive review. In 2021, Parisi was awarded with the Nobel prize in Physics also for this important discovery. Inspired by these successful examples in physics, more than 10 years ago, we had the idea to apply random strategies in socio-economic systems. In this paper, dedicated to our friend and colleague Alfredo Ferro, we review our work along this line of research, from the first papers on the Peter principle [5–7], where we investigated the efficiency in management of hierarchical organizations, to the next study about the role of sortition in politics [8,9], until the more recent results in financial markets [10–14] and in

LNCS 14070, pp. 186–222, 2024.
https://doi.org/10.1007/978-3-031-55248-9_10

the context of the 'science of success' [15–17]. We have presented these studies several times during the last editions of the Lipari Summer School in Complex Systems, so wonderfully organized by Alfredo. It is therefore our privilege and honor to summarize in this article the most important results.

2 The Peter Principle

Our first paper where we applied random strategies in order to improve the efficiency of an organization was published in 2010 [6], followed by a second one published in 2011 [7]. In this section we will review the main results of these two articles.

The Peter principle is an empirical finding due to Laurence Peter. In a famous book realized on the basis of several real examples taken from our daily life, Peter claimed the following statement: "In a hierarchical organization, every new member climbs the hierarchy until she reaches her level of incompetence" [5]. The book became a best seller in the 60s' and was translated into many languages. In 2009 we came across an Italian version of it and devised a simple hierarchical model to see if we could test it and eventually try to find a solution. This ended with our first paper where, by means of numerical simulations performed on a hierarchical group like that illustrated in Fig. 1, we verified the principle under certain circumstances: actually, if one promotes the best agent from the lower level and the agent's competence changes, since the new level requires a new competence, one ends with agents who finish their career with a minimal competence. In other words, we found that the Peter principle has strong numerical foundations. We were also able to quantify the efficiency of a simulated organization under different hypothesis and promotion strategies.

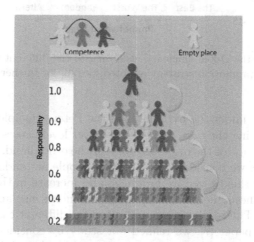

Fig. 1. The hierarchical organization considered in our first model. Different colors of the agents indicate a different competence distributed in a Gaussian way. Yellow agents indicate vacant positions. The vertical numbers close to the levels indicate the responsibility of the agents of that level. See [6] for further details. (Color figure online)

In our model we considered two mechanisms for transmitting the competence of a promoted agent from a level to the upper one: a) the common sense hypothesis (CS), according to which the agent maintains its competence in the new position, with a small random variation δ; b) the Peter hypothesis (PH), according to which the competence at the new level is independent from that at the previous level, that is, it is reassigned in a completely random way, and with the same distribution seen above.

For each of these two mechanisms we then examined four different promotion strategies, or four different ways to choose the agent to be promoted to a given vacant position from the lower level: the most competent (meritocratic strategy), the least competent, an agent chosen at random and an alternation of the first two strategies. Lastly, in order to assess the overall performance of the organization, we introduced a quantity called Global Efficiency (GE), calculated by adding the skills of the members level by level and multiplying these sums by a factor of responsibility that increases as we go up in the hierarchy, in order to take into account the fact that errors made at the highest levels are generally more burdensome for a company (see [6] for more details).

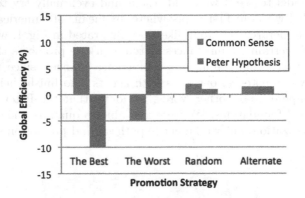

Fig. 2. The gain in the global efficiency according to the different working hypothesis considered and the promotion strategy adopted. See [6] for further details.

We found that, under the hypothesis of the Peter principle, the meritocratic strategy of promoting the best agent of the lower level causes a robust decrease in the global efficiency of the organization. On the other hand, when, inspired by physical examples, we considered to promote people at random from the lower level, we saw that this procedure produced and increase in the global efficiency under both hypothesis, i.e. that one of the common sense and that one of the Peter hypothesis. The results are summarized for different cases in Fig. 2.

In our second paper [7], we refined our study by considering different kind of organization structures. In particular we considered modular networks (hierarchical trees) still maintaining the pyramidal structure of the first model. We analyzed networks of different sizes, up to a number of agents equal to 1555.

We changed the time step of the promotion (one month instead of one year) and measured the Global Efficiency gain with respect to a meritocratic one. Also the percentage of random promotions to be adopted was explored, going from a 0% to a 100% of random promotions adopted.

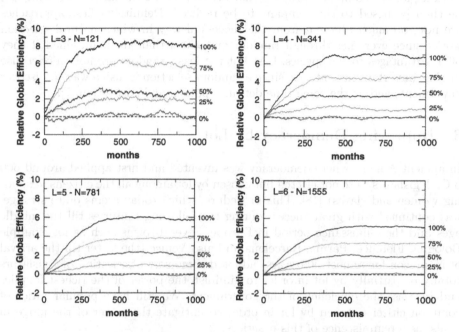

Fig. 3. The increase of Global Efficiency with respect to the meritocratic regime for the hierarchical organization of our second model as a function of time, for different number of levels and agents, and for increasing percentages of random promotions. See [7]

The numerical simulations performed with this new model did not change the essence of our final results. In fact, in any case we could see an increase of the global efficiency of the organization as illustrated in Fig. 3 for different sizes of the groups, different levels and different percentages of random promotions adopted. These results were also corroborated by further tests [7], carried out by introducing new characteristics to make our hierarchical tree model more realistic: variable skills of the agents with the age, non-linear increase in responsibility with the hierarchical level, mixed hypothesis for the transmission of skills from one level to the next, organizations of variable size over time. In all cases, our numerical results showed without any doubt that promoting at random, even with a small percentage of random promotions, one sees always an increase in the Global Efficiency of the organization. Therefore this conclusion is very robust and stable.

The fact that random promotions increase the efficiency of an organization may seem paradoxical at first sight, but on the other hand we have many real cases where it works. See for example the case of the SEMCO company in Brazil, that has grown a lot after adopting strategies similar to those we propose. Another good case is that of Google, which exploits the potentiality of random intuitions by allowing its employees to work on their own individual ideas while working for the company: projects that are considered the most interesting can be then proposed to the company to be realized. Definitely, these approaches are not too much different from the successful way in which natural selection works since ever, i.e. through random mutations that are maintained if they offer advantages to the species. Last but not least, random promotion offer also further important advantages, since a random selection is also a way to discover new talents and at the same time defeat nepotism and corruption [7].

3   Improving Democracy by Lot

In ancient Athens, where democracy was invented and first applied around 600 b.C., legislators were not elected but chosen by lot among all the citizens (excluding women and slaves) [18]. This procedure, which today seems quite strange, was continued with great success during the following centuries till the middle ages and the renaissance period in Europe. Several towns such as for example Bologna, Florence, Parma, Barcelona and also Venice (the latter till the arrival of Napoleon for almost 500 years) adopted some sort of selection of legislators totally or partially by lot in order to diminish the power of the richest families and increase the efficiency of the government. We still have popular juries of common citizens chosen by lot in order to mitigate the power of the judge in trials, as a reminiscence of this practices.

Along this line, in 2011 we decided to explore the theoretical possibility to improve the efficiency of a modern parliament by introducing a certain number of legislators chosen by lot [8], maintaining at the same time the presence of elected parliament members. In order to model the dynamics of a (simplified) parliament we considered the emergent behavior of a number of agents, who can only propose laws and vote them. Agents can have personal interests, like re-election or other benefits, and also a general interest for the society. They can belong to one of two parties or coalitions, thus being chosen through standard elections, or can be independent members chosen by lot.

Taking both a personal and a collective motivation into account, it is possible to represent individual legislators as points l_i (x, y) (with i = 1, ..., N) in a diagram (see Fig. 4), where we fixed arbitrarily the range of both axes in the interval $[-1,1]$ considering personal gain on the x-axis and social gain , i.e. the final outcome produced by law, on the y-axis. Each agent is described through her attitude to act for personal and general interest. A similar diagram was proposed by the economic historian Carlo M. Cipolla in 1976 [19], in order to characterize human population according to the attitude to promote personal or social interests. Each point in the Cipolla diagram represents the weighted average position of the actions of the correspondent person.

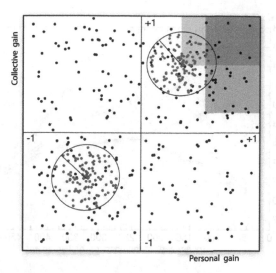

Fig. 4. The Cipolla diagram of our model of a parliament. The circles with red and green points represent two opposite coalitions, while the other point outside the circles represent independent legislators selected by lot with a uniform distribution. See [8, 19] (Color figure online)

An intuitive measure of the Parliament activity could be the number of accepted acts over the total ones proposed. But another important quantity is surely the average social welfare ensured by the accepted acts of the parliament. The latter can expressed for a single legislature L as

$$Y(L) = 1/N_{acc}\Sigma_{m=1}^{N_{acc}}y(a_m) \tag{1}$$

where N_{acc} is the number of accepted laws and y the social benefit of the accepted law a_m.

It is therefore convenient to define the global efficiency of a legislature L as the product of these two quantities i.e.

$$Eff(L) = N_{acc}(L)Y(L). \tag{2}$$

The latter is a real number included in the interval $[0, 100]$.

Then, in order to obtain a measure independent of the particular configuration of the parliament, the global efficiency has to be further averaged over many legislatures L, each one with the same number of proposals but with a different distribution of legislators and parties position in the Cipolla diagram.

We found that the efficiency of the parliament, defined as the product of the percentage of the accepted proposals times their overall social welfare, can be influenced by the introduction of a given number N_{ind} of *randomly selected* legislators, called 'independent' since we assume that they remain free from the influence of any party. These independent legislators will be represented as free points on the Cipolla diagram (Fig. 5).

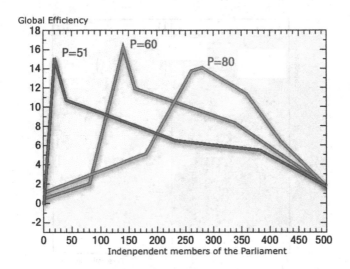

Fig. 5. Global Efficiency of a parliament vs the number of independent legislators (chosen by lot) for different percentage of the majority coalition. See [8] for further details.

The dynamics of our model is very simple. During a legislature L each legislator, independent or belonging to a party, can perform only two actions: (i) she can propose one or more acts of parliament, with a given personal and social advantage depending on her position on the diagram. (ii) she has to vote for or against a given proposal, depending on his/her *acceptance window*, i.e. a rectangular subset of the Cipolla diagram into which a proposed act has to fall in order to be accepted by the voter (whose position fixes the lower left corner of the window).

The main point is that, while each free legislator has her own acceptance window, so that her vote is independent from the others vote, all the legislators belonging to a party always vote by using *the same* acceptance window, whose lower left corner corresponds to the center of the circle of tolerance of their party. Furthermore, following the party discipline, any member of a party accepts *all* the proposals coming from any another member of the same party (see [8] for further details).

From our simulations, repeated numerous times by randomly varying the positions of independent parties and legislators in the Cipolla diagram, one can see that both a parliament without independent members (like the current ones) and a parliament in which all the members are independent (i.e. without parties) show very low efficiency, practically close to zero. In the first case because, the percentage of laws approved is high, but the social benefit insured is modest; in the second case, on the contrary, because only the laws with a very high collective advantage are approved, but consequently their number is extremely small. We therefore asked ourselves if, between these two extreme cases, there was an intermediate number of independent deputies who, with their presence, are able to maximize the efficiency of parliament. The simulations have confirmed that this optimal number, N_{ind}^*, exists, and we have also been able to derive the following analytical *Golden rule formula* for this optimal number

$$N_{ind}^* = \frac{2N - 4N \cdot (p/100) + 4}{1 - 4 \cdot (p/100)}, \tag{3}$$

where p is the percentage of the majority party. The Golden rule is in excellent agreement with the numerical results as shown in Fig. 6.

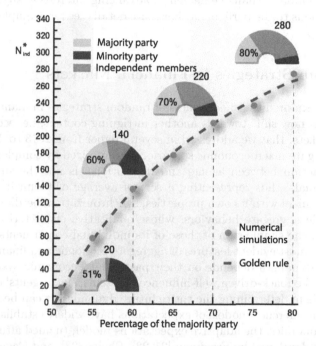

Fig. 6. The Golden rule: the optimal number of independent legislators which maximizes the parliament efficiency as a function of the percentage of the majority party or coalition. See [8] for further details.

These findings have been further corroborated in 2021 with another study where a one dimensional version of the original 2D model of parliament was presented and discussed [9]. In particular, through an analytical approach, we confirmed that, to overcome the detrimental effects of party discipline, one can beneficially move towards a parliament where independent, randomly selected legislators sit alongside elected members. It is interesting to notice that this is in line not only with the old sortition rules adopted in the past, but also with the most recent experiments of the last decades.

As a matter of fact, sortition tools are gradually making their way into the contemporary political scenario [9]. It is the case of the Irish Constitutional Convention of 2013 and, more recently, the cases of both the Observatorio de la Ciudad of Madrid (2018) and the permanent authority that will support the regional Parliament of German-speaking Community of Belgium (2019). In addition, there have been several recent successful examples in Europe and in other part of world where citizen assemblies, composed by lot, have worked very well, discussing and proposing laws which have then been approved by the parliaments. This has occurred for example in Ireland, Iceland, Poland, Belgium, Denmark and lately in France, where common citizens chosen by lot have discussed the climatic change emergence, advancing interesting suggestions and possible solutions to the parliament. For more details see for example the website democracyrd.org.

4   Random Strategies in Financial Markets

After having explored the effectiveness of random strategies in management and politics, let us now shift towards another intriguing context, i.e. economics and financial markets, that we addressed in several paper from 2013 to 2015 [10–14].

The aggregate macroeconomic scenario, characterized by complexity induced by the interaction between agents and institutions, is not the simple sum of single individuals, thus representing a sort of average of them; it is rather an emergent organism with its own properties, synchronization, herding, and asymmetric volatile aggregate behaviors, whose qualitative characteristics are very different from the simple sum of those of its individual constituents.

One of the most evident features of aggregate economic and financial systems is that their dynamics depends on their past. Thus, economic systems can be considered as feedback-driven, self-influenced systems, since agents' expectations play a key role in determining the entire future dynamics. It can be roughly said that two main reference models of expectations have widely established - among others - in economics: the adaptive expectations model, (named after Arrow and Nerlove [20]), developed by Friedman [21,22], Phelps [23], and Cagan [24] which assumes that the value of a variable is a weighted average of its past values; and the rational expectations model, whose birth dates back to contributions by Muth [25], Lucas [26], and Sargent-Wallace [27], which assumes that agents know exactly the entire model describing the economic system and, since they are endowed by perfect information, their forecast for any variable coincides with the objective prediction provided by theory.

Making predictions is a central issue in economic theory, for both real and monetary reasons. In financial markets, this problem is even more urgent, in terms of volatility effects on prospective individual earnings and global instability. The Efficient Market Hypothesis, which refers to the rational expectation framework, considers the information as the route to rational choices and to best predictions. Fama [28] defined financial efficiency depending on the existence of *perfect arbitrage*: the author suggests three forms of market efficiency - namely "weak", "semi-strong", and "strong" - according to the degree of completeness of the informative set. Inefficiency would then imply the existence of opportunities for unexploited profits that traders would immediately try to exploit. Other authors, such as Jensen [29] and Malkiel [30], also link the efficiency of the available information to the determination of assets' prices. Then, financial markets participants continuously seek to expand their informative set to choose the best strategy and this results in extreme variability and high volatility. The experience of severe and recurrent financial crises showed that mechanisms and trading strategies are not immune from failures. Also, the knowledge of available information (more or less incorporated in assets values) has proven not sufficient to provide a solid background against losses and dangerous investment choices. The interpretation of a fully working *perfect arbitrage* mechanism, without systematic forecasting errors, is not adequate to analyze financial markets: Cutler*et al.* [31], Engle [32], Mandelbrot [33, 34], Lux [35], and Mantegna and Stanley [47] just to mention some examples. The reason is quite intuitive: the hypothesis that information is available for everybody is not real. And it is not real even if in its semi-strong or weak versions. Many heterogeneous agent models have been introduced in the field of financial literature in order to describe what happens in true markets with different types of traders, each with different expectations, influencing each other by means of the consequences of their behaviors (some examples are: Brock [36, 37], Brock and Hommes [38], Chiarella [39], Chiarella and He [40], DeGrauwe *et al.* [41], Frankel and Froot [42], Lux [43], Wang [44], and Zeeman [45]). This approach, namely the "adaptive belief systems", tries to apply non-linearity and noise to financial market models in order to represent the highly complex behavior of markets.

Simon [46] explained that individuals assume their decision on the basis of a limited knowledge about their environment and thus face high search costs to obtain needed information. However, normally, they cannot gather all information they should. Thus, agents act on the basis of *bounded rationality*, which leads to significant biases in the expected utility maximisation that they pursue. This stream of literature gave birth to the development of further steps in the analysis of behavior and choice, culminating to relevant contributions of Thaler, Frank, Kahnemann and Tversky.

A multidisciplinary approach as revealed of great help in addressing the analysis of economic and financial complex systems: the adoption of powerful techniques from statistical physics and the use of agent-based models simulations in studying socio-economic phenomena have lead to innovative and robust results. In recent years, many physicists have started to investigate the complex dynamics of several phenomena beyond the field of physics and specifically referred to financial markets, thus creating the field of Econophysics [47–51]. The investigation of the role of random strategies in financial markets is the methodology to infer the role of noise in that complex system. In 2001, R. Wiseman explored the potentiality of random investments in a famous experiment, where a five-year old child, playing at random with shares of the London Stock Exchange, managed to contain losses better than a financial trader and an astrologist dur-

ing one year of turbulent market behaviour [52]. Similar results were obtained
also in other studies, by exploiting dartboard or monkeys [53,54]. Stimulated by
these findings, in the last years we started to investigate in detail the efficiency
of random trading with respect to standard technical strategies, both from an
individual point of view [10,11] and from a collective perspective [12], making
use of statistical analysis and agent based simulations.

4.1   The "micro" Perspective

Random investments can be studied from the point of view of the single trader
(micro level). Consider here three non interacting agents $A_i(i = 1, 2, 3)$, investing
(daily), for a long time period, in two stock markets (the Italian national stock
exchange, with the FTSE MIB All-Share index and the US stock market, with
the S&P 500 index) by adopting different trading strategies, in order to see which
strategy is more profitable, over long and short horizons.

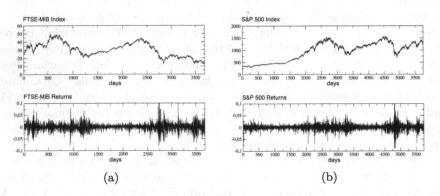

(a)   (b)

Fig. 7. Left panels: FTSE MIB All-Share index (from December, 31th 1997 to June,
29th 2012, for a total of $T = 3684$ days) and the corresponding series of returns. Right
panels: S&P 500 index (from September, 11th 1989 to June, 29th 2012, for a total of
$T = 5750$ days) and the corresponding series of returns.

In Fig. 7 we show the time series of the FTSE MIB index and that of the
S&P 500 index, and their corresponding 'returns' time series, whose standard
deviation in a given time window represents the volatility of the market in that
period, i.e. an indicator of the 'turbulent status' of the market.

We estimate the presence of correlations in a given time series by calculating
the time-dependent Hurst exponent through the so called 'detrending moving
average' (DMA) technique [55]. The DMA algorithm is based on the computation
of the following standard deviation $\sigma_{DMA}(n)$ as function of the size n of a time
window moving along a financial series F of length T:

$$\sigma_{DMA}(n) = \sqrt{\frac{1}{T - n} \sum_{j=n}^{T} [F_j - \tilde{F}_j(n)]^2}, \qquad (4)$$

where $\tilde{F}_j(n) = \frac{1}{n}\sum_{k=0}^{n-1} F_{j-k}$ is the average calculated in each time window of size n, while n is allowed to increase in the interval $[2, T/2]$. In general, the function $\sigma_{DMA}(n)$ exhibits a power-law behavior with an exponent H which is precisely the Hurst index of the time series F.

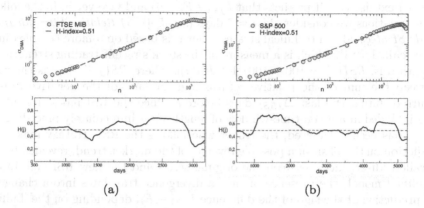

Fig. 8. Detrended analysis: power law behavior of the DMA standard deviation (upper panels) and time dependence of the Hurst index (lower panels) for the FTSE MIB (a) and the S&P 500 time series (b), respectively.

In the upper panels of Fig. 8 we show two log-log plots of the $\sigma_{DMA}(n)$, calculated over the complete FTSE MIB (a) and the S&P 500 time series (b), together with the corresponding power law fits $y \propto n^H$ (dashed lines): in both cases one observes a Hurst index very close to 0.5, indicating an absence of correlations on a large time scale. On the other hand, calculating the local value of the Hurst exponent day by day along the time series, significant oscillations around 0.5 seem to emerge. This is shown in the bottom panels of Fig. 8, where two sequences of Hurst exponent values $H(j)$ (solid lines) are obtained as function of time by considering subsets of the complete FTSE MIB (a) and the S&P 500 (b) series through sliding windows W_s of size T_s. These results suggest that correlations are important only on a local temporal scale, while they cancel out when averaging over long-term periods.

Simplifying with respect to the reality, the traders of our model have just to predict, day by day, the upward ('bullish') or downward ('bearish') movement of the index F_{j+1} on a given day with respect to the closing value F_j one day before: if the prediction is correct, we will say that they win, otherwise that they lose. We are interested only in comparing the percentage of wins of all the traders, which depend on their strategy of investment. We assume that they know only the past history of the index and that there isn't any influence from the market or from the other traders.

The three possible strategies, each one chosen by a given trader, are the following: *1) Random (RND) Strategy*, that is the simplest one, since the corresponding trader makes her 'bullish' (index increases) or 'bearish' (index decreases) prediction at the day j, for the next day $j+1$, completely at random (just tossing a coin). *2) Momentum (MOM) Strategy*, a technical strategy based on the so called 'momentum' $M(j)$, i.e. the difference between the value F_j and the value $F_{j-\Delta j_M}$, where Δj_M is a given trading interval (in days). Then, if $M(j) = F_j - F_{j-\Delta j_M} > 0$, the trader predicts an increase of the closing index for the next day (i.e. it predicts that $F_{j+1} - F_j > 0$) and vice-versa. In the following simulations we consider $\Delta j_M = 7$ days (see [56]). *3) Relative Strength Index (RSI) Strategy*, also a technical strategy, but it is based on a more complex indicator, called 'RSI', which is a measure of the stock's recent trading strength. Its definition is: $RSI(j) = 100 - 100/[1 + RS(j)]$, where $RS(j, \Delta j_{RSI})$ is the ratio between the sum of the positive returns and the sum of the negative returns occurred during the last Δj_{RSI} days before t. Once the RSI index, for all the days included in a given time-window of length T_{RSI} immediately preceding the day j, has been calculated, the trader who follows the RSI strategy makes her prediction on the basis of a possible reversal of the market trend, revealed by the so called 'divergence' between the original series and the new RSI one. In our simplified model, the presence of such a divergence translates into a change in the prediction of the sign of the difference $F_{j+1} - F_j$, depending on the 'bullish' or 'bearish' trend of the previous T_{RSI} days. In the following simulations we choose $\Delta j_{RSI} = T_{RSI} = 14$ days, since - again - this value is one of the most commonly used in RSI-based actual trading (see [56]).

In order to test the performance of the previous strategies, we divide each one of the two time series (FTSE MIB and S&P 500) into a sequence of N_w trading windows of equal size $T_w = T/N_w$ (in days) and we evaluate the average percentage of wins (with the corresponding standard deviation) of the three traders inside each window while they move along the series day by day, from $j = 0$ to $j = T$. In Fig. 9(a) and (b) we show the simulation results obtained for both the FTSE MIB and S&P 500 series. From top to bottom, we report the percentages of wins for the three strategies over all the windows and the corresponding standard deviations, averaged over 10 different runs (events) inside each window. It clearly appears that the random strategy shows, in comparison with the other strategies, a similar average performance in terms of wins, but with smaller fluctuations. The volatility of the two time series for each size of the trading windows is also reported at the bottom for completeness.

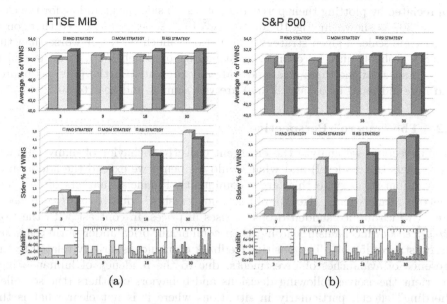

(a)   (b)

Fig. 9. (a) Results for the FTSE-MIB All-Share index series, divided into an increasing number of trading-windows of equal size $(3, 9, 18, 30)$, simulating different time scales. (b) Same analysis for results for the S&P 500 index series.

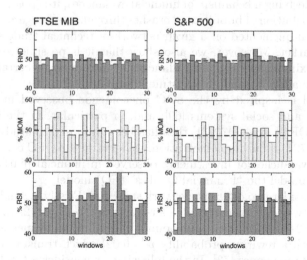

Fig. 10. The percentage of wins of the different strategies inside each time window - averaged over 10 different events - is reported, in the case $N_w = 30$, for the two markets considered.

Observing the local time behavior of the three traders, which can be better appreciated by plotting their percentage of wins inside each window for the case $N_w = 30$, as shown in Fig. 10 for the two time series, we can see that, on a small time scale, a given strategy may perform much better or worse than the others (probably just by chance, as suggested by Taleb [57]), but the global performances of the three strategies (already presented in the previous figures and here indicated by a dashed line) are very similar and near to 50%.

4.2   The "macro" Perspective

In this section, we analyze the emergent collective behavior of many interacting traders (macro level) and what would happen if we extend the adoption of the random strategy to this larger community. Financial markets often experience extreme events, like "bubbles" and "crashes", due to *positive feedback* effects which induce sudden drops or rises in prices, in contrast with the *negative feedback* mechanism leading to an equilibrium price under normal market circumstances [58,59]. The positive feedback dynamics is strictly related to the presence of avalanches of investments, due to the tendency of human beings to orient themselves following decisions and behaviors of others (the so called "herding" effect), particularly in situations where it is not clear what is the right thing to do [60]. Actually, such conditions are typical for financial markets, in particular during volatile periods. Remarkably, in this context, bubbles and crashes may reach any size and their probability distribution follows a power law behavior [34,47,61–67].

In this section we show that, assuming information cascades between agents [68] as the underlying mechanism of financial avalanches, it is possible to obtain a power law distribution of bubbles and crashes through a self-organizing criticality (SOC) model implemented on a given network of technical traders investing in a financial market. Moreover, we also show that it is possible to considerably reduce the maximum size of these avalanches by introducing a certain percentage of traders who adopt a random investment strategy.

Our model is inspired by the SOC phenomenon observed in many physical, biological and social systems [69], and, in particular, in the Olami-Feder-Christensen (OFC) model [70,71] that has been proposed to study earthquakes dynamics [64,72]. In our implementation, we identify the OFC earthquakes with the herding avalanches of investments observed in financial markets, therefore we called our model the "Financial Quakes" (FQ) model.

We will consider two different network structure. Firstly, a small-world (SW) undirected network of interacting traders (agents) A_i ($i = 1, 2, ..., N$), obtained from a regular $2D$ lattice (with open boundary conditions) by means of a rewiring procedure (with a rewiring probability $p = 0.02$) which transforms short range links into long range ones [73]. In the following we consider a total of $N = 1600$ agents, with an average degree $\langle k \rangle = 4$. Traders act as in the previous section, but now they have the possibility to interact, causing possible herding effects. For this purpose, we have imagined that, at each simulation time t, all the agents have a certain quantity of information $I_i(t)$ about the market considered. Initially, at $t = 0$, it assumes a random value in the interval $(0, I_{th})$, where $I_{th} = 1.0$ is an arbitrary threshold equal for all the traders. When the simulation starts, i.e. for $t > 0$, information may change due to two mechanisms: (1) a *global one*, in which due to external public information sources, all the variables $I_i(t)$ are simultaneously increased by a quantity δI_i, different for every agent. If, at a given time step t^*, the information $I_k(t^*)$ of one or more agents $\{A_k\}_{k=1,...,K}$ exceeds the threshold value I_{th}, these agents become "active" and take the decision of investing, i.e. they bet on the behavior of a given financial index value F_j compared to that one of the day before F_{j-1}; and (2) a *local* one, that depends on the topology of the network: as they invest on the market, all the active traders $\{A_k\}_{k=1,...,K}$ will also share their information with their neighbors according the following herding mechanism inspired by the stress propagation in the OFC model for earthquakes:

$$I_k > I_{th} \Rightarrow \begin{cases} I_k \to 0, \\ I_{nn} \to I_{nn} + \frac{\alpha}{N_{nn}} I_k \end{cases} \tag{5}$$

where "nn" denotes the set of nearest-neighbors of the agent A_k and N_{nn} is the number of her direct neighbors. The neighbors that, after receiving this *surplus* of local information, exceed their threshold, become active too and will also transfer their information to their neighbors, creating an avalanche of identical investments, i.e. what we call a financial quake. The parameter α in Eq. 5 controls the level of dissipation of the information during the dynamics ($\alpha = 1$ means no dissipation) and it is fundamental in order to drive the system in a SOC-like critical state. In analogy with the OFC model on a SW network [71, 73] we set here $\alpha = 0.84$, i.e. we consider some loss of information during the herding process. This value ensures the emergence of avalanches that can reach any size s, as shown in Fig. 11.

In a regular $2D$ lattice with small-world topology all the agents are equivalent, i.e. they all have, more or less, the same number of neighbors (four, on average). But it is interesting to investigate what would happen if the percentage of random traders is decreased further, but if, at the same time, their importance, in terms of connectivity within the network, is increased. Now we adopt another kind of network, an undirected scale-free (SF) network, i.e. an example of a network displaying a power-law distribution $p(k) \sim k^{-\gamma}$ in the node degree k. By using the preferential attachment growing procedure introduced in [74], we start from $m + 1$ all to all connected nodes and at each time step we add a new node with m links. These m links point to old nodes with probability $p_i = \frac{k_i}{\sum_j k_j}$, where k_i is the degree of the node i. This procedure allows a selection of the γ exponent of the power law scaling in the degree distribution with $\gamma = 3$ in the thermodynamic limit ($N \longrightarrow \infty$). In our simulations we adopt a SF network with $N = 1600$ traders and we consider two possibilities: (i) only technical RSI traders and (ii) a majority of technical RSI traders plus a small number N_H of random traders represented by the main hubs of the network (e.g. all the nodes with $k > 50$). Will these N_H hyper-connected random traders be able, alone, to reduce the size of financial quakes?

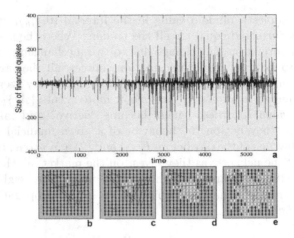

Fig. 11. (a) An example of time series of "financial quakes" for a SW network of traders investing in a given stock market (bottom panel). Both positive cascades ("bubbles") and negative avalanches ("crashes") are visible. (b-e) Time evolution of a single financial quake in the small-world network: starting from a single active trader (b), the herding-activated avalanche (in white) rapidly reaches different part of the network (c-d-e), helped by the presence of long range connections.

First of all, we should check if the dynamics of our financial quakes model on a scale-free network with only RSI traders is still able to reach a SOC-like critical state, as for the small world $2D$ lattice. We find that this system does show power law distributed avalanches of investments, providing that the parameter α in Eq. 5 is slightly increased with respect to the value adopted in the previous section. In particular, we use here $\alpha = 0.95$. Then, we perform two sets of simulations, with 10 different runs each, either in absence or in presence of the N_H hyper-connected random traders. In contrast to RSI ones, random traders are not activated by their neighbors, and they do not activate their neighbors, since a random trader has no specific information to transfer. In other words, random traders only receive the information δI_i from external sources ($\alpha = 0$) and so, they do not take part in the herding process, and are not involved in any financial quake. In Fig. 12 we report the probability distribution $P_N(s)$ of the absolute value of the size of the financial quakes, cumulated over 10 events, for the FTSE MIB and S&P 500 financial series, for both type of networks. For the SW, in the absence of random traders, i.e. with only RSI traders (circles), the distributions follow a well defined power law behavior. On the other hand, increasing the amount of random traders, in particular with percentages of 5% (squares) and 10% (triangles), the distributions tend to become exponential. For the SF, in the absence of random traders, i.e. with RSI traders only (gray circles), the distributions follow a well defined power law behavior. The introduction of just a few number of random traders (about 8 over 1600), corresponding with the hyper-connected nodes of the network (i.e. with $k > 50$), it is enough to dampen the avalanches reducing their maximum size to less than 40% (black circles). Therefore, comparing the results we find that with about only 8 hyper-connected

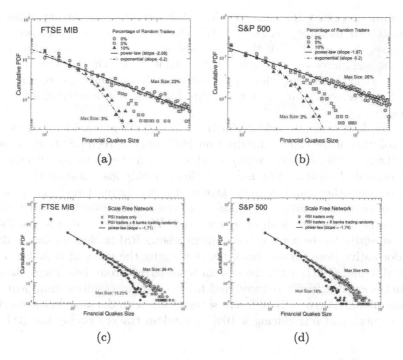

Fig. 12. Distributions of the absolute values of the size of herding avalanches occurring in the small-world (SW) (top panels) and in the scale-free (SF) (bottom panels) community of investors, with and without random traders, for both the FTSE MIB index (left panels) and the S&P 500 index (right panels).

random traders in the SF network, we have the same effect obtained with 2% (32) random traders in the SW. We can conclude that, compared to a small-world configuration of the network, with the scale-free topology a fewer number of random investors (provided that they are the mostly connected agents) are needed in order to dampen bubbles and crushes.

4.3   "Micro" Within "macro" Perspective

Now we are focusing on the personal gains or losses that the interacting agents experience during the whole trading period considered. We see how the interaction among traders, realized through different network topologies, affects the results found from the point of view of the wealth distribution. We refine the FQ dynamics by assigning at the beginning of each simulation exactly the same initial capital C of 1000 credits to all the traders, then we let them invest in the market (when established by the dynamical rules) according to the following prescriptions: the first bet of each agent does not modify her capital; if an agent wins thanks to a given bet, in the next investment he will bet a quantity δC of money equal to one half of her total capital C at that moment, i.e. $\delta C = 0.5C$ and if an agent loses due to an unsuccessful investment (for example after a neg-

ative financial quake), the next time he will invest only ten percent of her total
capital at that moment, i.e. $\delta C = 0.1C$. Due to these rules, after every financial quake, the capital of all the active agents involved in the herding-related
avalanche will increase or decrease by the quantity δC. On the other hand, the
wealth of random traders, who do not take part in avalanches, can change only
when they overcome their information threshold due to the external information
sources.

Firstly, for the small-world $2D$ lattice topology, if we consider a single-event
time evolution of the capital distribution $P(C)$ of $N = 1600$ RSI traders, investing in the S&P 500 market, we can see that, even if they start all with the same
initial capital of 1000 credits, many of them quickly lose most of their money
while, on the other hand, a small group of lucky agents largely increases the
capital until, at the end of the simulation, the resulting capital distribution is
a Pareto-like power law, with exponent -1.3. We verified that this distribution
is very sensitive to the herding mechanism among RSI traders: in fact, reducing
the informative flow among them, i.e. decreasing the value of α in Eq. 5 from
0.84 to 0.40 and then to 0.00, the system tends to exit from the critical-like state,
avalanches are drastically reduced and final capital inequalities disappear.

Cumulating data over 10 different simulations, with the same initial condition
about capital, and introducing a 10% of random traders among the RSI ones,

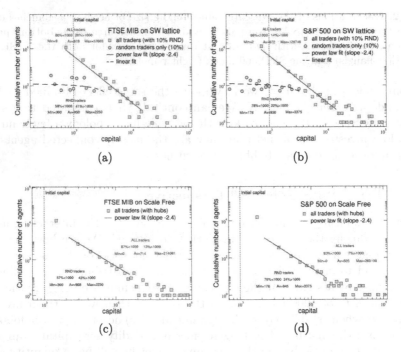

Fig. 13. Top Panels: Capital/wealth distribution for a small-world $2D$ lattice with a
10% of random traders over a total of $N = 1600$ RSI traders. Bottom Panels: Cap-
ital/wealth distribution for a scale-free network with, in average, 8 hyper-connected
random traders over a total of $N = 1600$ RSI traders.

we focus now on the final capital distribution and obtain the results presented in Fig. 13, for both the FTSE MIB and the S&P 500 time series and for both type of network.

In the top panels, corresponding to the SW with FTSE MIB series, the final global capital distribution shows a Pareto-like power law behavior, with exponent -2.4, but the partial distribution of the random traders only is completely different, staying almost constant (apart from the fluctuations) or, more precisely, decreasing linearly with an angular coefficient equal to 0.023. While, globally, 80% of traders have, in the end, a lower capital with respect to the initial one of 1000 credits, the same holds for only 59% of random traders. Moreover, the average final capital of all the traders is of 818 credits, against a higher average capital of random traders only, equal to 950 credits. This means that, on average, random trading seems more profitable with respect to the RSI one. On the other hand, the final range of capital is very different in the two cases: for all the traders, the final capital goes from a minimum of 9 credits to a maximum of 57665, while, for random traders only, it goes from 360 to 2250 credits. In other words, as expected, the random trading strategy seems also much less risky than the technical one. Similar results are shown for the S&P 500 time series. If we consider the bottom panels of Fig. 13, in which we have results for the scale-free network topology, we see that, summarizing, also for the scale-free network topology we find that the random strategy is, at the end of the game, more profitable and less risky. In this case, a very small number of hyper-connected random traders (hubs) (on average 0.5%, 8) results not only able to reduce financial avalanches (bubbles and crashes), but can also do this without great risks in terms of capital. On the other hand, if a small number of technical traders can gain very much (only about 0.9% of all the RSI traders end with more than 10000 credits for both the FTSE MIB and the S&P 500 series), the large majority of them, about 70%, lose more than the worst random traders. Furthermore, compering results, we found that the scale-free topology, with an unequal (power law distributed) connectivity, amplifies the inequalities in the final distribution of wealth: the richest traders in the SF network double the richest ones in the SW network, while only in the SF network we find a not negligible percentage of particularly unlucky traders - 4% for the FTSE MIB and even the 10% for the S&P 500 - that lose *all* their initial capital.

4.4   Conclusive Remarks and Policy Suggestions

We have reviewed some results from recent investigations about the positive role of randomness in socio-economic systems, mainly in financial markets.

We conducted our analysis both from a microeconomic and macroeconomic point of view (with two types of networks, $2D$ small-world and scale-free) and in both cases it was demonstrated that the presence of random investors reduces the risk and greatness of financial avalanches.

Therefore, as policy suggestion, stabilization of the financial market is possible if participants are told and convinced that they should not rely entirely on the signals they receive from other investors.

It remains to be investigated the dynamics of artificial markets with feed-back mechanisms, a natural step towards a self-regulating, participatory market society.

5   Talent, Luck and Success

In the previous sections we have shown that random strategies can be very efficient and successful in many different socioeconomic contexts. Although plausible and even evident to an in-depth analysis, this evidence often appears quite surprising at a first sight. Most part of the surprise probably comes from the common belief that individual success is mainly due to personal qualities such as talent, intelligence, skills, smartness, efforts, hard work or risk taking. And even if we are willing to admit that a certain degree of luck could also play a role in reaching significant achievements, it is rather common to underestimate the importance of external forces in individual successful stories.

It is clear that these assumptions are, on one hand, at the basis of the current, celebrated *meritocratic paradigm*, which affects not only the way our society grants work opportunities, fame and honors, but also the strategies adopted by Governments in assigning resources and funds to those who are considered the most deserving individuals. On the other hand, they appear to be in direct contrast with several studies which found that chance seems decisive in many aspect of our life. For example, just to name a few: roughly half of the variance in incomes across persons worldwide is explained only by their country of residence and by the income distribution within that country [75]; the probability of becoming a CEO is strongly influenced by your name or by your month of birth [76–78]; those with earlier surname initials are significantly more likely to receive tenure at top departments [79]; the innovative ideas are the results of a random walk in our brain network [80]; and even the probability of developing a cancer, maybe cutting a brilliant career, is mainly due to simple bad luck [81,82].

A macroscopic quantitative indication questioning the exclusive role of talent in reaching success can be also found by comparing the statistical distributions of these two observables among a population. Actually, while talent – as the other human qualities, like for example intelligence – exhibits the typical Gaussian distribution [83–86], the distribution of wealth – often considered a quantitative proxy of success – is very asymmetric, with a large majority of poor people and a very small number of billionaires. In particular, as originally discovered by Pareto [87], the wealth distribution (of a country, for example) shows a power law tail, where the 20% richest individuals own 80% of the total capital and the 80% own the remaining 20%. Such a large discrepancy suggests that something beyond talent has to be at work behind the scenes of success.

5.1   The TvL Model

In a 2018 paper [15] we supported the hypothesis that such an ingredient could be just randomness, possibly amplified by the complex feedback mechanisms of the socio-economic system, as for instance the *rich get richer* effect (also known in sociology as *Mathew effect*). With the help of a very simple agent-based model, called "Talent vs Luck" (TvL) model, we showed that a heavy-tailed distribution of capital in a population of individuals can spontaneously emerge in spite of the non heavy-tailed distribution of talent. The study was immediately noticed by the MIT Technology Review [88] and Scientific American [89], which wrote a couple of popular articles about it, so that in a few month it gained a lot of attention around the world reaching a very high Altmetric score [90].

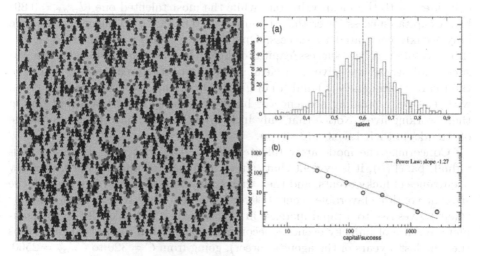

Fig. 14. Left Panel: An example of the TvL model's world, with $N = 1000$ agents exposed to 250 lucky events (green circles) and 250 unlucky events (red circles). Right Panels: (a) Normal distribution of talent, with $m_T = 0.6$ and $\sigma_T = 0.1$; (b) Final distribution of success/capital after a single simulation run: it can be fitted by a power-law with slope -1.27. (Color figure online)

In the TvL model, see Fig. 14, $N = 1000$ individuals are randomly placed in a virtual square world (left panel) and endowed with a fixed talent T, normally distributed in the interval $[0,1]$ (with mean $m_T = 0.6$ and standard deviation $\sigma_T = 0.1$, see right panel (a)), and with the same initial amount of capital $C(0) = 10$ (in arbitrary units). During a generic simulation run, all the (fixed) agents are exposed to the action of either lucky (green circles) or unlucky (red circles) events, which randomly move around the world and allow us to implement a multiplicative dynamics. Typically, $N_{ev} = 500$ events are present in the world, 50% lucky and 50% unlucky. At each time step t, if a lucky event occurs for a given agent (i.e. the position of the green circle coincides with the agent's position in the world), she doubles her capital with a probability equal to her talent; contrariwise, if an unlucky event occurs, capital is halved regardless of talent. At the end of a single simulation run (after 80 time steps, representing 40 years of simulated working life), despite the normal distribution of talent, the probability distribution of capital/success results clearly heavy-tailed, as shown in the right panel (b). More precisely, it can be fitted with a power-law with a negative exponent between 1 and 2, compatible with the Pareto law observed in the real world.

However, the TvL model allows us to go beyond this first result: in fact, as shown in Fig. 15(a), looking at the correlations between success and talent at the end of the same simulation run, it clearly emerges that richest individuals do not coincide with the most talented ones and vice-versa. More in detail, the most successful individual, with $C_{max} = 2560$, has a talent $T^* = 0.61$, basically coincident with the mean value 0.6, while the most talented one ($T_{max} = 0.89$) has a capital/success lower than 1 unit ($C = 0.625$). Thus, talent seems to be only weakly correlated to success. On the other hand, a further look at the time evolution of the success/capital of both the most successful and the less successful agents, can give us a clear indication on the great role of chance in their careers. In panels (b) and (c) of Fig. 15 these time evolutions are compared with the corresponding sequence of lucky or unlucky events occurred during their working lives. Notice that only lucky events agents have taken advantage of, thanks to their talent, are shown.

Concerning the moderately talented ($T^* = 0.61$) but very successful individual, panel (b), it is evident that, after a first part of working life with a low occurrence of lucky events, and then with an unchanged low level of capital, the sequence of three favorable events between 30 and 40 time steps (transformed by talent) gives rise to a rapid increase in capital, which – thanks also to the complete absence of unlucky events – becomes exponential in the last 10 time steps (i.e. the last 5 years of the agent's career), going from $C = 320$ to $C_{max} = 2560$. On the contrary, concerning the less successful individual, panel (c), it is evident that the impressive sequence of unlucky events in the second half of working life progressively reduces the capital/success, bringing it at its minimum final value $C = 0.00061$. It is interesting to notice that this poor agent had a talent $T = 0.74$, which was greater than that of the most successful agent, but without the possibility to express it due to the bud luck. Thus, in the competition between talent and luck, it seems that the latter wins hands down: even the greatest talent becomes useless against the fury of misfortune.

These findings are not a special case. Rather, they seem to be the rule for socio-economic systems endowed with multiplicative dynamics of capital: the maximum success never coincides with the maximum talent, and vice-versa. Moreover, such a misalignment between success and talent is disproportionate and highly nonlinear. In fact, the average capital of all people with talent $T > T^*$ in the previous simulation is $C \sim 20$: in other words, the capital/success of the most successful individual, who is moderately gifted, is 128 times greater than the average capital/success of people who are more talented than him. All these results, further confirmed in [15] by averaging over 100 simulations runs, should definitely convince the reader about the fundamental role of chance in our lives, a finding which – on the other hand – is also in agreement with our perception that

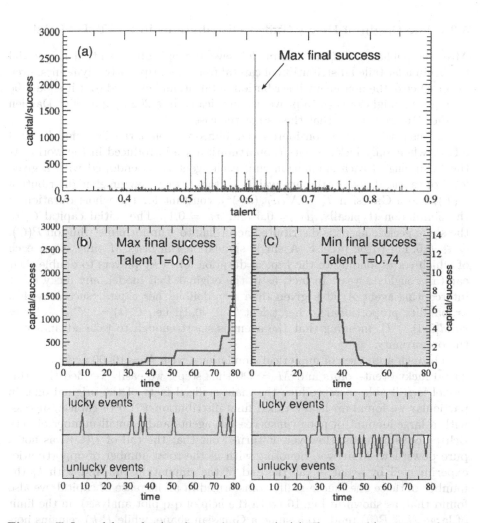

Fig. 15. TvL model results (single simulation run): (a) The capital/success is plotted as function of talent: the most successful agent, with $C_{max} = 2560$, has a talent only slightly greater than the mean value $m_T = 0.6$, while the most talented one has a capital/success lower than $C = 1$ unit, much less of the initial capital $C(0) = 10$; (b) Time evolution of success/capital for the most successful individual and (c) for the less successful one, compared with the corresponding sequences of lucky or unlucky events occurred during their working lives (80 semesters, i.e. 40 years) and reported in the corresponding bottom panels as upward or downwards spikes, respectively.

a great talent is not sufficient to guarantee a successful career and that, instead, less talented people are very often able to reach the top of success [91–93].

5.2 The Origin of Heavy Tailed Distributions in the TvL Model

After the publication of the TvL model, several people observed that it is trivial to obtain a fat tailed distribution of capital from a multiplicative dynamics, even in presence of the non-heavy distribution of talent, and claimed that it should be also quite straightforward to prove it analytically. In a 2020 paper with Damien Challet [16] we showed that this is not the case.

In this new study, we considered a simplified version of the TvL model, called STvL, where only lucky events (opportunities) are introduced in the world. At the beginning of each simulation run, each agent A_i is endowed with a given level of talent, drawn in the interval $[0, 1]$ from a known symmetric distribution $P(T)$, e.g. a Gaussian $T_i \sim \mathcal{N}(\mu_T, (\sigma_T)^2)$, constant for the whole duration of the simulation (typically, $\mu_T = 0.6$ and $\sigma_T = 0.1$). The initial capital C_i of the same generic agent is distributed according to a uniform distribution $P(C)$, $C \in [0.5, 1.5]$ to be precise. A single simulation run lasts M time steps, each of which corresponding to the typical duration for lucky players to double their capital, roughly a year. In fact, as in the original TvL model, any lucky event intercepting agent A_i at a given time step doubles her capital/success with a probability proportional to her talent $T_i \in [0, 1]$, i.e., $C_i(t) = 2C_i(t-1) \Leftrightarrow rand[0, 1] < T_i$, meaning that the agent is smart enough to take advantage of the opportunity.

Through a first set of numerical simulations, with $N = 10000$ agents, $N_{ev} = 20000$ lucky event-points and $M = 300$ time steps, we easily verified that this model is still able to reproduce the main stylized facts of the original one. In particular we found an heavy tailed final distribution $P(C)$ of capital/success, with a large amount of poor (unsuccessful) agents and a small number of very rich (successful) ones. However, it turned out that the tail of $P(C)$ was not a pure power-law. Moreover, denoting with n_i the total number of opportunities experimented by agent A_i at the end of the simulation run and with k_i the number of those ones successfully transformed in an increase of capital, we also found that, as shown in Fig. 16 (with the help of q-q plot analysis), in the limit of large N_{ev}, $P(n)$ tends to assume a Gaussian shape, while $P(k)$ remains not-Gaussian. Thus, we were interested in studying the relationship between the final distribution of capital $P(C)$ and the distributions $P(T)$, $P(n)$ and $P(k)$.

Let's suppose that, during a simulation, an agent experiences n lucky events, each of them transformed into a capital increase with probability T, for a total of k transformed events: it is easy to see that $P(k|n, T)$ is a binomial distribution $\mathcal{B}(n, T)(k)$. Assuming that all the agents have the same probability ρ to experience a lucky event in a given time step of the simulation, the number of lucky events of a given agent should also follow a binomial distribution, which can be approximated by a Gaussian distribution for large N: $P(n) \sim \mathcal{N}(\mu_n, \sigma_n^2)$ with $\mu_n = \rho M$ and $\sigma_n = M\rho(1 - \rho)$. Thus, if $P(k)$ is known, let us discuss how

$P(C)$ may gain its heavy tails. Using $C_i(T) = C_i(0)2^{k_i}$, assuming that $C_i(0) = 1$ $\forall i$, that C and k are continuous variables, and dropping the index i,

$$P(C) = P[k = \log_2 C]\frac{dk}{dC} \propto \frac{1}{C}P(\log_2 C).$$

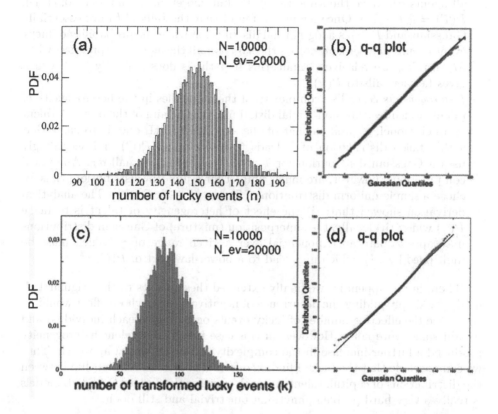

Fig. 16. STvL model with $N = 10000$ agents during a single run of $M = 300$ time steps and with $N_E = 20000$ lucky event points. Top Panels: (a) Probability distribution $P(n)$ for the number of lucky events; (b) The q-q plots shows a good agreement with the normal behavior. Bottom Panels: (c) Probability distribution $P(k)$ for the number of transformed lucky events; (d) The q-q plot of the same distribution shows a consistent deviation from the normal behavior.

Thus, for $P(C)$ to have a heavy tail, i.e., to decrease more slowly than any exponential, $P(k)$ must have a tail which decreases more slowly than e^{-e^k+k}. This is the case e.g. both for $P(k) \propto \exp(-\lambda k)$, which leads to $P(C) \propto C^{-\alpha}$ and $P(k) \sim \mathcal{N}(\mu_k, \sigma_k^2)$ which leads to a log-normal $P(C)$. In [16] we examined in detail the following specific cases:

– *Homogeneous case* $T = 1$. Suppose that all agents have the same talent $T = 1$. In this case $k = n$, $P(k) = P(n)$ is Gaussian and thus $P(C)$ is a pure log-normal distribution. In other words, the heavy tails of $P(C)$ are due to the combination of the stochastic nature of the number of lucky events and the multiplicative process which drives capital increases. While this case is relatively trivial, in the model $P(n) \neq P(k)$ when $T < 1$ for all agents.

- *Homogeneous case $0 < T < 1$, with constant T.* Let us assume now that all agents still have the same talent T', but chosen in the interval $(0,1)$, i.e. $P(T) = \delta(T - T')$. Once again, we found that the tails $P(k)$ are essentially Gaussian, and $P(C)$ is a log-normal distribution when the probability of lucky event occurrence per time step is the same for all the agents. In practice, when M and N_{ev} are relatively small, this hypothesis does not fully hold, which gives heavier tails to $P(n)$.

- *Heterogeneous case.* The complexity of the model lies in the heterogeneity of talent, which leads to non-trivial distributions. The aim of the original Talent vs Luck model, as well as that of the simplified STvL one, is to show that a thin-tailed distribution of T leads to heavy-tailed $P(C)$ and accordingly uses a Gaussian distribution for T, $\mathcal{N}(\mu_T, \sigma_T^2)$ with a small σ_T. Analytical computations, however, are much simpler for a uniform distribution, thus we chose a simple uniform distribution of T over $[T_0 - \frac{a}{2}, T_0 + \frac{a}{2}]$. The analytical derivation showed that: (i) the effect of heterogeneity of talent is to make $P(\gamma)$ wider; (ii) locally, the superposition (mixture) of Gaussian distributions may approximate an exponential over a given range of γ (which must be multiplied by M), which may lead to a power-law part of $P(C)$.

Then, in the appendix, we finally extended these results to the original TvL model by simply adding the occurrence of negative events, whose effect would be to reduce the effective number of lucky events occurring to each individual and to add some more noise. However, in this case the greater talent heterogeneity produced a further increase in the complexity of the analytical approach. Thus we can conclude that the task of finding a formal analytical relationship between the distributions of capital, talent and luck in either the TvL or the STvL models is really a very hard problem, anything but trivial and still open.

5.3   Success in Scientific Research and Funding Policies

Once established the heavy influence of random events on our everyday life and resized the role of talent in reaching success, in a 2019 paper [17], with Alfredo Ferro among the authors, we focused our attention on the importance of chance in scientific careers.

It is well known that, very often, researchers make unexpected and beneficial discoveries by chance, while they are looking for something else. Such a phenomenon is called "serendipity" [94,95] and there is a long anecdotal list of discoveries made just by lucky opportunities: from penicillin by Alexander Fleming to radioactivity by Marie Curie, from cosmic microwave background radiation by radio astronomers Arno Penzias and Robert Woodrow Wilson to the graphene by Andre Geim and Kostya Novoselov. This is further confirmed by the recent finding that scientists have the same chance along their career of publishing their biggest hit [96]. Therefore, many people think that curiosity-driven research should always be funded, because nobody can really know or predict where it can lead to [97]. However, the standard "meritocratic" strategy for assigning publicly-funded research grants still continues to pursuit excellence, based on an *ex-post* evaluation of merit without reference to real talent (which remains a hidden variable), and to drive out variety, even if this strategy seems destined to be loosing and inefficient [98–103].

In order to explore some of these features in the context of a real case study, in our 2019 paper we tried to reproduce with an agent-based model, inspired by the TvL model, the publication-citation dynamics of the physics research community, calibrating the control parameters through the information extracted from the American Physical Society (APS) data set. For the aims of our analysis we only considered the period $1980 - 2009$ and the articles of the $N = 7303$ authors who published their first paper between 1975 and 1985 and at least three papers between 1980 to 2009. Their total scientific production in this interval of 30 years consists of 89949 PACS classified articles, which received a total of 1329374 citations from the other articles in the data set.

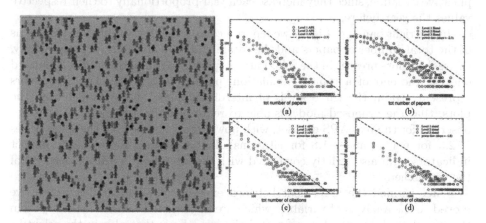

Fig. 17. Left Panel: The world of the publication-citation model inspired to the TvL one; for simplicity, only 500 agents are present, with 2000 event-points which here represent ideas triggering researches in one of the PACS classes (depending on the color). Right Panels: Distributions of the total number of papers published, during their entire careers, by the authors of the APS data set (a) and by the simulated ones (c), for the three groups with increasing levels of interdisciplinarity; The analogous distributions for the total number of citations are reported in panels (c) and (d).

In the left panel of Fig. 17 we show the 2D world where the $N = 7303$ simulated APS authors are randomly assigned to a position, fixed during each simulation. As in the TvL model, each author has a fixed talent $T_i \in [0, 1]$ (intelligence, skill, endurance, hard-working, ...), extracted from a Gaussian distribution with mean $m_T = 0.6$ and standard deviation $\sigma_T = 0.1$. Each simulation has a duration t_{max} of 30 years, with a time step t of 1 year. The agents are divided into the three groups of different color, according with the level of interdisciplinarity of their work (low-red, medium-green and high-blue, respectively). Instead of capital, the degree of success of an author is quantified, here, by the total number of papers published during her career and by the total number of received citations.

The virtual world also contains $N_{ev} = 2000$ event-points which, unlike the authors, randomly move around during the simulations. They are colored with different shades of magenta, one for each of the 10 PACS classes, and the relative abundance of points belonging to a given class is fixed in agreement with the information of the APS data set (also their total number N_{ev} was calibrated on the real data). Events represent random opportunities, ideas, encounters, intuitions, serendipitous events, etc., which can periodically occur to a given individual along her career, triggering a research line along one or more fields represented by the corresponding PACS class. Each agent can exploit the opportunities offered by the random occurrence of event-points by transforming them in publications that, in turn, can receive citations from other papers depending on the reputation of the author. Both publications and citations follow a multiplicative dynamic, since they increase each year proportionally to their respective values the year before.

In the right panels of Fig. 17 we report a comparison between the distributions of the total number of papers and the total number of citations for, respectively, the real authors present in the APS dataset, panels (a) and (c), and the simulated ones at the end of a typical simulation, panels (b) and (d). Different colors represent, again, the three levels of interdisciplinarity. It is evident from the plots, that the proposed model is able to reproduce the same kind of behavior observed for the real APS dataset, with power law tails showing the same slopes (-2.3 for papers and -1.8 for citations). In both cases, the scientific impact indicators seem also strictly correlated with the interdisciplinarity propensity of the researchers.

Once verified the agreement between experimental and modeled data, we moved to the analysis of variables which are impossible to observe directly from the real data. In particular, we tested the hypothesis that also in the scientific context individual talent is necessary but not sufficient to reach high levels of success, since external factors could still play a fundamental role. In Fig. 18 we plot the final number of papers (left column) and the final number of citations (right column) cumulated by each author belonging to the three interdisciplinarity groups during 10 simulation runs, each starting from a different realization of initial conditions, as a function of talent. It clearly appears that very talented scholars – for example those with a talent $T_i > 0.9$ – are very rarely the most successful ones, regardless the interdisciplinarity group they belong. Rather, their papers or citations score stays often quite low. On the other hand, scientists with a talent just above the mean – for example in the range $0.6 < T_i < 0.8$ – usually cumulate a considerable number of papers and citations. In other words, the most successful authors are almost always scientists with a medium-high level of talent, rather than the most talented ones. This happens because (i) talent

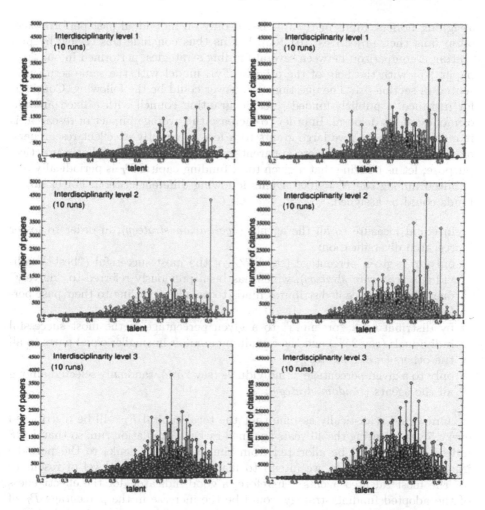

Fig. 18. Each circle in the figures represents the total number of papers (left column) or the total number of citations (right column) cumulated by each agent during a simulated run, reported as function of the corresponding talent. Each row represent a different interdisciplinarity group. These data, collected over 10 different simulation runs, indicate in a clear way that the most successful researchers are never the most talented ones.

needs lucky opportunities (chances, random meetings, serendipity) to exploits its potentialities, and (ii) very talented scientists are much less numerous than moderately talented ones (being the talent normally distributed in the population). Therefore, it is much easier to find a moderately gifted *and* lucky researcher than a very talented *and* lucky one.

These results further question the naively meritocratic assumptions still surviving in the context of scientific research and explain why the strategies assigning honors, funds or rewards based only on individual past performances often fails their objectives [104,105]. Let us thus conclude this review by presenting a comparison between several funding strategies, performed in our 2018 paper [15] with the help of the original TvL model with the same setup presented in section 5.1. The questions to answer could be the following. Consider, for instance, a publicly-funded research granting council with a fixed amount of money at its disposal. In order to increase the average impact of research, is it more effective to give large grants to a few apparently excellent researchers, or small grants to many more apparently ordinary researchers? For a practical purpose, let us imagine that a given total funding capital F_T is periodically distributed among the $N = 1000$ agents following different criteria. For example, funds could be assigned:

- in equal measure to all the agents (*egalitarian strategy*), in order to foster research diversification;
- only to a given percentage (say 25%) of the most successful ("best") individuals (*elitarian strategy*), which has been previously referred to "naively" meritocratic, since it distributes funds to people according to their past performance;
- by distributing a "premium" to a given percentage of the most successful individuals (say 25%) and the remaining amount in smaller equal parts to all the others (*mixed strategy*);
- only to a given percentage of individuals (say 25%), randomly selected among all the agents (*random strategy*);

One could realistically assume that the total capital F_T will be distributed every 5 years, during the 40 years spanned by each simulation run, so that $F_T/8$ units of capital will be allocated from time to time. Thanks to the periodic injection of these funds, we intend to maintain a minimum level of resources for the most talented agents. Therefore, a good indicator, for the effectiveness of the adopted funding strategy, could be the increase in the percentage P_T of individuals, with talent $T > m_T + \sigma_T$ (i.e. more than one standard deviation above the mean) and whose final success/capital is greater than the initial one, calculated with respect to the corresponding percentage in absence of funding. Let us call this indicator *efficiency index*, which can be estimated for the various strategies (averaging over 100 simulation runs) and normalized to its maximum value.

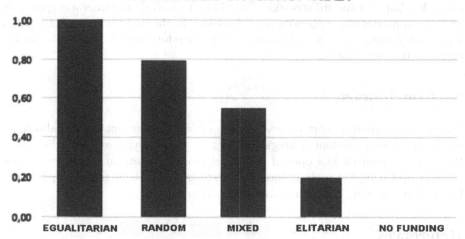

Fig. 19. A figure caption is always placed below the illustration. Please note that short captions are centered, while long ones are justified by the macro package automatically.

Looking at the histogram in Fig. 19, it is evident that, if the goal is to reward the most talented persons (thus increasing the global scientific production), it is much more convenient to distribute periodically an (even small) equal amounts of funds to all the agents (egalitarian strategy) rather than to give a greater capital only to a small percentage of them, selected through their level of success already reached at the moment of the distribution. Moreover, if – on one hand – the elitarian strategy, assigning all funds to the best 25% of the already successful individuals, is at the bottom of the ranking (thus reinforcing the thesis that this kind of approach is only apparently - i.e. naively - meritocratic), on the other hand assigning a "meritocratic" funding share to the 25% of the most successful individuals and distributing the remaining funds in equal measure to the rest of people (mixed strategy), gives back a better score in terms of efficiency index, even if the performance of this strategy is not able to overtake the egalitarian criterion. However, considering psychological factors (not modeled in the study), a mixed strategy could still be revalued with respect to the egalitarian one. Indeed, the premium reward - assigned to the more successful individuals - could induce all agents towards a greater commitment, while the equally distributed part would play a twofold role: at the individual level, it would act in fostering variety and providing unlucky talented people with new chances to express their potential, while feeding serendipity at the aggregate level, thus contributing to the progress of research and of the whole society.

Finally, it is also worthwhile to highlight the surprising high efficiency of the random strategy, which is second in the ranking. This is a further confirmation that, in complex social and economical contexts where chance plays a relevant role, the efficiency of alternative strategies based on random choices can easily overtake that of standard strategies based on the naively meritocratic approach. Such a counterintuitive phenomenon, discussed in the first sections of this paper for management, politics and finance, finds therefore new evidence also in the research funding context.

6   Conclusions

We have presented a short review of some of our most recent studies about the successful role of random strategies in socio-economic systems. We believe that this line of research has opened a fruitful cross-fertilization among statistical physics and complex socio-economical systems, suggesting useful applications for a more efficient, democratic and egalitarian society.

References

1. Metropolis, N., Ulam, S.: The Monte Carlo method. J. Am. Stat. Ass. **44**, 247 (1949)
2. Binder, K., Heermann, D.W.: Monte Carlo Simulation in Statistical Physics. Springer, Heidelberg (2010). https://doi.org/10.1007/978-3-642-03163-2
3. Benzi, R., Parisi, G., Sutera, A., Vulpiani, A.: Stochastic resonance in climate change. Tellus **34**, 10–16 (1982)
4. Gammaitoni, L., Hanggi, P., Jung, P., Marchesoni, F.: Stochastic resonance. Rev. Mod. Phys. **70**, 223 (1998)
5. Peter, L.J., Hull, R.: The Peter Principle: Why Things Always Go Wrong. William Morrow and Company, New York (1969)
6. Pluchino, A., Rapisarda, A., Garofalo, C.: The Peter principle revisited: a computational study. Phys. A **389**(3), 467 (2010)
7. Pluchino, A., Rapisarda, A., Garofalo, C.: Efficient promotion strategies in hierarchical organizations. Phys. A **390**(20), 3496 (2011)
8. Pluchino, A., Garofalo, C., Rapisarda, A., Spagano, S., Caserta, M.: Accidental politicians: how randomly selected legislators can improve parliament efficiency. Phys. A **390**(21), 3944 (2011)
9. Caserta, M., Pluchino, A., Rapisarda, A., Spagano, S.: Why lot: how sortition came to help representative democracy. Phys. A **565**, 125430 (2021)
10. Biondo, A.E., Pluchino, A., Rapisarda, A.: The beneficial role of random strategies in social and financial systems. J. Stat. Phys. **151**, 607–622 (2013)
11. Biondo, A.E., Pluchino, A., Rapisarda, A., Helbing, D.: Are random trading strategies more successful than technical ones. PLoS ONE **8**(7), e68344 (2013)
12. Biondo, A.E., Pluchino, A., Rapisarda, A., Helbing, D.: Reducing financial avalanches by random investments. Phys. Rev. E **88**, 06814 (2013)
13. Biondo, A.E., Pluchino, A., Rapisarda, A.: Micro and macro benefits of random investments in financial markets. Contemp. Phys. **55**(4), 318–334 (2014)
14. Biondo, A.E., Pluchino, A., Rapisarda, A.: Modelling financial markets by self-organized criticality. Phys. Rev. E **92**, 042814 (2015)
15. Pluchino, A., Biondo, A.E., Rapisarda, A.: Talent versus luck: the role of randomness in success and failure. Adv. Complex Syst. **21**(3,4), 1850014 (2018)

16. Challet, D., Pluchino, A., Biondo, A.E., Rapisarda, A.: The origins of extreme wealth inequality in the Talent versus Luck model. Adv. Complex Syst. **23**(2), 2050004 (2020)

17. Pluchino, A., et al.: Exploring the role of interdisciplinarity in Physics: Success, Talent and Luck. PLoS ONE **14**(6), e0218793 (2019)

18. Aristotle: The Athenian Constitution, Penguin (1984)

19. Cipolla, C.M.: The Basic Laws of Human Stupidity. The Mad Millers (1976)

20. Arrow, K.J., Nerlove, M.: Econometrica **26**, 297–305 (1958)

21. Friedman, M.: A Theory of the Consumption Function. Princeton University Press, Princeton (1956)

22. Friedman, M.: The role of monetary policy. Am. Econ. Rev. 1–17 (1968)

23. Phelps, E.: Phillips curve expectations of inflation, and output unemployment over time. Economica **34**(135), 254–281 (1967)

24. Cagan, P.: The monetary dynamics of hyperinflation. In: Friedman, M. (ed.) Studies in the Quantity Theory of Money. University of Chicago Press, Chicago (1956)

25. Muth, J.F.: Rational expectation and the theory of price movements. Econometrica **29**, 315–335 (1961)

26. Lucas, R.E.: Expectations and the neutrality of money. J. Econ. Theory **4**, 103–124 (1972)

27. Sargent, T.J., Wallace, N.: Rational expectations, the optimal monetary instrument, and the optimal money supply rule. J. Polit. Econ. **83**(2), 241–254 (1975)

28. Fama, E.F.: J. Financ. **25**, 383–423 (1970)

29. Jensen, M.: Some anomalous evidence regarding market efficiency. J. Financ. Econ. **6**, 95–101 (1978)

30. Malkiel, B.: Efficient market hypothesis. New Palgrave Dictionary of Money and Finance. Macmillan, London (1992)

31. Cutler, D.M., Poterba, J.M., Summers, L.H.: What moves stock prices? J. Portf. Manag., 4–12 April 1989

32. Engle, R.: Autoregressive conditional heteroscedasticity with estimates of the variance of UK inflation. Econometrica **50**, 987–1008 (1982)

33. Mandelbrot, B.B.: The variation of certain speculative prices. J. Bus. **36**, 394–419 (1963)

34. Mandelbrot, B.B.: Fractals and Scaling in Finance. Springer, New York (1997)

35. Lux, T.: The stable Paretian hypothesis and the frequency of large returns: an examination of major German stocks. Appl. Financ. Econ. **6**, 463–475 (1996)

36. Brock, W.A.: Pathways to randomness in the economy: emergent non-linearity and chaos in economics and finance. Estudios Económicos **8**, 3–55 (1993)

37. Brock, W.A.: Asset prices behavior in complex environments. In: Arthur, W.B., Durlauf, S.N., Lane, D.A. (eds.) The Economy as an Evolving Complex System II, pp. 385–423. Addison-Wesley, Reading (1997)

38. Brock, W.A., Hommes, C.H.: A rational route to randomness. Econometrica **65**, 1059–1095 (1997)

39. Chiarella, C.: The dynamics of speculative behavior. Ann. Oper. Res. **37**, 101–123 (1992)

40. Chiarella, C., He, T.: Heterogeneous beliefs, risk and learning in a simple asset pricing model. Comput. Econ. Spec. Issue Evol. Process. Econ. **19**(1), 95–132 (2002)

41. DeGrauwe, P., DeWachter, H., Embrechts, M.: Exchange Rate Theory. Chaotic Models of Foreign Exchange Markets. Blackwell, New York (1993)

42. Frankel, J.A., Froot, K.A.: Chartists, fundamentalists and the demand for dollars. Greek Econ. Rev. **10**, 49–102 (1988)

43. Lux, T.: Herd behavior, bubbles and crashes. Econ. J. **105**, 881–896 (1995)
44. Wang, J.: A model of competitive stock trading volume. J. Polit. Econ. **102**, 127–168 (1994)
45. Zeeman, E.C.: The unstable behavior of stock exchange. J. Math. Econ. **1**, 39–49 (1974)
46. Simon, H.A.: Models of Man. Wiley, New York (1957)
47. Mantegna, R.N., Stanley, H.E.: Introduction to Econophysics: Correlations and Complexity in Finance. Cambridge University Press, Cambridge (1999)
48. McCauley, J.L.: Dynamics of Markets: Econophysics and Finance (2007)
49. Bouchaud, J.-P., Potters, M.: Theory of Financial Risk and Derivative Pricing: From Statistical Physics to Risk Management. Cambridge University Press, Cambridge (2003)
50. Bass, T.A.: The Predictors: How a Band of Maverick Physicists Used Chaos Theory to Trade Their Way to a Fortune on Wall Street. Henry Holt & Company, New York (2000)
51. Sornette, D.: Why Stock Markets Crash: Critical Events in Complex Financial Systems (2004)
52. Wiseman, R.: Quirkology. Pan Macmillan, London (2007)
53. Porter, G.E.: The long term value of analysts' advice in the Wall Street Journal's Investment Dartboard Contest. J. Appl. Financ. **14** (2004)
54. Clare, A., Motson, N., Thomas, S.: An evaluation of alternative equity indices. Cass consulting report, March 2013. http://www.cassknowledge.com/research/article/evaluation-alternative-equity-indices-cass-knowledge
55. Carbone, A., Castelli, G., Stanley, H.E.: Time dependent Hurst exponent in financial time series. Phys. A **344**, 267–271 (2004)
56. Murphy, J.J.: Technical Analysis of the Financial Markets: A Comprehensive Guide to Trading Methods and Applications. New York Institute of Finance (1999)
57. Taleb, N.N.: Fooled by Randomness: The Hidden Role of Chance in the Markets and in Life. Random House, New York (2005)
58. Buchanan, M.: Forecast: What Physics, Meteorology, and the Natural Sciences Can Teach Us About Economics. Bloomsbury (2013)
59. Helbing, D., Kern, D.: Non-equilibrium price theories. Phys. A **287**, 259–268 (2000)
60. Helbing, D., Farkas, I., Vicsek, T.: Simulating dynamical features of escape panic. Nature **407**, 487–490 (2000)
61. Ghashghaie, S., Breymann, W., Peinke, J., Talkner, P., Dodge, Y.: Nature **381**, 767–770 (1996)
62. Farmer, J.D.: Ind. Corp. Change **11**(5), 895 (2002). https://doi.org/10.1093/icc/11.5.895
63. Bouchaud, J.-P., Potters, M.: Theory of financial risks: from statistical physics to risk management. Lavoisier (2004)
64. Sornette, D.: Why Stock Markets Crash: Critical Events in Complex Financial Systems. Princeton University Press (2003)
65. Helbing, D.: Globally networked risks and how to respond. Nature **497**, 51–59 (2013)
66. Krawiecki, A., Holyst, J.A., Helbing, D.: Volatility clustering and scaling for financial time series due to attractor bubbling. Phys. Rev. Lett. **89**, 158701 (2002)
67. Parisi, D.R., Sornette, D., Helbing, D.: Financial price dynamics and pedestrian counterflows: a comparison of statistical stylized facts. Phys. Rev. E **87**, 012804 (2013)

68. Bikhchandani, S., Hirshleifer, D., Welch, I.: Informational cascades and rational herding: an annotated bibliography and resource reference. Working paper, Anderson School of Management, UCLA (2008). www.info-cascades.info
69. Bak, P., Tang, C., Wiesenfeld, K.: Phys. Rev. Lett. **59**, 381 (1987)
70. Olami, Z., Feder, H.J.S., Christensen, K.: Phys. Rev. Lett. **68**, 1244 (1992)
71. Caruso, F., Pluchino, A., Latora, V., Vinciguerra, S., Rapisarda, A.: Analysis of self-organized criticality in the OFC model and in real earthquakes. Phys. Rev. E **75**, 055101(R) (2007)
72. Lillo, F., Mantegna, R.N.: Power-law relaxation in a complex system: Omori law after a financial market crash. Phys. Rev. E **68**, 016119 (2003)
73. Caruso, F., Latora, V., Pluchino, A., Rapisarda, A., Tadic, B.: Eur. Phys. J. B **50**, 243 (2006)
74. Barabasi, A.L., Albert, R.: Science **286**, 509 (1999)
75. Milanovic, B.: Global inequality of opportunity: how much of our income is determined by where we live? Rev. Econ. Stat. **97**(2), 452–60 (2015)
76. Du, Q., Gao, H., Levi, M.D.: The relative-age effect and career success: evidence from corporate CEOs. Econ. Lett. **117**(3), 660–662 (2012)
77. Deaner, R.O., Lowen, A., Cobley, S.: Born at the wrong time: selection bias in the NHL draft. PLoS ONE **8**(2), e57753 (2013)
78. Brooks, D.: The Social Animal. The Hidden Sources of Love, Character, and Achievement. Random House, 424 p. (2011)
79. Einav, L., Yariv, L.: What's in a surname? The effects of surname initials on academic success. J. Econ. Perspect. **20**(1), 175–188 (2006)
80. Iacopini, I., Milojevic, S., Latora, V.: Network dynamics of innovation processes. Phys. Rev. Lett. **120**, 048301 (2018)
81. Tomasetti, C., Li, L., Vogelstein, B.: Stem cell divisions, somatic mutations, cancer etiology, and cancer prevention. Science **355**, 1330–1334 (2017)
82. Newgreen, D.F., et al.: Differential clonal expansion in an invading cell population: clonal advantage or dumb luck? Cells Tissues Organs **203**, 105–113 (2017)
83. Wechsler, D.: The Measurement and Appraisal of Adult Intelligence, 4th edn. Williams and Witkins, Baltimore (1958)
84. Kaufman, A.S.: Assessing Adolescent and Adult Intelligence, 1st edn. Allyn and Bacon, Boston (1990)
85. Kaufman, A.S.: IQ Testing 101. Springer, New York (2009)
86. Stewart, J.: The distribution of talent. Marilyn Zurmuehlin Work. Papers Art Educ. **2**, 21–22 (1983)
87. Pareto, V.: Cours d'Economique Politique, vol. 2 (1897)
88. https://www.technologyreview.com/2018/03/01/144958/if-youre-so-smart-why-arent-you-rich-turns-out-its-just-chance/
89. https://blogs.scientificamerican.com/beautiful-minds/the-role-of-luck-in-life-success-is-far-greater-than-we-realized/
90. https://www.altmetric.com/details/33451664
91. Taleb, N.N.: Fooled by Randomness: The Hidden Role of Chance in Life and in the Markets. TEXERE, London (2001)
92. Taleb, N.N.: The Black Swan: The Impact of the Highly Improbable. Random House (2007)
93. Frank, R.H.: Success and Luck: Good Fortune and the Myth of Meritocracy. Princeton University Press, Princeton (2016)
94. Merton, R.K., Barber, E.: The Travels and Adventures of Serendipity. Princeton University Press, Princeton (2004)

95. Murayama, K., et al.: Management of science, serendipity, and research performance. Res. Policy **44**(4), 862–873 (2015)
96. Sinatra, R., Wang, D., Deville, P., Song, C., Barabási, A.-L.: Quantifying the evolution of individual scientific impact. Science **354**, 6312 (2016)
97. Flexner, A.: The Usefulness of Useless Knowledge. Princeton University Press, Princeton (2017)
98. Page, S.E.: The Diversity Bonus. How Great Teams Pay Off in the Knowledge Economy. Princeton University Press (2017)
99. Cimini, G., Gabrielli, A., Sylos Labini, F.: The scientific competitiveness of nations. PLoS ONE **9**(12), e113470 (2014). https://doi.org/10.1371/journal.pone.0113470
100. Curry, S.: Let's move beyond the rhetoric: it's time to change how we judge research. Nature **554**, 147 (2018)
101. Nicholson, J.M., Ioannidis, J.P.A.: Research grants: conform and be funded. Nature **492**, 34–36 (2012)
102. Bollen, J., Crandall, D., Junk, D., et al.: An efficient system to fund science: from proposal review to peer-to-peer distributions. Scientometrics **110**, 521–528 (2017)
103. Garner, H.R., McIver, L.J., Waitzkin, M.B.: Research funding: same work, twice the money? Nature **493**, 599–601 (2013)
104. Fortin, J.-M., Curr, D.J.: Big science vs. little science: how scientific impact scales with funding. PLoS ONE **8**(6), e65263 (2013)
105. Mongeon, P., Brodeur, C., Beaudry, C., et al.: Concentration of research funding leads to decreasing marginal returns. Res. Eval. **25**, 396–404 (2016)

Critical Density for Network Reconstruction

Andrea Gabrielli[1,2], Valentina Macchiati[3,4], and Diego Garlaschelli[4,5(✉)]

[1] Dipartimento di Ingegneria, Università degli Studi Roma Tre, Rome, Italy
[2] Enrico Fermi Research Center (CREF), Rome, Italy
[3] Scuola Normale Superiore, Pisa, Italy
[4] IMT School of Advanced Studies, Lucca, Italy
diego.garlaschelli@imtlucca.it
[5] Lorentz Institute for Theoretical Physics, University of Leiden, Leiden, The Netherlands

Abstract. The structure of many financial networks is protected by privacy and has to be inferred from aggregate observables. Here we consider one of the most successful network reconstruction methods, producing random graphs with desired link density and where the observed constraints (related to the market size of each node) are replicated as averages over the graph ensemble, but not in individual realizations. We show that there is a minimum critical link density below which the method exhibits an 'unreconstructability' phase where at least one of the constraints, while still reproduced on average, is far from its expected value in typical individual realizations. We establish the scaling of the critical density for various theoretical and empirical distributions of interbank assets and liabilities, showing that the threshold differs from the critical densities for the onset of the giant component and of the unique component in the graph. We also find that, while dense networks are always reconstructable, sparse networks are unreconstructable if their structure is homogeneous, while they can display a crossover to reconstructability if they have an appropriate core-periphery or heterogeneous structure. Since the reconstructability of interbank networks is related to market clearing, our results suggest that central bank interventions aimed at lowering the density of links should take network structure into account to avoid unintentional liquidity crises where the supply and demand of all financial institutions cannot be matched simultaneously.

Keywords: Network Reconstruction · Random Graphs · Financial Networks

1 Introduction

The interactions between the components of social, biological and economic systems are frequently unknown. In the case of financial networks, where nodes represent financial institutions (such as banks) and links represent credit relationships (such as loans), only aggregate exposures are observable due to confidentiality issues [1]. This means that only the total exposure of each node towards

© The Author(s), under exclusive license to Springer Nature Switzerland AG 2024
D. Cantone and A. Pulvirenti (Eds.): *From Computational Logic to Computational Biology*,
LNCS 14070, pp. 223–249, 2024.
https://doi.org/10.1007/978-3-031-55248-9_11

the aggregate of all other nodes is known. The field of *network reconstruction* is interested in devising methods that make the best use of the available partial information to infer the original network [2]. Among the probabilistic reconstruction techniques, particularly successful are those based on the maximum entropy principle [3]. By maximizing the entropy, which is an information-theoretic notion of uncertainty encoded in a probability distribution [4], these models generate the least biased distribution of random network configurations consistent with the constraints derived from the observed aggregate information. Clearly, the goodness of the reconstruction depends on the choice of the constraints: certain constraints are more informative, while others are less. Different implementations of the maximum entropy principle, originating from different choices of constraints, have been considered [2,5–8]. As a common aspect, these methods are *ensemble* ones as they generate not a single, but a whole set of random configurations that are compatible on average with the available data.

In this work we introduce the problem of *reconstructability* of the network under ensemble methods. Reconstructability is achieved when all the constraints, besides being reproduced on average, are also 'sufficiently close' to their expected value in individual typical realizations of the ensemble. We will first define this notion rigorously and then explore the role of the available empirical properties in making the network more or less easily reconstructable. Using both analytical calculations and empirical data, we study how the realized values of the constraints fluctuate around the empirical values as a function of network density. We find a minimum critical network density which is typically safely exceeded in the unrealistically dense regime, but not necessarily in the sparse regime of interest for real-world financial networks. If the network density is lower than the critical threshold, in typical configurations of the reconstructed networks the realized constraints are displaced away from the desired, observed values. These displacements are critical from the point of view of systemic risk. Indeed, if any of the many network-based models for financial shock propagation [9–14] (which are quite sensitive to the underlying topology of the network) is run on realizations of a network reconstruction method, a misalignment between the realized constraints and the empirical ones could severely bias the model-based estimation of systemic risk. Moreover, since reconstructability can be understood as related to decentralized market clearing (where all constraints are met simultaneously via purely pairwise matchings), the lack thereof might have adverse implications in terms of illiquidity of the interbank system. Therefore the possibility of a crossover to unreconstructability suggests that regulatory authorities should monitor the density of interbank relationships with an increased awareness.

The rest of the paper is structured as follows: Sect. 2 introduces the general formalism and method, Sect. 3 illustrates some reference cases that are analytically tractable, Sect. 4 makes a technical check of the role played by self-loops to warrant the validity of the general approach, Sect. 5 looks at empirical data from the Bankscope dataset and uses them to extrapolate the results to realistic regimes, and finally Sect. 6 provides final remarks and conclusions.

2   General Setting

2.1   The Reconstruction Method

Links in financial networks correspond to exposures, e.g. they indicate the amount of money that a bank has borrowed from another bank in a certain time window. Individual links are unobservable, while the sum of the outgoing link weights and the sum of the incoming link weights for each node are both observable, as they represent the total interbank assets and the total interbank liabilities that can be derived from the balance sheet of the corresponding institution. While these relationships are clearly directional, for simplicity in this paper we will frame the problem within the context of undirected networks, where we interpret each link as representing the bilateral exposure (the sum of the exposure in both directions) between two nodes. This does not change the essence of the phenomenology we address in this paper, i.e. the identification of a regime where the margins of the empirical network cannot be properly replicated in individual reconstructed networks, even if they are still replicated on average.

We denote the weights of the links of the (unobserved) original network as $\{w_{ij}^*\}_{ij}$ (where $w_{ij}^* = w_{ji}^* > 0$ denotes the amount of the bilateral exposure between nodes i and j, while $w_{ij}^* = 0$ denotes the absence of a link) and we arrange these weights into a $N \times N$ symmetric matrix $\mathbf{W}^*$. We assume that, while we cannot observe $\mathbf{W}^*$, we can observe its margins, i.e. the so-called *strength*

$$s_i^* \equiv \sum_{j=1}^{N} w_{ij}^* \tag{1}$$

of each node i ($i = 1, N$). This quantity represents the sum of the interbank assets and the interbank liabilities for node i. Let us denote with $\vec{s}^*$ the N-dimensional vector of entries $\{s_i^*\}$, i.e. the (observable) *strength sequence* of the (unobservable) empirical network.

Clearly, inferring $\mathbf{W}^*$ from $\vec{s}^*$ with certainty is impossible, and this is the main drawback of deterministic reconstruction methods that identify a single possible solution to the reconstruction problem [2]. By contrast, probabilistic methods look for a solution, given $\vec{s}^*$, in terms of a probability distribution $P(\mathbf{A})$ over the set $\{\mathbf{A}\}$ of all $N \times N$ symmetric binary adjacency matrices (where the entry a_{ij} of one such matrix is $a_{ij} = 1$ if a link between nodes i and j is present, and $a_{ij} = 0$ otherwise). This ensures that the unobserved adjacency matrix $\mathbf{A}^*$ corresponding to the unobserved weighted matrix $\mathbf{W}^*$ (where $a_{ij}^* = 1$ if $w_{ij}^* > 0$ and $a_{ij}^* = 0$ if $w_{ij}^* = 0$) is a member of this ensemble and is therefore assigned a non-zero probability $P(\mathbf{A}^*)$. Given each generated adjacency matrix, a procedure to assign a weight to each realized link must also be designed [2]. This step turns the random ensemble of binary matrices into a random ensemble of weighted matrices where the entry w_{ij} of a generic matrix $\mathbf{W}$ is a random variable, not to be confused with the deterministic (unknown) value w_{ij}^*. Similarly, the strength $s_i = \sum_{j=1}^{N} w_{ij}$ is a random variable, not to be confused with the deterministic (known) value s_i^*, and finally the total link weight $W = \sum_i \sum_{j<i} w_{ij} = \sum_{i=1}^{N} s_i/2$ is a random variable, not to be confused with the total (known) weight $W^* = \sum_{i=1}^{N} s_i^*/2$.

In particular, we consider a successful reconstruction method, proposed in various variants [7,15,16], where the probability $P(\mathbf{A})$ is factorized over pairs of nodes, i.e. the edges of the graph are assumed to be independent Bernoulli random variables. Specifically, a link between node i and node j is established with probability

$$p_{ij}(z) = \frac{z \, s_i^* \, s_j^*}{1 + z \, s_i^* \, s_j^*}, \tag{2}$$

where z is the only free parameter (since $\bar{s}^*$ is known) and is chosen in order to tune the resulting expected link density

$$d(z) = \frac{2}{N(N-1)} \sum_{i=1}^{N} \sum_{j<i} p_{ij}(z) \tag{3}$$

of the reconstructed networks. The specific functional form of the connection probability in Eq. (2) derives from a maximum-entropy construction where, morally, one enforces the degree (i.e. the number of links) of each node as a constraint to be met as an ensemble average [3]. In such a construction, the combined quantity $x_i \equiv \sqrt{z} s_i^*$ in Eq. (2) is in principle unrelated to the strength, as it is technically a transformed Lagrange multiplier required to enforce the degree of node i. However, since the degree itself is typically not observable in financial networks, the quantity x_i cannot be determined from the data. The core of the 'fitness ansatz' is the observation that, for a few networks whose structure has been analysed and for which the value of x_i has been calculated from the empirical degrees, this value has been found to display a strong linear correlation with the empirical strength s_i^* of the corresponding node [7]. This linear correlation suggests that, for networks with unobservable topology, the undetermined value of x_i can be assumed to be proportional (by a factor $\sqrt{z}$) to the observable strength s_i^*, thereby giving rise to Eq. (2). In this way, the only free parameter is z and its effect is that of controlling for the overall link density. Now, the empirical density is also not necessarily known, however the method allows for the exploration of a range of realistic densities as a function of z, based for instance on published analyses of networks of the same type for which the empirical density has been documented. Indeed, several real financial networks are found to be sparse, which means that their empirical density scales as the inverse of N [10]. This implies that a choice for z could be $z = z_{\text{sparse}}$ where z_{sparse} is such that

$$d(z_{\text{sparse}}) \simeq \frac{k}{N}, \quad k > 1 \tag{4}$$

with k finite. Clearly, the value z_{sparse} realizing the above condition depends on the entire strength sequence, and in general on the strength distribution when $N \to \infty$.

Once generated, each binary adjacency matrix $\mathbf{A}$ drawn from the ensemble is 'dressed' with link weights. In the simplest specifications of the model [7, 15], the link weights are deterministic functions of the observable margins: if the link is realized, the random variable w_{ij} is assigned a value $\frac{s_i^* s_j^*}{2W^* p_{ij}(z)}$, otherwise it is given a zero value. This means that w_{ij} is a Bernoulli random variable given by

$$w_{ij} = \begin{cases} \dfrac{s_i^* s_j^*}{2W^* p_{ij}(z)} & \text{with probability} \quad p_{ij}(z), \\ 0 & \text{with probability} \quad 1 - p_{ij}(z). \end{cases} \tag{5}$$

A variant of this approach where the (conditional) weights on the realized links are placed not deterministically, but following a second random process resulting in exponentially distributed weights, with expected value given again by $\frac{s_i^* s_j^*}{2W^* p_{ij}(z)}$, has also been considered [16]. While this approach is found to be superior from the point of view of the reconstruction of the unobserved weight matrix (as $\mathbf{W}^*$ is always generated with positive likelihood), our main focus here is the (simpler) reconstruction of the margins $\vec{s}^*$ of $\mathbf{W}^*$ in typical realizations, for which we can use the specification given by Eq. (5) without loss of generality (exponentially distributed weights around the values considered here do not change the essence of our results).

Equation (5) ensures that the (unconditional) expected weight of the link connecting nodes i and j equals

$$\langle w_{ij} \rangle = p_{ij}(z) \frac{s_i^* s_j^*}{2W^* p_{ij}(z)} = \frac{s_i^* s_j^*}{2W^*}, \tag{6}$$

so that the expected strength $\langle s_i \rangle$ of each node i equals the corresponding observed strength s_i^* (which is a prerequisite for a successful reconstruction):

$$\langle s_i \rangle = \sum_{j=1}^{N} \langle w_{ij} \rangle = \sum_{j=1}^{N} \frac{s_i^* s_j^*}{2W^*} = s_i^* \quad \forall i. \tag{7}$$

Note that, in the above summations, it is crucial that j takes also the value i and that the 'diagonal' expected value is equal to

$$\langle w_{ii} \rangle = \frac{(s_i^*)^2}{2W^*}, \tag{8}$$

i.e. it must be described by Eq. (6) (with $i = j$) just like any other 'non-diagonal' expected weight. This means that, actually, the ensemble of binary adjacency matrices should allow for self-loops: the diagonal entry a_{ii} should also be a Bernoulli random variable with probability $p_{ii}(z)$ given by Eq. (2) with $i = j$, and similarly the entry w_{ii} should also follow Eq. (5) with $i = j$. Therefore, even if the empirical (unobserved) matrices $\mathbf{A}^*$ and $\mathbf{W}^*$ have no self-loops (as it makes no sense to say that a bank lends or borrows from itself), the method needs the generation of self-loops with appropriate probabilities and weights in order to ensure that Eq. (7) holds and that all strengths are exactly replicated on average. What is important for practical purposes is that the self-loops in the reconstructed ensemble retain a negligible expected weight, so that their existence does not cause any relevant difference with respect to their absence in the real matrix $\mathbf{W}^*$. In Sect. 4 we will check that this condition is met for the cases of relevance for real-world financial networks, and for the moment assume that self-loops can be safely added to the model as they will not play a crucial role.

In what follows, we are interested in studying various properties of the model as a function of N, and eventually in the thermodynamic limit $N \to \infty$. We assume that, as N increases, the average empirical strength $\overline{s^*} = N^{-1} \sum_{i=1}^{N} s_i^*$ remains finite. This accounts for the fact that, irrespective of how many banks enter the market, each bank retains a finite value of assets and liabilities. Again, this regime is consistent with the sparse regime typically observed for real-world networks. We therefore choose units such that

$$\overline{s^*} = 1, \qquad 2W^* = N\overline{s^*} = N, \tag{9}$$

i.e. divide each empirical node strength by the average strength over all nodes. Later, we will consider different empirical distributions of the node strengths with average value given by $\overline{s^*} = 1$. Besides its simplicity, this choice of units has the advantage that, by construction, the average value of the connection probability $p_{ij}(z)$ given by Eq. (2) over all pairs of nodes, which is precisely the link density given by Eq. (3), is of order $z/(1+z)$. This implies that, in order to realize the sparsity condition in Eq. (4), a necessary condition is

$$z_{\text{sparse}} \to 0^+ \quad \text{for} \quad N \to +\infty. \tag{10}$$

The necessary condition, i.e. the specific speed with which z_{sparse} has to decay as N grows, depends on the particular strength distribution, as we will discuss for explicit examples later.

2.2 Transition from Reconstructability to Unreconstructability

The above model ensures that the expected value of each constraint matches the corresponding observed value exactly, i.e. $\langle \vec{s} \rangle = \vec{s}^*$, however it does not ensure that all constraints can be met in each realization of the network. This is normal for any canonical ensemble where the constraints are by construction allowed to fluctuate around their expected values [3] and is not undesirable, as long as all constraints are *close enough* to their observed values in a *typical* realization of the network. This means that, for the network to be satisfactorily reconstructed, a necessary condition is that the *relative fluctuation* (defined as the ratio of the standard deviation to the expected value) of the strength of each node vanishes sufficiently fast in the limit of large N. In other words, we want to avoid the undesired situation where the typical realizations violate some of the constraints by an unacceptable amount, even though the expected value of each of them still coincides with the desired, observed value.

For the unobserved network $\mathbf{W}^*$ to be reconstructable from the observed strength sequence $\vec{s}^*$ we therefore require that, for each node i, the relative strength fluctuations

$$\delta_i(z) \equiv \frac{\sqrt{v_i(z)}}{\langle s_i \rangle} = \frac{\sqrt{v_i(z)}}{s_i^*} \tag{11}$$

decay at least as fast as $1/\sqrt{N^\beta}$, where $v_i(z) \equiv \mathrm{Var}[s_i]$ is the variance of s_i and $\beta > 0$ is some desired exponent. The 'canonical' case is $\beta = 1$, although one might in principle allow for more general scenarios. As we now show, this requirement implies that z should be larger than a critical value z_c, i.e. that the expected network density $d(z)$ defined in Eq. (3) exceeds a critical threshold $d_c \equiv d(z_c)$.

Note that if $z = +\infty$ then $d(+\infty) = 1$ and the model has a deterministic outcome producing only one fully connected network where all strengths are replicated exactly, with zero variance $v(+\infty) = 0$. At the opposite extreme, if $z = 0$ then $d(0) = 0$ and the model is again deterministic, but the output is now a single, empty network where none of the strengths are replicated as they are all zero with zero variance $v(0) = 0$. The critical value z_c we are looking for is an intermediate value separating a reconstructable phase from an unreconstructable one, corresponding to a critical scaling for the largest relative fluctuation (as we want all node strengths to be satisfactorily replicated) given by

$$\max_i \{\delta_i(z_c)\} = \max_i \frac{\sqrt{v_i(z_c)}}{s_i^*} \equiv \sqrt{cN^{-\beta}} \tag{12}$$

where $c > 0$ is a finite constant. Since we want $\max_i \{\delta_i(z_c)\} < 1$ (so that the standard deviation of the strength does not exceed the expected strength) for all values of $\beta > 0$ including values arbitrarily close to 0, we also require $c < 1$.

To identify z_c we first compute the variance of the weight w_{ij} as

$$\mathrm{Var}[w_{ij}] = \langle w_{ij}^2 \rangle - \langle w_{ij} \rangle^2$$

$$= p_{ij}(z) \frac{(s_i^* s_j^*)^2}{(2W^*)^2 \, p_{ij}^2(z)} - \langle w_{ij} \rangle^2$$

$$= \left(\frac{s_i^* s_j^*}{2W^*} \right)^2 \left(\frac{1}{p_{ij}(z)} - 1 \right)$$

$$= \left(\frac{s_i^* s_j^*}{2W^*} \right)^2 \frac{1}{z \, s_i^* s_j^*}$$

$$= \frac{s_i^* s_j^*}{(2W^*)^2 \, z}$$

$$= \frac{s_i^* s_j^*}{N^2 \, z}. \tag{13}$$

Next, using the independence of different edges in the graph, we compute the variance $v_i(z)$ of the strength s_i as

$$v_i(z) = \sum_{j=1}^{N} \text{Var}[w_{ij}] = \frac{s_i^*}{2W^* z} = \frac{s_i^*}{Nz} \tag{14}$$

and the resulting relative fluctuations as

$$\delta_i(z) \equiv \frac{\sqrt{s_i^*/Nz}}{s_i^*} = \frac{1}{\sqrt{Nz\,s_i^*}}. \tag{15}$$

Clearly, the largest fluctuation is attained by the node with minimum strength:

$$\max_i\{\delta_i(z)\} = \frac{1}{\sqrt{Nz\,\min_i\{s_i^*\}}}. \tag{16}$$

Now, imposing that $\max_i\{\delta_i(z_c)\}$ equals the critical expression in Eq. (12) implies that the critical value for z is

$$z_c = \frac{N^{\beta-1}}{c\,\min_i\{s_i^*\}}, \tag{17}$$

which is essentially driven by the statistics of the minimum strength. This is our general result. To ensure that all node strengths are replicated satisfactorily in a typical single realization of the model, we need $z > z_c$. We are particularly interested in determining whether, in the realistic sparse regime $z = z_{\text{sparse}}$ given by Eq. (4), the network is reconstructable, i.e. whether

$$z_{\text{sparse}} > z_c. \tag{18}$$

Note that the above requirement, combined with the necessary sparsity condition in Eq. (10), implies another necessary condition:

$$z_c \to 0^+ \quad \text{for} \quad N \to +\infty. \tag{19}$$

In what follows, we are going to consider specific theoretical and empirical cases, corresponding to different distributions $f(s^*)$ of the node strengths.

3   Specific Cases

3.1   Homogeneous Networks

As a first, trivial example, we consider the case of equal empirical strengths $s_i^* = 1$ for all i, i.e. $f(s^*) = \delta(s^* - 1)$. Inserting this specification into Eq. (2), it is clear that the underlying binary network reduces to an Erdős-Rényi (ER) random graph with homogeneous connection probability

$$p_{ij}(z) = \frac{z}{1+z} \equiv p(z) \qquad \forall i, j. \tag{20}$$

Equation (5) indicates that, when realized, each link gets the same conditional weight $w_{ij} = N^{-1}p^{-1}(z) = (1+z)/(Nz)$, so that the expected unconditional weight in Eq. (6) is $\langle w_{ij} \rangle = N^{-1}$ for each pair of nodes i, j.

Now, since in this particular case $p(z)$ coincides with the expected link density $d(z)$, Eq. (4) implies that, to have a sparse network, we need $z = z_{\text{sparse}}$ with

$$z_{\text{sparse}} \simeq \frac{k}{N}, \quad k > 1 \tag{21}$$

so that $d(z_{\text{sparse}}) = \frac{z_{\text{sparse}}}{1+z_{\text{sparse}}} \simeq z_{\text{sparse}} = k/N$ as required. Equation (21) sets the specific way in which the sparsity condition in Eq. (10) is realized in this completely homogeneous case. Note that the corresponding relative fluctuations of the strength are asymptotically constant:

$$\delta_i(z_{\text{sparse}}) = \frac{1}{\sqrt{N z_{\text{sparse}} s_i^*}} \simeq \frac{1}{\sqrt{k s_i^*}} = \frac{1}{\sqrt{k}} \quad \forall i \tag{22}$$

and hence do not vanish in the sparse case. The critical value z_c for the reconstructability found using Eq. (17) is

$$z_c = \frac{1}{c} N^{\beta-1}, \quad 0 < \beta < 1, \tag{23}$$

where we have enforced the additional requirement $\beta < 1$ to realize the necessary condition given in Eq. (19). Indeed, the 'canonical' case $\beta = 1$ (relative fluctuations vanishing like $1/\sqrt{N}$) would yield a finite threshold $z_c = c^{-1} > 1$ and a finite critical density $d(z_c) = p_c = (1+c)^{-1} > 1/2$. In this case, in the reconstructability phase $z > z_c$ the network would necessarily be dense. Similarly, the case $\beta > 1$ would yield $z_c = +\infty$ and $p_c = 1$, so asymptotically the network would be a complete graph.

Therefore, in the only possible case $0 < \beta < 1$ for the sparse regime, when N is large we have $z_c \ll 1$ and hence $p_c \equiv p(z_c) = z_c/(1+z_c) \simeq z_c = c^{-1}N^{\beta-1}$. Importantly, p_c vanishes more slowly than the critical threshold $p_{\text{gcc}} \simeq 1/N$ associated with the onset of the giant connected component (gcc) in the ER graph and also more slowly than the critical threshold $p_{\text{ucc}} \simeq \log N/N$ associated with the onset of an overall connectivity in the ER graph (unique connected component, ucc). This clarifies that the reconstructability transition considered here is different from both transitions in the underlying ER model. In particular, to ensure reconstructability z need be larger than the values required for the entire network to have a giant and even a unique connected component.

The above results suggest that, at least when all strengths are (nearly) equal to each other, one should necessarily set $0 < \beta < 1$ (hence deviating from the canonical choice for the scaling of the relative fluctuations) if the network has to simultaneously be reconstructable and have a vanishing density $d(z) \to 0$. In any case, $\beta > 0$ implies that z_c in Eq. (23) is asymptotically always much larger than the value z_{sparse} in Eq. (21), which is required to have a realistically sparse graph with density scaling as in Eq. (4). Therefore the condition in Eq. (18) is always violated, except possibly for small values of N. This means that real-world networks, if they were simultaneously large, sparse and homogeneous, could not be reconstructed with the approach considered here. Indeed we see that only for dense and homogeneous networks, i.e. finite $d(z)$, it is possible to achieve the reconstructability of the network, with $\beta = 1$. Finally, $\beta > 1$ is uninteresting as it leads to a fully connected network.

3.2   Core-Periphery Structure

We now consider a less trivial setting where we introduce some heterogeneity in the network in the form of a core-periphery structure, where nodes in the core have a larger strength and nodes in the periphery have a smaller strength. The presence of a core-periphery structure has been documented in real financial networks [17–20]. For simplicity we assume that all nodes in the core are still homogeneous, i.e. they all have the same value s_c^* of the expected strength, and the same goes for all nodes in the periphery, i.e. they all have the same expected strength s_p^*, with clearly $s_p^* < s_c^*$. We place a fraction q of the N nodes in the core, i.e. the core has $N_c = qN$ nodes and the periphery has $N_p = (1 - q)N$ nodes. Then the distribution of empirical strengths is

$$f(s^*) = (1 - q)\,\delta(s^* - s_p^*) + q\,\delta(s^* - s_c^*) \tag{24}$$

and, setting $\overline{s^*} = 1$, we get

$$\overline{s^*} = (1 - q)\,s_p^* + q\,s_c^* \equiv 1 \tag{25}$$

where $s_p^* < 1 < s_c^*$. Inverting,

$$q = \frac{1 - s_p^*}{s_c^* - s_p^*}. \tag{26}$$

Since $\min_i\{s_i^*\} = s_p^*$, Eq. (17) indicates that the critical density is achieved by the condition

$$z_c = \frac{1}{c\,s_p^*}\,N^{\beta-1}, \quad \beta, c > 0 \tag{27}$$

whose asymptotic behaviour essentially depends on how s_p^* is chosen to scale as N grows.

As in the previous example, we need to compare the above z_c with the value z_{sparse} given by Eq. (4). To this end, we first compute the expected density defined in Eq. (3) as

$$d(z) = d_{cc}(z) + d_{pp}(z) + d_{cp}(z), \tag{28}$$

where $d_{cc}(z)$, $d_{pp}(z)$ and $d_{cp}(z)$ represent the core-core, periphery-periphery and core-periphery contributions given by

$$d_{cc}(z) = \frac{N_c(N_c - 1)}{N(N - 1)}\,\frac{z\,(s_c^*)^2}{1 + z\,(s_c^*)^2}, \tag{29}$$

$$d_{pp}(z) = \frac{N_p(N_p - 1)}{N(N - 1)} \frac{z\,(s_p^*)^2}{1 + z\,(s_p^*)^2},$$ (30)

$$d_{cp}(z) = \frac{2N_c\,N_p}{N(N - 1)} \frac{z\,s_c^*\,s_p^*}{1 + z\,s_c^*\,s_p^*},$$ (31)

respectively. Next, in order to look for z_{sparse} given by Eq. (4), we need to specify how N_c, N_p, s_c^* and s_p^* scale with N.

If both N_c and N_p grow linearly in N and if both s_c^* and s_p^* remain finite (with $s_p^* < 1 < s_c^*$), then it is easy to see that the three terms in Eqs. (29), (30) and (31) are all of the same order and the implied z_{sparse} is inversely proportional to N, as in the previous example. Again, this means $z_{\text{sparse}} < z_c$ and the network is unreconstructable in the sparse regime.

To better exploit the potential of the core-periphery model, we therefore consider a highly concentrated (or 'condensed') case where, as N increases, only a finite number N_c of nodes remain in the core, while the size $N_p = N - N_c$ of the periphery grows extensively in N. We also assume that the strength of peripheral nodes decreases with N as

$$s_p^* \simeq \alpha\,N^{-\eta}, \quad \alpha, \eta > 0.$$ (32)

Since $q = N_c/N$, Eq. (26) implies that asymptotically

$$s_c^* \simeq q^{-1} = N/N_c.$$ (33)

Inserted into Eqs. (29), (30) and (31), the above specifications imply that $d_{cc}(z)$ is subleading with respect to $d_{pp}(z)$ and $d_{cp}(z)$. Therefore we can look for z_{sparse} by setting $d_{pp}(z_{\text{sparse}}) + d_{cp}(z_{\text{sparse}}) \simeq k/N$. Taken separately, the requirement $d_{pp}(z_{\text{sparse}}) \simeq k_1 N^{-1}$ for some $k_1 > 0$ implies

$$z_{\text{sparse}} \simeq \frac{k_1}{N(s_p^*)^2} \simeq \frac{k_1}{\alpha^2} N^{2\eta - 1}, \quad 0 < \eta < 1/2,$$ (34)

where we have enforced the extra condition $\eta < 1/2$ to ensure that the exponent $2\eta - 1$ of N is negative, in order to meet the necessary condition given by Eq. (10). When inserted into $d_{cp}(z)$, Eq. (34) implies $d_{cp}(z_{\text{sparse}}) \simeq k_2/N$ with $k_2 = 2N_c$, showing that $d_{pp}(z_{\text{sparse}})$ and $d_{cp}(z_{\text{sparse}})$ are of the same order and both contribute to the desired scaling $d(z_{\text{sparse}}) \simeq k/N$ with $k = k_1 + k_2$.

The above scaling of z_{sparse} has to be compared with the reconstructability threshold z_c in Eq. (27) which, for the chosen behaviour of s_p^*, reads

$$z_c = \frac{1}{c\,\alpha}\,N^{\beta - 1 + \eta}, \quad \alpha, \beta, c > 0, \quad 0 < \eta < 1/2.$$ (35)

Comparing Eqs. (34) and (35), we see that the reconstructability condition $z_{\text{sparse}} > z_c$ is met asymptotically (for large N) if

$$0 < \beta < \eta,$$ (36)

while it is asymptotically violated if $0 < \eta < \beta$. Indeed the largest relative fluctuations in the sparse regime, following Eq. (16), are given by

$$\max_i\{\delta_i(z_{\text{sparse}})\} = \frac{1}{\sqrt{N}\, z_{\text{sparse}}\, s_p^*} \simeq \sqrt{\frac{s_p^*}{k_1}} \simeq \sqrt{\frac{\alpha}{k_1}}\, N^{-\eta/2} \qquad (37)$$

and they remain within the critical value given by Eq. (12) (with the identification $c = \alpha/k_1$) if the condition in Eq. (36) is met. So, for any chosen $\beta > 0$, there is a critical exponent $\eta_\beta \equiv \beta$ such that for $\eta > \eta_\beta$ the network is reconstructable while for $\eta < \eta_\beta$ the network is unreconstructable. Since $0 < \eta < 1/2$, we see that reconstructability is possible only if $0 < \beta < 1/2$. So, also in this case, the canonical decay $\beta = 1$ of the relative fluctuations cannot be achieved in the sparse regime.

The core-periphery model considered here is a simple, yet important example showing that a sufficiently heterogeneous network structure, by acting on the statistics of the minimum strength, can imply the presence of two phases (reconstructability and unreconstructability) in the sparse regime. Contrary to what naive intuition may suggest by looking at Eq. (17), i.e. that a decreasing (with N) value of $\min_i\{s_i^*\}$ could make the reconstructability only more difficult by increasing the value of z_c, actually the net effect might be the opposite if at the same time z_{sparse} increases as well, sufficiently fast. This circumstance, which is impossible in the completely homogeneous case discussed in Sect. 3.1, is precisely what happens in Eq. (34) due to the presence of $(s_p^*)^2$, rather than s_p^* itself, in the denominator.

3.3 Arbitrary Strength Distribution

We now generalize the calculation of the critical reconstructability threshold z_c to the case of an arbitrary empirical strength distribution. To calculate $\min_{i=1,N}\{s_i^*\}$ as a function of N, we assume that the N values $\{s_i^*\}_{i=1}^N$ of the empirical strength are realized via sampling N times (in an i.i.d. manner) from some probability density function $f(s)$. Then, a simple argument from Extreme Value Theory indicates that the typical value of the realized minimum strength $s_{\min}^*(N) = \min_{i=1,N}\{s_i^*\}$ is such that the expected number of nodes with strength $s^* \leq s_{\min}^*$ is of order one, or equivalently

$$\frac{1}{N} \simeq \int_0^{s_{\min}^*(N)} f(s)\, ds. \qquad (38)$$

For a given choice of $f(s)$, inverting the above equation produces the sought-for scaling of $s_{\min}^*(N)$ with N.

We are mainly interested in the behaviour of $f(s)$ for low values $s \to 0^+$ of the strength, because that is the behaviour that determines the statistics of the minimum strength and the behaviour of the integral in Eq. (38). In particular we consider the following behaviour

$$f(s) \to_{s\to 0^+} a\, s^{\psi-1}, \quad a, \psi > 0. \qquad (39)$$

Inserting this into Eq. (38) we get

$$\frac{1}{N} \simeq a \int_0^{s_{\min}(N)} s^{\psi-1}\, ds = \frac{a}{\psi}\, s_{\min}^{\psi}(N) \tag{40}$$

and, inverting,

$$s_{\min}(N) \simeq \left(\frac{\psi}{a}\right)^{1/\psi} N^{-1/\psi} \tag{41}$$

This implies

$$z_c = \frac{(a/\psi)^{1/\psi}}{c}\, N^{\beta-1+1/\psi} \tag{42}$$

where, to realize the condition in Eq. (19), we need $\beta-1+1/\psi < 0$ or equivalently

$$\psi > (1-\beta)^{-1}. \tag{43}$$

A comparison with Eq. (35) highlights that $1/\psi$ has an effect on z_c similar to the one that η has in the core-periphery model. Indeed, if in general we have a scaling of the form

$$s_{\min}(N) \simeq m\, N^{-\mu} \qquad m > 0,\ \mu \geq 0, \tag{44}$$

then Eq. (17) implies

$$z_c \simeq \frac{1}{cm}\, N^{\beta-1+\mu} \tag{45}$$

where, to realize the condition in Eq. (19), we need

$$0 < \mu < 1 - \beta. \tag{46}$$

In the examples considered so far, $\mu \equiv 0$ (with $m \equiv 1$) for the homogeneous case discussed in Sect. 3.1 (confirming that the above inequality cannot be realized), while $\mu \equiv \eta$ with $0 < \eta < 1/2$ for the core-periphery model discussed in Sect. 3.2 (for which we need $0 < \beta < 1/2$), and finally $\mu \equiv 1/\psi$ for the general case discussed in this Section, for which Eq. (43) has to hold. In order to check for reconstructability, one should of course calculate also z_{sparse}, which requires the full knowledge of $f(s)$ and has to be evaluated for the specific case at hand.

4   The Role of Induced Self-loops

As anticipated, we now show that, even though the probabilistic reconstruction model considered here generates self-loops, the role of the latter in the reconstructability problem is negligible for the typical regimes of relevance for real-world networks. We recall that, in our units such that $\overline{s^*} = 1$, the total weight of all links in the empirical network is $W^* = \sum_{i=1}^{N} s_i^*/2 = N/2$ and its expected value in the model replicates the empirical value perfectly, i.e. $\langle W \rangle = W^*$. However the model, differently from the real network, produces self-loops, each with an expected weight $\langle w_{ii} \rangle$ given by Eq. (8). The resulting expected total weight of self-loops can be calculated as

$$\langle W_{\mathrm{SL}} \rangle = \sum_{i=1}^{N} \langle w_{ii} \rangle = \sum_{i=1}^{N} \frac{(s_i^*)^2}{2W^*} = \frac{N\overline{s^{*2}}}{2W^*} = \overline{s^{*2}} \qquad (47)$$

where $\overline{s^{*2}} = \sum_{i=1}^{N}(s_i^*)^2/N$ is the empirical second moment of the node strengths. If we require that the weight of all self-loops is negligible with respect to the expected weight $\langle W \rangle$ of all links (including self-loops), we should impose

$$\frac{\langle W_{\mathrm{SL}} \rangle}{\langle W \rangle} = \frac{2\overline{s^{*2}}}{N} = o(1) \qquad \Longrightarrow \qquad \overline{s^{*2}} = o(N) \qquad (48)$$

Note that Eq. (48) essentially sets a bound on the second moment, hence on the right tail, of the strength distribution $f(s)$. This should be combined with the previous bounds on the left tail.

We now show that the above condition is satisfied in the regimes that are typical for real-world networks. In particular, if we assume realistic strength distributions that are observed in empirical financial networks, we should mainly consider log-normal distributions and distributions with a power-law tail decay $f(s) \simeq b\,s^{-\gamma}$ (with $b > 0$) for large s. If the second moment of $f(s)$ is finite (as for the log-normal distribution and the power-law distribution with $\gamma > 3$), we can automatically conclude that Eq. (48) is verified. If the second moment of $f(s)$ diverges (as for the power-law distribution with $\gamma \leq 3$), we should more carefully look at how fast this occurs as N grows and check whether Eq. (48) is still satisfied. We can do so using an Extreme Value Theory argument analogous to the one in Eq. (38): we can first estimate how the realized maximum strength $s_{\max}^*(N)$ (out of an i.i.d. sample of N values) scales as a function of N, and then use $s_{\max}^*(N)$ to establish how the realized empirical second moment $\overline{s^{*2}}$ scales with N. Considering the non-trivial case $f(s) \simeq b\,s^{-\gamma}$ for large s, we estimate $s_{\max}^*(N)$ in analogy with Eq. (38) as

$$\frac{1}{N} \simeq \int_{s_{\max}^*(N)}^{+\infty} f(s)\,ds \simeq b \int_{s_{\max}^*(N)}^{+\infty} s^{-\gamma}\,ds, \qquad (49)$$

which leads to (note that $\gamma - 1 > 0$)

$$s_{\max}^*(N) \simeq \left(\frac{b}{\gamma - 1} \right)^{\frac{1}{\gamma-1}} N^{\frac{1}{\gamma-1}}. \qquad (50)$$

We can then estimate the scaling of $\overline{s^{*2}}(N)$ as follows:

$$\overline{s^{*2}}(N) = \int_0^{s^*_{\max}(N)} f(s)\, s^2 \, ds$$

$$\simeq b \int_0^{s^*_{\max}(N)} s^{2-\gamma} \, ds$$

$$\simeq \frac{b}{\gamma - 3} \left[s^*_{\max}(N) \right]^{3-\gamma}$$

$$\simeq \frac{b}{\gamma - 3} \left(\frac{b}{\gamma - 1} \right)^{\frac{3-\gamma}{\gamma - 1}} N^{\frac{3-\gamma}{\gamma - 1}}. \tag{51}$$

To ensure that self-loops contribute only negligibly as stated in Eq. (48), we need to require $(3 - \gamma)/(\gamma - 1) < 1$, which simply boils down to $\gamma > 2$. Importantly, this means that empirical strength distributions that have an asymptotically diverging second moment are still viable, provided they have an asymptotically finite first moment. This is indeed what is observed in most real-world financial networks, where the strength distribution typically has either a log-normal form or a power-law tail with an exponent in the range $2 < \gamma < 3$. So all our calculations in the previous sections are justified in the regimes of relevance for real-world networks.

The same results can be derived more rigorously by analysing the large N convergence of the probability distribution of the random variable W_{SL} (whose possible realized values will be denoted as w_{SL}). Let us assume that the probability distribution $f(s)$ of node strengths does not depend on N, in agreement with the assumption of sparse networks with $2W^* = N$. Let us consider the PDF $g(u)$ of the variable $u = s^2$:

$$g(u) = \frac{f(\sqrt{u})}{2\sqrt{u}}.$$

The PDF $p(w_{\mathrm{SL}})$ of W_{SL} is related to $g(u)$ by

$$p(w_{\mathrm{SL}}) = \int_0^\infty \cdots \int_0^\infty \left[\prod_{i=1}^N du_i \, g(u_i) \right] \delta \left(w_{\mathrm{SL}} - \sum_{i=1}^N \frac{u_i}{N} \right)$$

Consequently, its characteristic function is

$$\hat{p}(t) = \int_0^\infty dw_{\mathrm{SL}} \, p(w_{\mathrm{SL}}) e^{-t w_{\mathrm{SL}}} = \left[\int_0^\infty du \, g(u) e^{-tu/N} \right]^N. \tag{52}$$

We have to distinguish two fundamental cases: (i) u has a finite mean value (i.e. s has a finite variance); (ii) u has diverging mean value (i.e. s has an infinite variance).

$-\ \overline{u} = \overline{s^2} < \infty$: in this case at small enough t we have

$$\int_0^\infty du \, g(u) e^{-tu/N} = 1 - \frac{\overline{u}}{N} t + o(t/N). \tag{53}$$

Using this expression in Eq. (52) and taking the large N limit we get

$$\hat{p}(t) = e^{-\bar{u}t}. \tag{54}$$

which is the Laplace transform of $p(w_{\text{SL}}) = \delta(w_{\text{SL}} - \bar{u})$. Using Eq. (47) we can therefore conclude that in this case the relative weight of self-loops is $\langle W_{\text{SL}} \rangle / \langle W \rangle = O(1/N)$, in accordance with Eq. (48). This confirms that log-normal distributions or power-law distributions with finite variance ($\gamma > 3$) are always acceptable.

- $\bar{u} = \overline{s^2} \to \infty$: in this case we consider again the case $f(s) \simeq b\, s^{-\gamma}$ at large s with $1 < \gamma \leq 3$ (note that $\gamma > 1$ is required in order to guarantee the integrability of the PDF). Consequently $g(u) \simeq (b/2)\, u^{-(\gamma+1)/2}$ at large u (note that $\gamma > 1$ implies $(\gamma+1)/2 > 1$ too). By the properties of the Laplace transform we now have that, at small t,

$$\int_0^\infty du\, g(u) e^{-tu/N} = 1 - A \frac{|t|^{(\gamma-1)/2}}{N^{(\gamma-1)/2}} + o[(|t|/N)^{(\gamma-1)/2}] \tag{55}$$

for some finite A. If we now use this expression in Eq. (52) at large but finite N we have

$$\hat{p}(t) \simeq e^{-AN^{(3-\gamma)/2}|t|^{(\gamma-1)/2}} \tag{56}$$

which implies the typical value $W_{\text{SL}} \sim N^{(3-\gamma)/(\gamma-1)}$. Recalling Eq. (47), this result coincides with Eq. (51). Indeed Eq. (56) implies $p(w_{\text{SL}}) \sim w_{\text{SL}}^{-(\gamma+1)/2}$ for $w_{\text{SL}} \gg N^{(3-\gamma)/(\gamma-1)}$ and, in order to have a total measure of order 1 for the variable W_{SL}, we have to go up to values of order $N^{(3-\gamma)/(\gamma-1)}$. From Eq. (48) we see that the weight of self-loops can be neglected if $(3 - \gamma)/(\gamma - 1) < 1$ i.e. $2 < \gamma < 3$. This coincides with the 'realistic' regime identified above.

We can summarize the discussion above by saying that self-loops can be neglected with respect to the rest of the connections if the empirical strength distribution has a finite variance (as in log-normal distributions or power-law distributions with $\gamma > 3$) or if it has a power-law tail with $2 < \gamma < 3$ (such that the mean is finite but the variance diverges). Realistic situations typically fall into one of those two cases. The next Section provides an empirical example that confirms this picture.

5   Reconstructed Networks from Bankscope Data

5.1   Dataset

The Bureau Van Dijk *Bankscope* database contains information on banks' balance sheets and aggregate exposures. Our dataset consists of a subset of $N = 119$ anonymized European banks that were publicly traded between 2006 and 2013 [21]. For each of the N banks and each year t in the data, we have

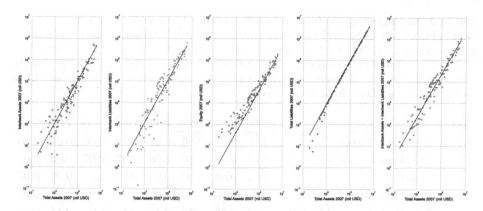

Fig. 1. Scatter plot and simple linear regression of the type $y_{i,k}(t) = a_k(t) \cdot x_i(t)$ of the Interbank Assets ($k = 1$), Interbank Liabilities ($k = 2$), Equity ($k = 3$), Total Liabilities ($k = 4$), and Interbank Assets + Liabilities ($k = 5$) of each bank i ($i = 1, 119$) in year $t = 2007$, versus the total assets $x_i(t)$ of the same bank in the same year. Data from Bankscope, reported in mil USD.

access to the yearly values of total interbank assets, total interbank liabilities, total assets, total liabilities, and equity.

To characterize the data, we start by noticing that all these bank-specific variables are strongly linearly correlated, as the plots in Fig. 1 illustrate for year 2007. This observation holds for all the years in the data. Indeed, in Table 1 we report, for each year t in the data, the fitted proportionality coefficient $a_k(t)$ and the corresponding coefficient of determination $R_k^2(t)$ of a simple linear regression of the type $y_{i,k}(t) = a_k(t) \cdot x_i(t)$, where the independent variable $x_i(t)$ is always the total assets of bank i in year t and the dependent variable $y_{i,k}(t)$ is the Interbank Assets ($k = 1$), Interbank Liabilities ($k = 2$), Equity ($k = 3$), Total Liabilities ($k = 4$), and Interbank Assets+Liabilities ($k = 5$) of the same bank in the same year.

The approximate linearity of all the quantities allows us to proceed with an undirected description of the data, in line with our discussion so far. We define the strength of node i as

$$s_i^* = A_i + L_i \tag{57}$$

where A_i and L_i are the total interbank assets and interbank liabilities of bank i, respectively. From now on, we limit ourselves to year 2007 and rescale the yearly strengths to the average strength so that $\overline{s^*} = 1$, in line with our choice of units so far.

The set of empirical strengths $\{s_i^*\}_{i=1}^{119}$ represents the starting point of our analysis and its empirical cumulative distribution function (CDF) is shown in Fig. 2a. Since many of our results refer to the behavior of z_c and z_{sparse} as N grows, we also construct synthetic replicas of the dataset for increasing values of N. To achieve this, we first fit a log-normal distribution

Table 1. Coefficient $a_k(t)$ (top) and coefficient of determination $R_k^2(t)$ (bottom) of a simple linear regression of the type $y_{i,k}(t) = a_k(t) \cdot x_i(t)$ (plus zero-mean i.i.d. additive noise), where $x_i(t)$ is the total assets of bank i in year t (for $t = 2006, \dots, 2013$) and $y_{i,k}(t)$ is the Interbank Assets ($k = 1$), Interbank Liabilities ($k = 2$), Equity ($k = 3$), Total Liabilities ($k = 4$), and Interbank Assets + Liabilities ($k = 5$) of the same bank in the same year. Data from Bankscope.

Year t		2006	2007	2008	2009	2010	2011	2012	2013
$a_k(t)$	Interbank Assets	0.11	0.11	0.06	0.06	0.06	0.06	0.05	0.06
	Interbank Liabilities	0.12	0.12	0.10	0.08	0.07	0.07	0.06	0.05
	Equity	0.04	0.04	0.03	0.05	0.05	0.05	0.05	0.05
	Total Liabilities	0.96	0.96	0.97	0.95	0.95	0.95	0.95	0.95
	Interbank Assets+Liabilities	0.23	0.23	0.16	0.15	0.13	0.13	0.11	0.11
$R_k^2(t)$	Interbank Assets	0.69	0.74	0.46	0.53	0.47	0.46	0.41	0.44
	Interbank Liabilities	0.87	0.88	0.86	0.90	0.86	0.80	0.71	0.65
	Equity	0.87	0.89	0.88	0.92	0.93	0.93	0.92	0.93
	Total Liabilities	0.99	0.99	0.99	0.99	0.99	0.99	0.99	0.99
	Interbank Assets+Liabilities	0.85	0.89	0.78	0.83	0.73	0.69	0.62	0.59

$$f(s) = \frac{1}{s\,\sigma\sqrt{2\pi}} \exp\left(-\frac{(\ln(s) - \mu)^2}{2\sigma^2}\right), \quad s > 0 \tag{58}$$

to the empirical distribution of the $N = 119$ node strengths and then use it to sample any desired number N of i.i.d. strength values $\{s_i^*\}_{i=1}^N$ from it. Note that the theoretical mean value of the log-normal distribution is

$$\bar{s} = \int_0^{+\infty} f(s)\,s\,ds = \exp\left(\mu + \frac{\sigma^2}{2}\right), \tag{59}$$

therefore, to ensure an expected unit mean value $\bar{s}^* = 1$ for the sampled strengths, we set $\mu \equiv -\sigma^2/2$, leaving out only the free parameter σ. When fitted to the data, the latter gets the value $\sigma = 2.28$. The resulting theoretical CDF of the strengths is

$$F(s) = \int_0^s f(x)\,dx = \frac{1}{2}\left[1 + \mathrm{erf}\left(\frac{\ln s + \sigma^2/2}{\sigma\sqrt{2}}\right)\right], \quad s > 0. \tag{60}$$

Figure 2a shows the good agreement between the empirical CDF of the 119 empirical Bankscope strengths and the CDFs of the corresponding N synthetic values (for increasing N) sampled from the fitted log-normal distribution with CDF given by Eq. (60). Note that the accordance with the log-normal distribution automatically ensures that the induced self-loops in the network reconstruction method do not represent any problem, as discussed in Sect. 4. We will

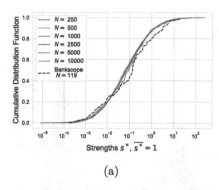
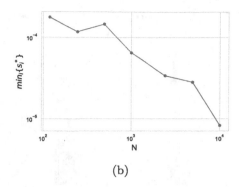

(a)   (b)

Fig. 2. Empirical and synthetic data, using the 2007 Bankscope dataset as a reference. (a) Cumulative distribution function of the rescaled node strengths for the Bankscope data ($N = 119$) and for the synthetic data (for increasing values of N) drawn from a log-normal distribution fitted to empirical data (with $\sigma = 2.28$ and $\mu = -\sigma^2/2$). (b) Minimum realized strength $s^*_{\min}(N)$ as a function of the number N of nodes. Again, the point at $N = 119$ is the Bankscope dataset.

confirm this result explicitly later on. We also report, in Fig. 2b, the value of the realized minimum strength $s^*_{\min}(N) = \min_{i=1,N}\{s^*_i\}$ as a function of N. As in our examples considered in Sects. 3.2 and 3.3, $s^*_{\min}(N)$ decreases with N, realizing the necessary condition given in Eq. (10) to have a sparse network. The behaviour of $s^*_{\min}(N)$ will also affect the critical value z_c given by Eq. (17) for the reconstructability of the network.

In our following analysis, we will also consider a completely homogeneous benchmark where, for a given set of N banks, each bank i is assigned exactly the same strength $s^*_i = 1$, as in our discussion in Sect. 3.1. This benchmark (in which clearly $s^*_{\min} = 1$ independently of N) will serve as a reference to emphasize the role of bank heterogeneity in the reconstructed networks.

5.2   Network Reconstruction

We now apply the method described in Sect. 2.1 to the sets of empirical and synthetic strengths to test our theoretical results and check for possible transitions from reconstructability to unreconstructability. We are interested in the sparse regime where the link density is given by $d(z_{\text{sparse}}) \simeq k/N$ as in Eq. (4), where k is the average node degree in the network. For a given strength distribution, we explore various numbers of nodes, $N = \{119, 250, 500, 1000, \ldots, 10000\}$, and two values of the average degree, $k = \{50, 100\}$. For each pair of N and k, we consider an ensemble of 1000 realizations of weighted undirected networks. We are interested in determining whether typical realizations produce the desired strengths with vanishing relative fluctuations, i.e., whether Eq. (12) is verified.

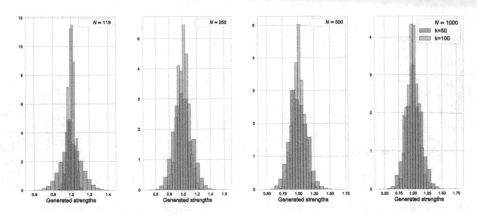

Fig. 3. Distribution of the reconstructed strengths in the case of identical nodes (homogeneous networks), across all the nodes and all the 1000 network realizations in the ensemble. The distribution is peaked around the expected value $\overline{s^*} = 1$. Different colors correspond to different choices of the constant k, while each subplot corresponds to a different number N of nodes.

Identical Strengths. We start from the homogeneous benchmark where all banks are identical and all strengths are therefore equal. As discussed in Sect. 3.1, the underlying binary network reduces to an Erdős-Rényi random graph with homogeneous connection probability.

In Fig. 3 we report, for different values of N and k, the distribution of the realized strengths in the networks sampled from the ensemble, across all the N nodes and all the 1000 network realizations. The results confirm a distribution peaked around the expected value $\overline{s^*} = 1$. However, in order to investigate whether this distribution is 'narrow enough' in order to achieve the desired reconstructability of the network, we have to look more closely at the relative fluctuations $\delta_i(z_{\text{sparse}})$ for each node i.

This is what we illustrate in Fig. 4a, where for one randomly chosen node (note that all nodes are statistically equivalent in the homogeneous case) we compare the theoretical value $\delta_i^{\text{theo}}(z_{\text{sparse}})$ calculated in Eq. (15) with the sample fluctuations $\delta_i^{\text{sample}}(z_{\text{sparse}})$ measured using the sample variance across the 1000 realizations of the network, for different values of N and k. We confirm that sample and theoretical values are in excellent agreement. In Fig. 4b we also confirm that, after a transient trend that increases with N, the relative fluctuations approach the asymptotically constant value $\delta_i(z_{\text{sparse}}) = 1/\sqrt{k}$ for all i, as expected from Eq. (22), and hence they do not vanish. Indeed, as discussed in Sect. 3.1, the relative fluctuations for homogeneous networks decay only in the dense regime.

In Fig. 5 we compare the behaviour of $z_{\text{sparse}}(N)$, obtained by inverting Eq. (4) (for $k = 50$ and $k = 100$), with the critical $z_c(N)$ given by Eq. (23), with the identification $c = 1/(k \min_i\{s_i^*\}) = 1/k$ that would correspond to the

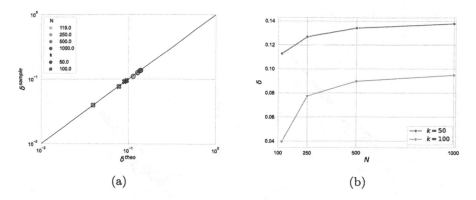

(a)   (b)

Fig. 4. Relative strength fluctuations $\delta_i(z_{\text{sparse}})$ in the case of identical nodes (homogeneous networks), for a randomly chosen node i. (a) The theoretical value $\delta_i^{\text{theo}}(z_{\text{sparse}})$ given by Eq. (15) matches the realized value $\delta_i^{\text{sample}}(z_{\text{sparse}})$ calculated across 1000 sampled networks, for different values of N and k. (b) Realized $\delta_i^{\text{sample}}(z_{\text{sparse}})$ versus the number N of nodes, which asymptotically approaches the theoretical value $1/\sqrt{k}$.

baseline asymptotic behaviour for $\delta(z_{\text{sparse}})$ when $\beta \to 0^+$. The figure confirms that, asymptotically, the network is unreconstructable ($z_c > z_{\text{sparse}}$) for all the values $0 < \beta < 1$ allowed by Eq. (23). Only for small $\beta \to 0^+$ and moderate values of N the network can be found transiently in the reconstructability phase.

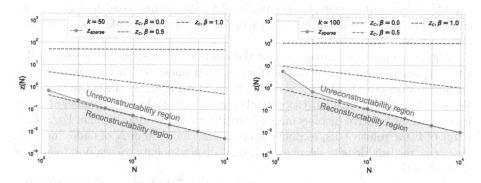

Fig. 5. Comparison between $z_{\text{sparse}}(N)$ and $z_c(N) \simeq c^{-1}N^{\beta-1}$ with $c = 1/k$, for different values of β and N in the case of homogeneous networks. Each subplot corresponds to a different choice of the constant k. The shaded area is the region $z(N) < z_{\text{sparse}}(N)$: the network is reconstructable whenever $z_c(N)$ crosses that region. This occurs only for $\beta \to 0^+$ and small N, while for large N the network is always in the unreconstructability regime.

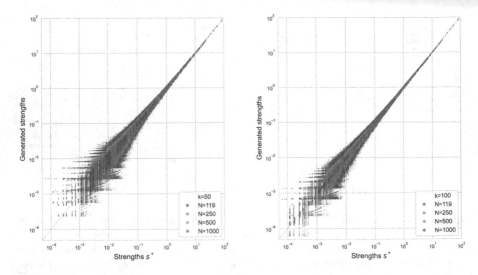

Fig. 6. Empirical strengths s^* versus the generated ones. Each dot figures out the value of the generated strength of a certain node and a certain network between the 1000 realizations in the ensemble versus its empirical strength. Each subplot corresponds to a different choice of the constant k and different colors correspond to a different number of nodes N.

Bankscope Strengths. We then consider the heterogeneous case where the strengths of banks are either taken from the Bankscope data or sampled from the fitted distribution with CDF given by Eq. (60), as discussed in Sect. 5.1. We enforce the same values of density as in the previous example (sparse regime) and vary N and k as before.

Figure 6 reports a scatter plot of the realized node strength of each node i (in a typical realization out of the 1000 sampled ones) versus the corresponding empirical strength s_i^*, which is also the expected value across the entire ensemble. We see that, in line with our theoretical calculations, nodes with smaller strength feature larger relative displacements from the expected identity line. For a fixed value of k, the relative displacement increases as the number N of nodes increases (hence as the link density decreases). Larger displacements correspond to a worse reconstruction.

To quantify the above effect rigorously, in Fig. 7 we focus on the relative fluctuations $\delta_i(z_{\text{sparse}})$ and, in analogy with Fig. 3a, compare the sample fluctuations $\delta_i^{\text{sample}}(z_{\text{sparse}})$ (measured using the sample variation across 1000 realizations of the network) with the theoretical value $\delta_i^{\text{theo}}(z_{\text{sparse}})$ calculated in Eq. (15), for different values of N and k. Since in this case the banks' strengths vary greatly in size, we report in different subplots the fluctuations for different representative nodes, each corresponding to a certain quantile of the strength distribution. Again, we confirm the good accordance between the sampled fluctuations and our theoretical calculations, for all quantiles.

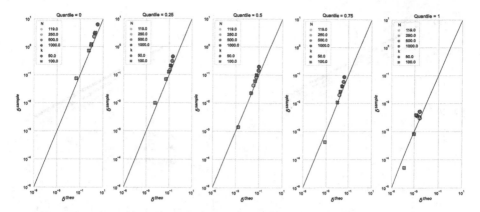

Fig. 7. Relative strength fluctuations $\delta_i(z_\text{sparse})$ in the reconstructed networks based on the Bankscope data (heterogeneous networks). The theoretical value $\delta_i^\text{theo}(z_\text{sparse})$ given by Eq. (15) is compared with the realized value $\delta_i^\text{sample}(z_\text{sparse})$ calculated across 1000 sampled networks, for different values of N and k. Each subplot shows the fluctuations for a representative node that corresponds to the indicated quantile of the strength distribution.

As a related check, Fig. 8 shows the sample relative fluctuation $\delta_i^\text{sample}(z_\text{sparse})$ versus the empirical strength s_i^*. For fixed k and N, $\delta_i^\text{sample}(z_\text{sparse})$ is proportional to $1/\sqrt{s_i^*}$, in accordance with the expectation from Eq. (15). Moreover we see that, for fixed s_i^*, the dependence of $\delta_i^\text{sample}(z_\text{sparse})$ on N is not suppressed. This is different from the behaviour we found in Eq. (21) for the homogeneous case, where the dependence of $\delta_i^\text{sample}(z_\text{sparse})$ on N is exactly cancelled out by the expression $z_\text{sparse} \simeq k/N$ given by Eq. (22), which in this case evidently does not hold (at least for the considered range of values for N). In particular, here we see that $\delta_i^\text{sample}(z_\text{sparse})$ increases with N, a behaviour that is confirmed in Fig. 9.

In Fig. 10 we compare $z_\text{sparse}(N)$ and $z_c(N) \sim c^{-1}N^{\beta-1}$ with $c^{-1} = k\,\min_i\{s_i^*\}$, which again would correspond to the baseline decay of the relative fluctuations, for different values of β and N. We find the presence of two regimes (reconstructability and unreconstructability), depending on the combination of values of β and N. So, in this case the reconstructability depends sensibly on the details of the strength distribution and on the size of the network. Even if certain values of β lead asymptotically to unreconstructability, we find that for finite but realistically large N the system may still be in the reconstructability regime. In other words, depending on the chosen values for k and β, the networks are reconstructable up to a critical number of nodes whose value increases as β decreases (hence as the decay of the largest relative fluctuations of the strength slows down).

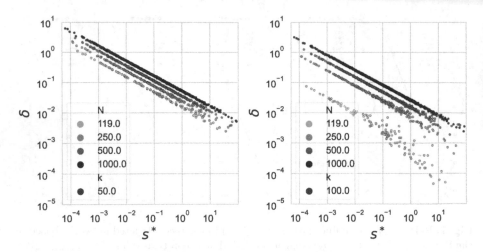

Fig. 8. Sample relative fluctuation $\delta_i^{\mathrm{sample}}(z_{\mathrm{sparse}})$ versus the empirical strength s_i^*. Each subplot corresponds to a different value of k, and different colors correspond to a different number N of nodes.

Self-loops Contribution. We finally confirm that, as expected, the effect of the induced self-loops in the reconstruction is negligible. This is shown in Table 2 where we report the fraction of weight associated to self-loops, confirming that the condition in Eq. (48) is met, as the fraction decreases for increasing N.

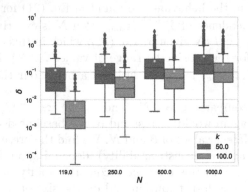

Fig. 9. Boxplot representing the distribution of the sample relative fluctuations $\delta_i^{\mathrm{sample}}(z_{\mathrm{sparse}})$ over nodes, as a function of the number N of nodes. Yellow dots represent the mean. Different colors correspond to different values of k. (Color figure online)

Table 2. Fraction $\langle W_{SL}\rangle/\langle W\rangle$ of total link weight associated to the self-loops induced by the reconstruction method, for different numbers of nodes N in the case of homogeneous (equal values) and heterogeneous (from Bankscope data) strength distributions.

N	Homogeneous networks	Heterogeneous networks
119	0.008	0.061
250	0.004	0.073
500	0.002	0.049
1000	0.001	0.026

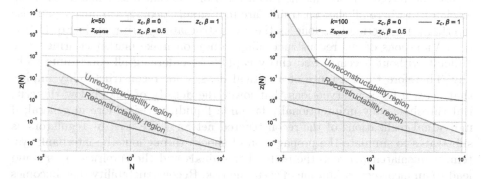

Fig. 10. Comparison between $z_{\mathrm{sparse}}(N)$ and $z_c(N) \simeq c^{-1}N^{\beta-1}$ with $c = 1/(k \min_i\{s_i^*\})$, for different values of β and N in the case of networks with strengths based on the Bankscope data. Each subplot corresponds to a different choice of the constant k. The shaded area is the region $z(N) < z_{\mathrm{sparse}}(N)$, and the network is reconstructable whenever $z_c(N)$ crosses that region. Whether this occurs depends strongly on β: for $\beta \to 0^+$ the network is reconstructable for all values of N, for $\beta = 1$ the network is always unreconstructable, while for intermediate values of β the reconstructability depends on N.

6   Conclusions

In this paper we focused on the reconstruction of financial networks from aggregate constraints (node strengths, representing assets and liabilities) and introduced the concept of *reconstructability*, occurring when the constraints, besides being reproduced on average, are also close to their expected value in individual typical realizations of the ensemble. We considered different situations arising in the sparse regime, first from a theoretical point of view and then by generating networks from real-world strength distributions. In the homogeneous case of equal strengths, we found that simultaneously sparse and large networks are always unreconstructable. By contrast, if an appropriate degree of heterogeneity is introduced (specifically, a core-periphery structure), we found that the system can be in one of two regimes (reconstructability and unreconstructability), depending on the asymptotic decay of the minimum strength. In general, the behaviour of the minimum strength plays a crucial role in the reconstruction. Using data from the Bankscope dataset, and extrapolating to a larger number of nodes, we found again the presence of two regimes, which additionally depend sensibly on the number of nodes and on the details of the strength distribution.

It should be noted that the independence of different links captured by the reconstruction method discussed in Sect. 2.1, together with the fact that both the connection probability p_{ij} in Eq. (2) and the link weight w_{ij} in Eq. (5) are determined by purely local properties (s_i^* and s_j^*), indicates that it is possible to use the reconstruction method as a form of a decentralized market clearing mechanism where all the constraints are met simultaneously via purely pairwise interbank interactions, as long as the reconstructability criterion in Eq. (18) is met. Violations of the reconstructability condition make market clearing more difficult without a centralized entity capable of enforcing all constraints globally. Therefore the existence of a critical density for reconstructability implies that central bank interventions that lower the density of links below the critical threshold may unintentionally favour a liquidity crisis. More in general, if individual realizations of the reconstructed networks are used by regulators as substrates to simulate the propagation of shocks throughout the interbank system, a mismatch between the realized marginals and the empirical ones could lead to an incorrect estimation of systemic risk. Reconstructability then becomes an important criterion to be met in order to avoid the resulting bias.

Our result suggest that network reconstructability is an important aspect of probabilistic reconstruction techniques and deserves further study, including its generalization to directed and more complicated network structures.

Acknowledgements. DG is supported by the Dutch Econophysics Foundation (Stichting Econophysics, Leiden). This work is supported by the European Union - NextGenerationEU - National Recovery and Resilience Plan (Piano Nazionale di Ripresa e Resilienza, PNRR), project 'SoBigData.it - Strengthening the Italian RI for Social Mining and Big Data Analytics' - Grant IR0000013 (n. 3264, 28/12/2021). We also acknowledge the PRO3 project "Network Analysis of Economic and Financial Resilience" by the IMT School of Advanced Studies Lucca, the Scuola Normale Superiore in Pisa and the Sant'Anna School of Advanced Studies in Pisa.

References

1. Bardoscia, M., et al.: The physics of financial networks. Nat. Rev. Phys. **3**(7), 490–507 (2021)
2. Squartini, T., Caldarelli, G., Cimini, G., Gabrielli, A., Garlaschelli, D.: Reconstruction methods for networks: the case of economic and financial systems. Phys. Rep. **757**, 1–47 (2018)
3. Squartini, T., Garlaschelli, D.: Maximum-Entropy Networks: Pattern Detection, Network Reconstruction and Graph Combinatorics. Springer, Cham (2017). https://doi.org/10.1007/978-3-319-69438-2

4. Cover, T.M., Thomas, J.A.: Elements of Information Theory. Wiley-Interscience, New York (2006)
5. Mastrandrea, R., Squartini, T., Fagiolo, G., Garlaschelli, D.: Enhanced reconstruction of weighted networks from strengths and degrees. New J. Phys. **16**(4), 043022 (2014)
6. Garlaschelli, D., Loffredo, M.I.: Maximum likelihood: extracting unbiased information from complex networks. Phys. Rev. E **78**(1), 015101 (2008)
7. Cimini, G., Squartini, T., Garlaschelli, D., Gabrielli, A.: Systemic risk analysis on reconstructed economic and financial networks. Sci. Rep. **5**(1), 1–12 (2015)
8. Squartini, T., Cimini, G., Gabrielli, A., Garlaschelli, D.: Network reconstruction via density sampling. Appl. Netw. Sci. **2**(1), 1–13 (2017)
9. Bardoscia, M., Battiston, S., Caccioli, F., Caldarelli, G.: DebtRank: a microscopic foundation for shock propagation. PLoS ONE **10**(6), e0130406 (2015)
10. Barucca, P., et al.: Network valuation in financial systems. Math. Financ. **30**(4), 1181–1204 (2020)
11. Caccioli, F., Barucca, P., Kobayashi, T.: Network models of financial systemic risk: a review. J. Comput. Soc. Sci. **1**(1), 81–114 (2018)
12. Eisenberg, L., Noe, T.H.: Systemic risk in financial systems. Manage. Sci. **47**(2), 236–249 (2001)
13. Macchiati, V., Brandi, G., Di Matteo, T., Paolotti, D., Caldarelli, G., Cimini, G.: Systemic liquidity contagion in the European interbank market. J. Econ. Interac. Coord. **17**(2), 443–474 (2022)
14. Ramadiah, A., et al.: Network sensitivity of systemic risk. J. Netw. Theory Financ. **5**(3), 53–72 (2020)
15. Cimini, G., Squartini, T., Gabrielli, A., Garlaschelli, D.: Estimating topological properties of weighted networks from limited information. Phys. Rev. E **92**(4), 040802 (2015)
16. Parisi, F., Squartini, T., Garlaschelli, D.: A faster horse on a safer trail: generalized inference for the efficient reconstruction of weighted networks. New J. Phys. **22**(5), 053053 (2020)
17. Alves, I., et al.: The structure and resilience of the European interbank market. ESRB: Occasional Paper Series, March 2013
18. Barucca, P., Lillo, F.: Disentangling bipartite and core-periphery structure in financial networks. Chaos Solitons Fractals **88**, 244–253 (2016)
19. Fricke, D., Lux, T.: Core-periphery structure in the overnight money market: evidence from the e-mid trading platform. Comput. Econ. **45**(3), 359–395 (2015)
20. Van Lelyveld, I., et al.: Finding the core: network structure in interbank markets. J. Bank. Financ. **49**, 27–40 (2014)
21. Battiston, S., Caldarelli, G., D'Errico, M., Gurciullo, S.: Leveraging the network: a stress-test framework based on DebtRank. Stat. Risk Model. **33**(3-4), 117–138 (2016)

Motif Finding Algorithms: A Performance Comparison

Emanuele Martorana[1], Roberto Grasso[2], Giovanni Micale[2],
Salvatore Alaimo[2], Dennis Shasha[3], Rosalba Giugno[4],
and Alfredo Pulvirenti[2]($\boxtimes$)

[1] Istituto Oncologico del Mediterraneo, Viagrande, CT, Italy
[2] Department of Clinical and Experimental Medicine, University of Catania,
Catania, CT, Italy
alfredo.pulvirenti@unict.it
[3] Courant Institute of Mathematical Sciences, New York University,
New York, NY, USA
[4] University of Verona, Verona, VR, Italy

Abstract. Network motifs are subgraphs of a network that occur more
frequently than expected, according to some reasonable null model. They
represent building blocks of complex systems such as genetic interaction
networks or social networks and may reveal intriguing typical but perhaps unexpected relationships between interacting entities. The identification of network motif is a time consuming task since it subsumes the
subgraph matching problem. Several algorithms have been proposed in
the literature. In this paper we aim to review the motif finding problem
through a systematic comparison of state-of-the-art algorithms on both
real and artificial networks of different sizes. We aim to provide readers
a complete overview of the performance of the various tools. As far as
we know, this is the most comprehensive experimental review of motif
finding algorithms to date, with respect both to the number of compared
tools and to the variety and size of networks used for the experiments.

Keywords: Network motifs · Network motifs search · Network motifs
significance · Network motifs tools comparison

1 Introduction

Complex systems are characterized by a set of entities (e.g. people, molecules or
objects) that can interact each other in many ways (e.g. physically, chemically).
Graphs (or networks) are a natural way to represent such complex systems.
Networks are characterized by the presence of recurrent structural patterns,
called motifs, that may have a functional role and can explain some network
properties emerging at a meso-scale level. Thus, network motifs can be seen as
the basic building blocks of complex networks that work together for an higher-
level process. Network motifs were originally defined as recurrent and statistically

© The Author(s), under exclusive license to Springer Nature Switzerland AG 2024
D. Cantone and A. Pulvirenti (Eds.): *From Computational Logic to Computational Biology*,
LNCS 14070, pp. 250–267, 2024.
https://doi.org/10.1007/978-3-031-55248-9_12

significant patterns of interaction, whose frequency in the network is higher than expected with respect to a null model [25]. The null model consists of an ensemble of random networks sharing some characteristics of the input network (e.g. the degree distribution).

Most network motif discovery algorithms i) compute the frequency F of the motif in the input network, ii) generate a set of random networks from the ensemble, iii) compute the frequency of the motif in each random network, iv) estimate a p-value of statistical significance of the motif by comparing F to the average frequency of the motif in the set of random networks. Motifs with p-values lower than a fixed threshold (e.g. 0.05) are considered as significantly over-represented in the input network. The most expensive task in motif enumeration is the frequency counting both in the input and in each random network. The counting problem is related to subgraph isomorphism, which is known to be an NP-complete problem [9]. This makes the frequency counting problem computationally expensive in large networks even for a single network. For this reason, many motif search algorithms are limited to motifs with few nodes (from 3 to 8 nodes).

Here we propose a comprehensive review of existent software for motifs analysis. Though there are many reviews on network motifs, some of them describe only the existing approaches and algorithms for network motif search [11,29,35,50,51] and some of them make a limited experimental comparison of such methods, considering only small networks [31,43,48]. In this paper, we describe different approaches and make a widespread comparison between these algorithms by using a dataset of medium and large artificial and real networks of different sizes. For each tool we briefly describe the algorithms proposed for the enumeration, classification and detection of motifs. In the last section we compare their execution time for motifs of up to 8 nodes and recommend the best software solution according to the size of motifs to seek.

2   Motif Finding Problem

2.1   Preliminary Definitions

Here we formally define the concepts of graph, subgraph, subgraph isomorphism and network motifs.

Graph. A *graph* G is a pair (V, E), where V is the set of vertices (also called nodes) and $E \subseteq V \times V$ is the set of edges. Let Σ_V and Σ_E be two set of labels. A graph is *labeled* if there are two functions, $l_V : V \to \Sigma_V$ and $l_V : E \to \Sigma_E$, that assigns labels to vertices and edges, respectively. A graph is *undirected* if $\forall (u, v) \in E(G)$ we have $(v, u) \in E(G)$, otherwise it is *directed*.

Subgraph. A *subgraph* S of a graph G is a graph such that $V(S) \subseteq V(G)$ and $E(S) \subseteq E(G)$. A subgraph S is *induced* if $\forall u, v \in V(S) : \nexists (u, v) \in E(G) \Rightarrow (u, v) \notin E(S)$, otherwise S is *non-induced*.

Subgraph Isomorphism Problem. Given a query graph Q and a target graph T, the *subgraph isomorphism problem* consists in finding an injective function $f : V(Q) \to V(G)$, called *mapping*, such that the following conditions hold:

1. each vertex of Q is mapped to a unique vertex of T;
2. $\forall v \in V(Q) \Rightarrow l_V(v) = l_V(f(v))$;
3. $\forall (u, v) \in E(Q) \Rightarrow (f(u), f(v)) \in E(T) \land l_E(u, v) = l_E(f(u), f(v))$.

If $V(Q) = \{q_1, q_2, ..., q_k\}$, the subgraph S of T formed by nodes $f(q_1), f(q_2), ..., f(q_k)$ and all edges between such nodes is called a *non-induced occurrence* of Q in T. The subgraph S of T formed by nodes $f(q_1), f(q_2), ..., f(q_k)$ and edges $(f(q_i), f(q_j))$ such that $(q_i, q_j) \in Q \, \forall \, 1 \leq i, j \leq k$ is called an *induced occurrence* of Q in T. The induced frequency F of Q is the number of induced occurrences of Q in T. A similar definition holds for the non-induced frequency of Q.

Network Motif. According to the definition proposed by Milo et al. [25], *network motifs* are patterns of inter-connections occurring in complex networks significantly more than in similar randomized networks. This means that a motif is a subgraph which is statistically over-represented. A more-formal definition was proposed by Ribeiro et al. [29]. An induced subgraph S of a graph G is called a *network motif* when for a given set of parameters $\{P, U, D, N\}$ and an ensemble of N randomized networks:

1. $p(\overline{F}_{rand}(S) > F_{real}(S)) \leq P$
2. $F_{real}(S) \geq U$
3. $F_{real}(S) - \overline{F}_{rand}(S) > D \times \overline{F}_{rand}(S)$

where i) $F_{real}(S)$ is the frequency of a motif in the real network, ii) $\overline{F}_{rand}(S)$ is the average frequency in all randomized networks, iii) P is a probability threshold determined with the *Z-score* (it is described in Sect. 2.2), iv) U is a cutoff value for $F_{real}(S)$, v) D is a proportional threshold that ensures the minimum difference between $F_{real}(S)$ and $\overline{F}_{rand}(S)$. Randomized networks can be built according to any random model (see [10] for a comprehensive review of graph random models). In the context of motif search, randomized variants are graphs that preserve some features of the input network, typically the degree distribution. Examples of such models are the Fixed Degree Distribution (FDD) model [27] the Expected Degree Distribution (EDD) model [7,30] or the more general analytical maximum-likelihood method [42].

2.2   Measures of Statistical Significance

Statistical significance, especially compared with randomized networks, is one of the most important features for network motifs [49]. Several measures for statistical significance of motifs have been proposed. Here we briefly describe the most important ones.

Frequency is the simplest measure. In order to be a motif, a subgraph must have a frequency in the real network greater than the mean frequency in the set of randomized networks. There are several strategies for the subgraph frequency computation depending on how much overlap is permitted the among different embeddings [8,12,39,40] to count all the several embeddings. In the first embedding counting rule, called F_1, motif occurrences should be different at least by a node or an edge. In the second embedding counting rule, called F_2, motif occurrences should not have edges in common. In the third and last counting rule, called F_3, motif occurrences should not have edges or nodes in common.

Another important measure is the ratio between the frequency of a subgraph G in the real network and the mean frequency of G in randomized networks. The statistical significance of a subgraph G compared to that in randomized networks can be estimated with the *Z-score* [25],

$$Z(G) = \frac{F_{real}(G) - \overline{F}_{rand}(G)}{\sigma_{F_{rand}(G)}}$$

where $F_{real}(G)$ is the frequency of G in the real network, $\overline{F}_{rand}(G)$ is the mean frequency of G in randomized networks and $\sigma_{F_{rand}(G)}$ is the standard deviation of the frequency of G in randomized networks.

A statistical measure similar to the *Z-score* is the *abundance* [8] of a subgraph G,

$$\Delta(G) = \frac{F_{real}(G) - \overline{F}_{rand}(G)}{F_{real}(G) + \overline{F}_{rand}(G) + \epsilon}$$

where ϵ is used to avoid the over-growth of Δ when the frequency of G is low both in the original network and in randomized networks.

Finally, the *concentration* [25] of a subgraph G is defined as the ratio between the frequency of G in the network and the sum of the frequencies of all the subgraphs of the network having as many nodes as G.

$$C(G) = \frac{F(G)}{\sum_{S \subseteq N:|V(G)|=|V(S)|} F(S)}$$

Analytical models have been proposed to approximate the number of occurrences of a particular subgraph s based on a null hypothesis model (e.g., uniform or fractal) [20,23,32,37]. Those approximations and their confidence intervals support the calculation of the significance of the actually found number of occurrences of s. We assume the use of such approaches, but this paper focuses on the discovery of the number of instances of a given subgraph.

2.3 Approaches

In what follows we briefly describe the main approaches for motif counting using the classification proposed by Ribeiro et al. [29]. The two main approaches for motif counting are the enumeration and the analytic approach. Each of these approaches can be further divided into other sub-categories. Enumeration approaches counts and categorizes all subgraph occurrences. Network centric methods search all possible subgraphs with n-nodes. Algorithms that follow this strategy are FANMOD [46,47], Kavosh [13] and mFinder [25]. Motif centric methods use a a query-graph and search all occurrences of a fixed topology in an input network. An algorithm that follows this strategy is NeMo [16]. Finally, encapsulation methods merge the enumeration and categorization steps together using common topological features of subgraphs or pre-computing some information about subgraphs to avoid repeated computation of isomorphism. Some algorithms that follow this strategy are gtrieScanner [36], NetMODE [19], QuateXelero [14] and FaSE [34]. On the other hand, analytic approaches avoid enumerating all subgraphs in a graph. Matrix based methods try to relate the frequency of each subgraph with the frequency of other subgraphs of the same or smaller size. Based on this relations, we can build a matrix of linear equation that can be solved using linear algebra methods. Decomposition methods decompose each subgraph in several smaller patterns of graph properties, like common neighbors, or triangles that touch two vertices. Algorithms that follow this strategy are Acc-Motif [22] and PGD [2,3].

3   Motif Finding Tools

In this section we briefly describe the motif finding algorithms that we are comparing in our review, explaining for each tool:

- the type of network that can be analyzed (e.g. directed, undirected, labeled or unlabeled);
- the characteristics of motif (e.g. induced, non-induced and restriction on motif size);
- the approach followed and the algorithm designed for motif counting;
- the algorithm used for calculating statistical significance.

   Table 1 summarizes the features of each algorithm. We believe that all this information, combined with the extensive experimental evaluation presented in Sect. 4 can guide the reader to choose the most suitable algorithm for his or her problem.

Table 1. Features of the algorithms covered in this survey

Algorithm	Year	Parallel	Directed	Induced	Approach	Type	Max Motif Size
FANMOD [46,47]	2006	✗	✓	Induced	Enum	Network-centric	≤8
Kavosh [13]	2009	✗	✓	Induced	Enum	Network-centric	∞
mFinder [25]	2002	✗	✓	Induced	Enum	Network-centric	∞
MAVisto [41]	2005	✗	✓	Non-induced	Enum	Network-centric	∞
gtrieScanner [36]	2010	✗	✓	Induced	Enum	Encapsulation	∞
NetMODE [19]	2012	✗	✓	Induced	Enum	Encapsulation	≤6
QuateXelero [14]	2013	✗	✓	Induced	Enum	Encapsulation	∞
FaSE [34]	2013	✗	✓	Induced	Enum	Encapsulation	∞
Acc-Motif [22]	2012	✗	✓	Induced	Analytic	Decomposition	≤6
NeMo [16]	2011	✗	✓	Non-induced	Enum	Motif-centric	∞
PGD [2,3]	2016	✓	✓	Induced	Analytic	Decomposition	≤4

3.1   FANMOD

FANMOD [46,47] is a motif finding algorithm which can retrieve induced subgraphs on both directed and undirected unlabeled or labeled networks, where labels can be specified on nodes and/or edges. The tool has a graphical interface and does sampling to find motifs in big networks. The algorithm can detect motifs with up to eight nodes. FANMOD consists of three steps: enumeration, classification and motif detection. The first step relies on the RAND-ESU [45] algorithm to improve the efficiency of subgraph enumeration. The algorithm is based on a full enumeration methodology that takes each vertex of the graph and creates a subgraph by iterative extension steps, where node neighbors are selected with some constraints. An a-priori defined probability of randomly skipping an extension step during full enumeration guarantees that sampling is unbiased. When this skip probability is equal to zero, RAND-ESU performs a complete enumeration task. The second step, i.e. determining isomorphic classes of motifs, is done using Nauty [21], a powerful algorithm for finding isomorphic graphs by canonical labeling. Finally, motif detection is performed through a classical permutation test.

3.2 Kavosh

Kavosh [13] is a tool for finding induced unlabeled motifs of any size and can handle both directed and undirected networks. Kavosh consists of four steps: enumeration, classification, random graph generation and motif identification. The first task aims at counting all k-size subgraphs, where k is the size of the motif. For each node in the graph, different trees are built following the neighbor relationship. Then, subgraphs are extracted using the composition operation of an integer combined with revolving door algorithm. Once all subgraphs are enumerated, they are classified into isomorphic classes using Nauty [21]. Motif significance is computed through a permutation test. Random graphs are generated through iterated vertex rewiring that preserves degree distribution.

3.3 mFinder

mFinder [25] can find unlabeled induced topologies on directed or undirected networks. The algorithm implements both exhaustive enumeration and subgraph sampling for finding motifs. The exhaustive enumeration performs a complete search of network motifs by iteratively extracting one edge and all relatively subgraphs with k nodes containing that edge. The sampling process consists of repeatedly picking one edge at random and performing a random walk until we obtain a k-size subgraph. Although this sampling approach enables approximate motif estimation in big graphs, it is biased due to different sampling probabilities for the same subgraph shape. mFinder implements different random network generation methods, such as switching edges, matching and "go to the winners" algorithms [24]. The switching method swaps the endpoint vertices of two randomly selected edges repeatedly and provides additional randomization options such as the possibility of preserving triads, layers and clustering coefficient of nodes.

3.4   MAVisto

MAVisto [41] is one of the few available tools to find non-induced motifs. The algorithm works on both directed and undirected labeled and unlabeled graphs. The tool is based on the Gravisto visualization toolkit [5] and offers a user interface for editing graphs, analyzing motifs and browsing results. MAVisto can count the number of occurrences of subgraphs of a given size according to three different definitions of frequencies F_1, F_2 and F_3 illustrated in Subsect. 2.2. MAVisto uses a frequent pattern finder algorithm for motif search [17]. The algorithm is based on a technique called generating parent, which builds a tree by extending patterns of previous levels. Branches of the tree represent only patterns supported by the network, i.e. all patterns obtained by extending a motif with all incident edges within the graph. Starting from one-edge topologies all matches over the network are found. Then, MAVisto performs one-edge extensions of patterns at the previous level until the desired size of patterns is obtained. Thanks to the downward closure property of F_2 and F_3 frequencies, branches representing patterns with lower frequencies are cut off. Finally, motif significance is computed through a classical permutation test.

3.5   gtrieScanner

gtrieScanner [36] is a motif finding tool for counting induced subgraphs in unlabeled directed and undirected networks. The algorithm performs subgraph census using a novel data structure called GTrie (Graph reTRIEval). gtrieScanner follows an encapsulation approach, which consists of searching a set of topologies of a given size by visiting the target graph only once. GTrie data structure is similar to the prefix trees for sequences. Each vertex of the tree is a graph G and the children of an internal vertex contains all the non-isomorphic graphs that can be obtained from G adding one edge in all possible ways. Leaves of the tree contain all motifs of a certain size that must be searched in the target graph. To avoid counting occurrences more than once, vertices of the tree are enriched by breaking symmetry conditions among motif nodes. The census is performed by recursively visiting all the nodes of the GTrie starting from the root and searching the corresponding graph of the GTrie vertex in the target graph. However, when we visit an internal vertex of the GTrie, the search starts from the occurrences of the subgraphs of the GTrie's ancestor nodes and this makes the census very efficient. Motif significance is computed by using a permutation test. In [26] authors proposed glabTrie, an extension of gtrieScanner to deal with labeled networks.

3.6   NetMODE

NetMODE [19] is a network motif detection tool based on Kavosh. It works on directed and undirected unlabeled induced subgraphs up to 6 nodes. This constraint is needed to apply the authors' strategy to the graph isomorphism problem. The main differences between NetMODE and Kavosh concern canonical labeling, random graphs generation and subgraph enumeration. In contrast to Kavosh, where canonical labels are provided through a runtime isomorphism evaluation using Nauty [21], NetMODE stores canonical labels for topologies up to 5 nodes in memory for an efficient comparison while processing input or random graphs. For motifs with 6 nodes the previous approach is infeasible, so NetMODE uses a reconstruction conjecture. Given a k-size subgraph S, a unique signature called deck(S) is built, using $(k-1)$-size subgraphs obtained by removing one vertex from S. Then, two graphs H and G are isomorphic if deck(H) = deck(G). Random graph generation is based on a switching method, where the endpoints of two random edges are swapped repeatedly. Parallelism is applied only for the census in random graphs while the algorithm works sequentially in the target network. Motif significance is expressed by z-score or p-value through a permutation test.

3.7 QuateXelero

QuateXelero [14] is a motif detection tool for unlabeled and induced subgraphs on both directed and undirected networks. QuateXelero uses a similar procedure to the one used by FANMOD for enumerating k-size subgraph, employing a quaternary tree. Edges of the tree can be labeled with symbols $\{-1, 0, 1, 2\}$. Symbol '-1' denotes a one-way connection from a node to its children, '0' denotes no edges between these vertices, '1' represents a reversed one-way connection, while '2' indicates a two-edge connection in two distinct directions between them. Concatenation of symbols represents a string of edges used to drive search into the quaternary tree in order to indicate whether the same topology has already been founded. Labels are also used to manage quaternary tree generation: if a symbol of a searched input string does not exist, a new node will be created and connected to the previous one through a new edge labeled with the searched symbol. Reaching a leaf means that k nodes of the graph are connected. If a leaf does not exist, a new node is created and Nauty algorithm [21] is run in order to determine its related non-isomorphic class. If it exists, previous calculations are exploited avoiding the computation of canonical labeling. In both cases the occurrences count of the reached leaf is incremented. If two subgraphs fall in the same leaf, they are isomorphic. However, two or more leaves of the quaternary tree may have the same isomorphic class. Motif significance is performed through a classic permutation test, but enumeration in random graphs is done avoiding path generation for a subgraph that does not exist in original network.

3.8 FaSE

FaSE (FAst Subgraph Enumeration) [34] is a subgraph census tool which performs only the enumeration and classification steps. It works with directed or undirected networks. It detects unlabeled and induced subgraphs using a network-centric approach. FaSE tries to improve the efficiency of the subgraph frequency problem by avoiding redundant isomorphic tests. For this purpose, authors use a modified G-Trie structure [36]. In the enumeration phase, FaSE allows any approaches that incrementally grows the set of connected nodes to create a topology with one-by-one node insertion from the neighborhood of a "root" node, such as the ESU algorithm [45]. Classification is performed by encapsulating isomorphic information in a customized G-Trie tree, where each node represents a partial subgraph and edges are labeled to trace connections in a deterministic way. Frequencies are stored in tree nodes and incremented when FaSE reaches a leaf during the enumeration step. Since two different leaves may represent the same subgraph, an isomorphic test over all leaves is needed.

3.9   Acc-Motif

Acc-Motif [22] is a motif-finding tool which works on both directed and undirected networks. It finds induced and unlabeled topologies. The motif detection problem is formalized by the authors as a motif-k problem, which consists in counting k-size induced subgraphs by grouping them into isomorphic classes. In order to enumerate and classify subgraphs, Acc-Motif exploits a set of linear relations to retrieve the frequencies of all k-nodes motifs starting from $(k-2)$-nodes topologies. Given a list of $(k-2)$-nodes subgraphs and a precomputed set of variables, Acc-Motif computes an histogram of frequencies H_k of k-nodes motifs. The precomputed variables are:

- R, a vector representing the number of times that an induced subgraph p must be counted;
- ID, a vector which uniquely identifies all topologies of a given size;
- P, a matrix encoding connections between triplets (X, L, M) and a vector ID, where X is a base pattern of G, while L and M are two extension set of nodes;
- Variants of P, named $P^{\rightarrow}$ and $P^{\leftrightarrow}$, encoding directed or bidirectional edges between the two sets L and M, respectively.

Histogram of frequencies for k-nodes motifs are computed also in random graphs. Topologies are considered motifs if their frequencies in the original graph is significantly higher than the average one measured in random graphs.

3.10   NeMo

NeMo [16] is a motif detection tool, implemented as a package in R. NeMo works on directed and undirected graphs. It deals with non-induced unlabeled subgraphs. In contrast with all other motif finding tools, NeMo performs motif detection through the analytical computation of p-values of motifs, avoiding the generation of a sample of random graphs and the estimation of motif frequencies in each random graph. NeMo addresses both a complete subgraph enumeration and a partial subgraph counting. In the first case, the algorithm performs enumeration by sorting graph and motif adjacency matrices according to the indices of connected nodes. This trick allows a pruning of the search tree during backtracking. The partial counting approach is less time-consuming than the full enumeration approach because it increments only the counter of the corresponding motif for each occurrence found.

3.11   PGD: Parallel Graphlet Decomposition

PGD [2,3] is a graphlet counting tool which can work in parallel for motifs of 3 and 4 nodes. PGD analyzes undirected networks and can search for both connected and unconnected induced motifs. The algorithm combines counts of a few subgraphs by using a combinatorial approach that returns exact counts of other topologies in constant time. Given a graph $G = (V, E)$, PGD scans each edge $e(a, b) \in E$ and looks at all nodes connected to it, i.e. its neighborhood, denoted as $N(e)$. The idea of the algorithm is to map a high-dimensional problem, such as motif count over big networks, into a lower-dimensional one. For instance, for a 3-node clique with nodes a, b and c, triangles can be represented as the number of overlapping nodes in the set $N(a) \cap N(b)$. For a 3-node star with nodes a, b and c, PGD counts all nodes c which make a 2-star topology connected to a or b. Likewise, all 4-nodes topologies can be decomposed into four 3-node graphlets by deleting one node each time. Total counts of 4 or more nodes topologies can be obtained from lower-dimensional subgraph counts in constant time, by using equations describing the relationship between their counts.

4   Experimental Analysis

We performed an experimental comparison of the motif search algorithms described in the previous section. We restricted our analysis to induced motif tools, because the induced definition of motifs is the most accepted one and most network analysis framework relies on this definition.

All experiments were run on a workstation consisting of an Intel i7-3770 quad-core multithreading processor with a frequency of 3,40 GHz per core and 8 GB RAM. Two different O.S. were used to ensure tools compatibility: Microsoft Windows 10 PRO and Ubuntu 17.04.

We considered undirected artificial networks generated using the Barabási-Albert model [4] and directed and undirected real networks of different size. Each tool was tested to find motifs having two to eight nodes. For each algorithm we computed the running time to complete the task.

4.1   Dataset Description

Our benchmark dataset consists of eight real and 20 artificial networks. Networks can be directed or undirected and have different sizes. In what follows we briefly describe their features:

barabási
A set of 20 undirected networks built using the Barabási-Albert model [4] which generates scale-free network based on the preferential attachment principle. Networks in this set have from 1,000 to 20,000 nodes and from 1,000 to 200,000 edges.

hpylo
A protein-protein interaction network of the Helicobacter pylori (first generation) organism taken from DIP [38], with 706 nodes that represent protein and 1,353 undirected edges which encode physical interactions between proteins.

roget
An undirected network taken from Roget's thesaurus, a list of synonyms or verbal equivalents. It is formed by 1,011 nodes representing categories and 3,648 edges between them if there exists a relation between categories or their words [15].

openflights
An undirected network extracted from the Openflights airport database of flight connections between airports (http://openflights.org) [28]. The network contains 2,940 nodes that represent airports and 15,677 edges which are flight connections.

ppiHuman
A protein-protein interaction network of human, taken from the HPRD database, with 9,507 nodes representing proteins and 37,054 undirected edges which are physical interactions between proteins [33].

polblogs
A directed network of hyperlinks between weblogs on presidential elections of United States in 2005 [1]. The network contains 1,224 nodes representing blogs and 19,022 edges representing links between blogs.

neuralWorm
A neural network modeling chemical and electrical synapses between neurons in Caenorhabditis elegans [44]. It is formed by 279 nodes and 2,287 directed edges.

dblp
A directed network of citations of publications taken from the database DBLP, which is the online reference for research papers in computer science [18]. The network has 12,591 nodes which represent publications and 49,744 edges that are citations between them.

foldoc
A directed network linking words contained in the Foldoc dictionary of computer science terms (http://foldoc.org) [6]. It is formed by 13,357 nodes that are terms and 120,238 directed edges. A directed edge between two nodes A and B exists iff B appear in the definition of term A.

4.2   Results

In Tables 2, 3 and 4 we show the running times for artificial scale-free network built using Barabási-Albert model [4] while Tables 5 and 6 shows execution times for directed and undirected real networks, respectively. In each of these tables the first column contains some statistics about the network (name, number of vertices and edges), while the first row contains the names of the analyzed tools. The tables list the execution time of each algorithm (in the format hh:mm:ss) for each motif size and input network. Special characters are used in the table to denote failure conditions, namely:

– *blank*, denoting a software which is unable to enumerate topologies of a generic size k,
– /, indicating that a test for motifs of size k has not been performed since the algorithm took more than 24 h to retrieve motifs of size $k - 1$;
– *OOM* (Out Of Memory), indicating that a tool requires more than 8 GB to complete a task.

Table 2. Running times (in the format hh:mm:ss) of the algorithms covered in this survey for all artificial networks with 1,000 nodes, denoted as B (#*nodes*, #*edges*).

Networks	Tools									
		mFinder	Kavosh	FANMOD	gTrieScanner	NetMODE	FaSe	QuateXelero	accMotif	PGD
B (1000V, 1000E)	3	<00:00:01	<00:00:01	<00:00:01	<00:00:01	<00:00:01	<00:00:01	<00:00:01	<00:00:01	<00:00:01
	4	00:00:01	<00:00:01	<00:00:01	<00:00:01	<00:00:01	<00:00:01	<00:00:01	<00:00:01	<00:00:01
	5	00:00:33	<00:00:01	<00:00:01	<00:00:01	00:00:01	<00:00:01	<00:00:01	<00:00:01	
	6	00:39:07	00:00:11	00:00:12	00:00:08	00:00:02	<00:00:01	<00:00:01	00:00:05	
	7	>24:00:00	00:03:12	00:03:04	00:02:10		00:00:03	00:00:03		
	8	/	01:13:30	00:54:04	00:44:45		00:00:44	00:00:40		
B (1000V, 2000E)	3	<00:00:01	<00:00:01	<00:00:01	<00:00:01	<00:00:01	<00:00:01	<00:00:01	<00:00:01	<00:00:01
	4	00:00:21	<00:00:01	<00:00:01	<00:00:01	<00:00:01	<00:00:01	<00:00:01	<00:00:01	<00:00:01
	5	00:18:54	00:00:10	00:00:08	00:00:08	00:00:02	<00:00:01	<00:00:01	<00:00:01	
	6	>24:00:00	00:05:50	00:04:12	00:03:17	00:00:22	00:00:05	00:00:04	00:00:32	
	7	/	02:41:48	01:59:35	01:31:03		00:02:07	00:01:48		
	8	/	>24:00:00	>24:00:00	>24:00:00		01:16:48	00:56:43		
B (1000V, 5000E)	3	00:00:03	<00:00:01	<00:00:01	<00:00:01	<00:00:01	<00:00:01	<00:00:01	<00:00:01	<00:00:01
	4	00:05:02	00:00:04	00:00:03	00:00:03	<00:00:01	<00:00:01	<00:00:01	<00:00:01	<00:00:01
	5	>24:00:00	00:04:20	00:02:12	00:02:26	00:00:07	00:00:05	00:00:04	00:00:04	
	6	/	03:33:35	02:10:12	02:09:23	00:15:14	00:03:54	00:03:04	00:10:33	
	7	/	>24:00:00	>24:00:00	>24:00:00		04:42:50	04:08:15		
	8	/	/	/	/		>24:00:00	>24:00:00		
B (1000V, 10000E)	3	00:00:24	<00:00:01	<00:00:01	<00:00:01	<00:00:01	<00:00:01	<00:00:01	<00:00:01	<00:00:01
	4	00:21:05	00:00:22	00:00:19	00:00:17	<00:00:01	<00:00:01	<00:00:01	<00:00:01	<00:00:01
	5	>24:00:00	00:38:11	00:28:19	00:25:09	00:00:48	00:00:53	00:00:34	00:00:27	
	6	/	>24:00:00	>24:00:00	>24:00:00	03:18:30	01:12:25	00:50:59	01:51:05	
	7	/	/	/	/		>24:00:00	>24:00:00		
	8	/	/	/	/		/	/		

Table 3. Running times (in the format hh:mm:ss) of the algorithms covered in this survey for all artificial networks with 5,000 nodes, denoted as $B\,(\#nodes, \#edges)$.

Networks		mFinder	Kavosh	FANMOD	gTrieScanner	NetMODE	FaSe	QuateXelero	accMotif	PGD
B (5000V, 5000E)	3	00:00:01	<00:00:01	<00:00:01	<00:00:01	<00:00:01	<00:00:01	<00:00:01	<00:00:01	<00:00:01
	4	00:00:19	<00:00:01	<00:00:01	<00:00:01	<00:00:01	<00:00:01	<00:00:01	<00:00:01	<00:00:01
	5	00:17:35	00:00:15	00:00:13	00:00:08	00:00:01	<00:00:01	<00:00:01	<00:00:01	
	6	>24:00:00	00:06:00	00:07:07	00:04:22	00:00:18	00:00:08	**00:00:07**	00:00:39	
	7	/	02:11:59	03:00:37	01:43:56		00:02:59	**00:02:35**		
	8	/	>24:00:00	>24:00:00	>24:00:00		01:00:32	**00:59:07**		
B (5000V, 10000E)	3	00:00:04	<00:00:01	<00:00:01	<00:00:01	<00:00:01	<00:00:01	<00:00:01	<00:00:01	<00:00:01
	4	00:03:34	<00:00:01	00:00:05	00:00:04	<00:00:01	<00:00:01	<00:00:01	<00:00:01	<00:00:01
	5	>24:00:00	00:00:06	00:05:26	00:05:09	00:00:09	00:00:13	00:00:10	**00:00:05**	
	6	/	**00:06:30**	05:56:17	06:12:28	00:30:18	00:11:18	00:09:28	00:15:37	
	7	/	07:37:15	>24:00:00	>24:00:00		10:03:51	**06:55:56**		
	8	/	**>24:00:00**	/	/		>24:00:00	>24:00:00		
B (5000V, 25000E)	3	00:00:15	<00:00:01	<00:00:01	<00:00:01	<00:00:01	<00:00:01	<00:00:01	<00:00:01	<00:00:01
	4	00:50:28	00:00:51	00:00:44	00:00:47	00:00:02	00:00:03	00:00:02	<00:00:01	<00:00:01
	5	>24:00:00	01:54:32	01:34:15	01:35:31	00:02:37	00:03:21	00:02:52	**00:00:28**	
	6	/	>24:00:00	>24:00:00	>24:00:00	15:03:23	05:12:30	**03:48:11**	04:23:15	
	7	/	/	/	/		>24:00:00	>24:00:00		
	8	/	/	/	/		/	/		
B (5000V, 50000E)	3	00:00:60	00:00:02	<00:00:01	00:00:02	<00:00:01	<00:00:01	<00:00:01	<00:00:01	<00:00:01
	4	05:54:07	00:04:29	00:04:47	00:04:07	00:00:08	00:00:15	00:00:10	00:00:01	<00:00:01
	5	>24:00:00	12:39:09	10:27:46	09:13:45	00:20:23	00:39:47	00:19:20	**00:05:03**	
	6	/	**>24:00:00**	**>24:00:00**	**>24:00:00**	>24:00:00	>24:00:00	>24:00:00	>24:00:00	
	7	/	/	/	/		/	/		
	8	/	/	/	/		/	/		

Table 4. Running times (in the format hh:mm:ss) of the algorithms covered in this survey for all artificial networks with 20,000 nodes, denoted as $B\,(\#nodes, \#edges)$.

Networks		mFinder	Kavosh	FANMOD	gTrieScanner	NetMODE	FaSe	QuateXelero	accMotif	PGD
B (20000V, 20000E)	3	00:00:05	<00:00:01	<00:00:01	<00:00:01	<00:00:01	<00:00:01	<00:00:01	<00:00:01	<00:00:01
	4	00:09:58	00:00:14	00:00:12	00:00:11	<00:00:01	<00:00:01	<00:00:01	<00:00:01	<00:00:01
	5	>24:00:00	00:23:30	00:22:55	00:18:09	00:00:28	00:00:39	00:00:18	**00:00:06**	
	6	/	>24:00:00	>24:00:00	>24:00:00	01:41:50	00:55:34	**00:26:56**	00:48:47	
	7	/	/	/	/		>24:00:00	>24:00:00		
	8	/	/	/	/		/	/		
B (20000V, 40000E)	3	00:00:22	<00:00:01	<00:00:01	<00:00:01	<00:00:01	<00:00:01	<00:00:01	<00:00:01	<00:00:01
	4	01:13:11	00:00:59	00:00:49	00:00:43	00:00:01	00:00:04	00:00:02	<00:00:01	<00:00:01
	5	>24:00:00	02:21:46	01:58:12	01:36:45	00:02:36	00:07:22	00:03:11	**00:00:26**	
	6	/	>24:00:00	>24:00:00	>24:00:00	17:58:53	11:59:33	**02:59:25**	06:25:20	
	7	/	/	>24:00:00	/		>24:00:00	>24:00:00		
	8	/	/	/	/		/	/		
B (20000V, 100000E)	3	00:01:39	00:00:03	00:00:02	00:00:03	<00:00:01	<00:00:01	<00:00:01	<00:00:01	<00:00:01
	4	09:20:43	00:05:14	00:07:12	00:06:03	00:00:15	00:00:28	00:00:12	00:00:03	<00:00:01
	5	>24:00:00	16:00:35	23:42:09	12:19:20	00:28:30	01:00:35	00:28:54	**00:05:43**	
	6	/	**>24:00:00**	**>24:00:00**	**>24:00:00**	>24:00:00	>24:00:00	>24:00:00	>24:00:00	
	7	/	/	/	/		/	/		
	8	/	/	/	/		/	/		
B (20000V, 200000E)	3	00:08:18	00:00:09	00:00:09	00:00:10	<00:00:01	00:00:01	<00:00:01	<00:00:01	<00:00:01
	4	>24:00:00	00:29:59	00:34:13	00:27:48	00:01:35	00:03:37	00:01:17	00:00:15	<00:00:01
	5	/	>24:00:00	>24:00:00	>24:00:00	05:18:04	08:47:09	02:37:18	**00:55:54**	
	6	/	/	/	/	>24:00:00	>24:00:00	>24:00:00	>24:00:00	
	7	/	/	/	/		/	/		
	8	/	/	/	/		/	/		

Table 5. Running times (in the format hh:mm:ss) of the algorithms covered in this survey for all real directed networks, denoted as *network* (*#nodes*, *#edges*).

Networks		Tools								
		mFinder	Kavosh	FANMOD	gTrieScanner	NetMODE	FaSe	QuateXelero	accMotif	PGD
neuralworm (279V, 2287E)	3	<00:00:01	<00:00:01	<00:00:01	<00:00:01	<00:00:01	<00:00:01	<00:00:01	<00:00:01	<00:00:01
	4	00:01:22	00:00:01	00:00:02	<00:00:01	<00:00:01	<00:00:01	<00:00:01	<00:00:01	<00:00:01
	5	>24:00:00	00:00:52	00:01:22	00:00:47	00:00:03	00:00:01	00:00:02	00:00:04	
	6	/	00:45:33	00:43:18	00:32:45	00:04:05	00:00:55	00:01:07	00:05:47	
	7	/	OOM	>24:00:00	OOM		00:31:32	OOM		
	8	/		/			12:47:48			
polblogs (1224V, 19022E)	3	00:00:31	<00:00:01	<00:00:01	<00:00:01	<00:00:01	<00:00:01	<00:00:01	<00:00:01	<00:00:01
	4	03:30:02	00:00:01	00:01:19	00:01:03	00:00:03	00:00:03	00:00:02	00:00:02	<00:00:01
	5	>24:00:00	02:17:29	02:33:50	02:19:55	00:05:34	00:06:01	00:03:34	00:06:33	
	6	/	>24:00:00	>24:00:00	>24:00:00	>24:00:00	00:00:55	OOM	19:11:53	
	7	/	/	/	/		08:35:19			
	8	/	/	/	/		>24:00:00			
dblp (12591V, 49744E)	3	00:01:45	<00:00:01	00:00:01	00:00:01	<00:00:01	<00:00:01	<00:00:01	<00:00:01	<00:00:01
	4	07:39:11	00:03:25	00:04:28	00:02:53	00:00:12	00:00:15	00:00:11	00:00:01	<00:00:01
	5	>24:00:00	14:09:35	>24:00:00	10:03:21	00:26:09	00:42:57	00:29:47	00:02:14	
	6	/	>24:00:00	/	>24:00:00	>24:00:00	>24:00:00	>24:00:00	07:10:51	
	7	/	/	/	/		/	/		
	8	/	/	/	/		/	/		
ppiHuman (9507V, 37054E)	3	00:01:31	<00:00:01	<00:00:01	<00:00:01	<00:00:01	<00:00:01	<00:00:01	<00:00:01	<00:00:01
	4	02:19:48	00:00:50	00:01:19	00:01:20	00:00:03	00:00:05	00:00:02	00:00:03	<00:00:01
	5	>24:00:00	02:08:37	02:02:34	01:50:10	00:04:29	00:06:29	00:03:26	00:04:04	
	6	/	>24:00:00	>24:00:00	>24:00:00	>24:00:00	07:52:14	05:09:06	13:52:31	
	7	/	/	/	/		>24:00:00	>24:00:00		
	8	/	/	/	/		/			

Table 6. Running times (in the format hh:mm:ss) of the algorithms covered in this survey for all real undirected networks, denoted as *network* (*#nodes*, *#edges*).

Networks		Tools								
		mFinder	Kavosh	FANMOD	gTrieScanner	NetMODE	FaSe	QuateXelero	accMotif	PGD
Hpylo (706V, 1353E)	3	<00:00:01	<00:00:01	<00:00:01	<00:00:01	<00:00:01	<00:00:01	<00:00:01	<00:00:01	<00:00:01
	4	00:00:07	<00:00:01	<00:00:01	<00:00:01	<00:00:01	<00:00:01	<00:00:01	<00:00:01	<00:00:01
	5	00:05:09	00:00:03	00:00:02	00:00:04	00:00:01	<00:00:01	<00:00:01	<00:00:01	
	6	>24:00:00	00:01:23	00:01:23	00:01:15	00:00:12	00:00:02	00:00:01	00:00:18	
	7	/	00:37:35	00:34:17	00:27:23		00:00:31	00:00:29		
	8	/	14:22:18	13:57:03	10:12:34		00:10:21	00.09.04		
	9	/	>24:00:00	/	>24:00:00		03:20:21	02:51:21		
roget (1011V, 3648E)	3	00:00:01	<00:00:01	<00:00:01	<00:00:01	<00:00:01	<00:00:01	<00:00:01	<00:00:01	<00:00:01
	4	00:00:14	<00:00:01	<00:00:01	<00:00:01	<00:00:01	<00:00:01	<00:00:01	<00:00:01	<00:00:01
	5	00:08:39	00:00:06	00:00:05	00:00:05	00:00:01	<00:00:01	<00:00:01	00:00:01	
	6	>24:00:00	00:02:04	00:01:45	00:01:45	00:00:16	00:00:03	00:00:03	00:00:34	
	7	/	00:54:01	00:32:21	00:27:55	/	00:01:05	00:00:55	/	
	8	/	21:26:55	11:10:20	09:44:30	/	00:21:01	00:20:00		
openflights (2940V, 15677E)	3	00:00:09	<00:00:01	<00:00:01	<00:00:01	<00:00:01	<00:00:01	<00:00:01	<00:00:01	<00:00:01
	4	00:29:58	00:00:59	00:00:49	00:00:36	00:00:02	00:00:02	00:00:02	00:00:04	<00:00:01
	5	>24:00:00	01:24:02	01:14:48	01:02:05	00:02:34	00:02:32	00:02:03	00:07:41	
	6	/	>24:00:00	>24:00:00	>24:00:00	10:25:12	03:24:41	02:21:45	>24:00:00	
	7	/	/	/	/	/	>24:00:00	/	/	
	8	/	/	/	/	/	/	/		
foldoc (13357V, 120238E)	3	00:03:16	00:00:01	<00:00:01	00:00:02	<00:00:01	<00:00:01	<00:00:01	<00:00:01	<00:00:01
	4	09:34:12	00:03:03	00:03:13	00:03:55	00:00:13	00:00:13	00:00:08	00:00:04	<00:00:01
	5	>24:00:00	15:10:51	18:24:14	17:13:24	00:29:37	00:31:40	00:25:20	00:09:30	
	6	/	>24:00:00	>24:00:00	>24:00:00	>24:00:00	>24:00:00	>24:00:00	>24:00:00	
	7	/	/	/	/	/	/	/	/	
	8	/	/	/	/	/	/	/	/	

The experimental results suggest that *acc-Motif, QuateXelero* and *NetMODE* take the least time to count the number of occurrences of motifs up to 6 nodes. For motifs having 7 or 8 nodes, the fastest algorithms are *QuateXelero* and *FaSe*. As network density increases, the running time of almost all tools tends to grow exponentially and, at a certain point, the algorithms fail to complete the task before the timeout. *mFinder* already gets stuck with 4-nodes motifs in networks with at least 10,000 nodes and 100,000 edges. *Kavosh, gTrieScanner* and *FANMOD* fail to complete before the timeout starting from motifs with 5 nodes and networks with 10,000 nodes and 100,000 edges. The others tools begin to suffer with 6-node topologies and networks having at least 5,000 nodes and 50,000 edges. *PGD* has a different behaviour w.r.t. the other tools because it is by far the fastest model for counting motifs of 3 and 4-nodes and its performance is not affected by network's size or density. However, the algorithm cannot handle bigger motif sizes. So, *PGD* is suitable to count very small motifs even in large networks formed by millions of nodes and edges.

Note that our analysis focused only on the time performance of algorithms to find the number of instances of individual motifs. The analysis of significance of such identified motifs will use the analytical models mentioned above [20,23, 32,37].

5   Perspectives and Future Work

Network analysis through motif discovery aims to identify subgraphs which are statistically significantly over-represented in a network. A common approach to evaluate statistical significance consists in calculating the motif's frequency both in the input network and in a set of hundreds or thousands of randomized variants of the network and finally comparing frequencies to estimate a p-value of significance. This process is time-consuming and makes motif search hard, even for motifs of moderate size. In this paper we reviewed the main approaches for the motif counting problem and the metrics used for evaluating a motif's statistical significance. We briefly described the most popular motif search tools and conducted an extensive experimental evaluation using both real and artificial networks of different sizes. Our work is intended to be a guide for the user in order to choose the best tool for motif analysis depending on the size of the network and the size of motifs.

In the future we plan to extend our experimental analysis, by including tools that handle non-induced graphs [23,26], labeled graphs as well as parallel motif search tools.

References

1. Adamic, L.A., Glance, N.: The political blogosphere and the 2004 U.S. election: divided they blog. In: ACM, pp. 36–43 (2005)
2. Ahmed, N., Neville, J., Rossi, R., Duffield, N.: Efficient graphlet counting for large networks. In: ICDM, pp. 1–10 (2015)

3. Ahmed, N., Neville, J., Rossi, R., Duffield, N.: Fast parallel graphlet counting for large networks. Technical report, arXiv:1506.04322 (2016)
4. Albert, R., Barabási, A.L.: Statistical mechanics of complex networks. Rev. Mod. Phys. **74**(47), 47–97 (2002)
5. Bachmaier, C., Brandenburg, F.J., Forster, M., Holleis, P., Raitner, M.: *Gravisto*: graph visualization toolkit. In: Pach, J. (ed.) GD 2004. LNCS, vol. 3383, pp. 502–503. Springer, Heidelberg (2005). https://doi.org/10.1007/978-3-540-31843-9_52
6. Batagelj, V., Mrvar, M., Zavesnik, M.: Network analysis of dictionaries. In: Language Technologies, pp. 135–142 (2002)
7. Chung, F., Lu, L.: The average distances in random graphs with given expected degrees. Proc. Natl. Acad. Sci. **99**(25), 15879–15882 (2002)
8. Ciriello, G., Guerra, C.: A review on models and algorithms for motif discovery in protein-protein interaction networks. Brief. Funct. Genomic. Proteomic. **7**(2), 147–56 (2008)
9. Cook, S.A.: The complexity of theorem-proving procedures. In: Proceedings of the 3rd ACM Symposium on Theory of Computing, pp. 151–158 (1971)
10. Drobyshevskiy, M., Turdakov, D.: Random graph modeling: a survey of the concepts. ACM Comput. Surv. **52**(6), 1–36 (2019)
11. Jain, D., Patgiri, R.: Network motifs: a survey. In: Advances in Computing and Data Sciences, ICACDS 2019. Communications in Computer and Information Science, vol. 1046 (2019)
12. Jazayeri, A., Yang, C.: Motif discovery algorithms in static and temporal networks: a survey. J. Complex Netw. **8**, 1–38 (2020)
13. Kashani, Z.R.M., Ahrabian, H., Elahi, E., Nowzari-Dalini, A., Ansari, E.S., et al.: Kavosh: a new algorithm for finding network motifs. BMC Bioinform. **10**, 3–8 (2009)
14. Khakabimamaghani, S., Sharafuddin, I., Dichter, N., Koch, I., Masoudi-Nejad, A.: QuateXelero: an accelerated exact network motif detection algorithm. PLoS ONE **8**(7), e68073 (2013)
15. Knuth, D.E.: The Stanford GraphBase: A Platform for Combinatorial Computing. ACM Press (1993)
16. Koskas, M., Grasseau, G., Birmelé, E., Schbath, S., Robin, S.: NeMo: fast count of network motifs. In: MARAMI 2011: 2. Conférence sur les Modèles et l'Analyse des Réseaux: Approches Mathématiques et Informatique (2011)
17. Kuramochi, M., Karypis, G.: Finding frequent patterns in a large sparse graph. Data Min. Knowl. Disc. **11**, 243–271 (2004)
18. Ley, M.: The DBLP computer science bibliography: evolution, research issues, perspectives. In: Proceedings of the International Symposium on String Processing and Information Retrieval, vol. 2476, pp. 1–10 (2002)
19. Li, X., Stones, D.S., Wang, H., Deng, H., Liu, X., et al.: NetMODE: network motif detection without Nauty. PLoS ONE **7**(12), e50093 (2012)
20. Martorana, E., Micale, G., Ferro, A., Pulvirenti, A.: Establish the expected number of injective motifs on unlabeled graphs through analytical models. Complex Netw. Appl. **8**, 255–267 (2020)
21. McKay, B.D., Piperno, A.: Practical graph isomorphism II. J. Symb. Comput. **60**, 94–112 (2014)
22. Meira, L.A.A., Máximo, V.R., Fazenda, A.L., Conceição, A.F.: Acc-Motif: accelerated network motif detection. Trans. Comput. Biol. Bioinform. **11**(5), 853–862 (2014)

23. Micale, G., Giugno, R., Ferro, A., Mongiovì, M., Shasha, D., Pulvirenti, A.: Fast analytical methods for finding significant labeled graph motifs. Data Min. Knowl. Disc. **32**(2), 504–531 (2018)
24. Milo, R., Kashtan, N., Itzkovitz, S., Newman, M.E.J., Alon, U.: On the uniform generation of random graphs with prescribed degree sequences. Technical report, 0312028, arXiv (2004)
25. Milo, R., Shen-Orr, S., Itzkovitz, S., et al.: Network motifs: simple building blocks of complex networks. Science **298**(5594), 824–827 (2002)
26. Mongiovi, M., Micale, G., Ferro, A., Giugno, R., Pulvirenti, A., Shasha, D.: gLab-Trie: a data structure for motif discovery with constraints. In: Fletcher, G., Hidders, J., Larriba-Pey, J. (eds.) Graph Data Management. DSA, pp. 71–95. Springer, Cham (2018). https://doi.org/10.1007/978-3-319-96193-4_3
27. Newman, M.E.J., Strogatz, S.H., Watts, D.J.: Random graphs with arbitrary degree distributions and their applications. Phys. Rev. E **64**, 026118 (2001)
28. Opsahl, T.: Why anchorage is not (that) important: binary ties and sample selection. Technical report, Tore Opsahl (2011). http://toreopsahl.com/2011/08/12
29. Ribeiro, P., Silva, F., Kaiser, M.: Strategies for network motifs discovery. In: e-Science 2009 – 5th IEEE International Conference on e-Science, pp. 80–87 (2009)
30. Park, J., Newman, M.: The origin of degree correlations in the internet and other networks. Phys. Rev. E **68**, 026112 (2003)
31. Patra, S., Mohapatra, A.: Review of tools and algorithms for network motif discovery in biological networks. IET Syst. Biol. **14**(4), 171–189 (2020)
32. Picard, F., Daudin, J.J., Koskas, M., et al.: Assessing the exceptionality of network motifs. J. Comput. Biol. **15**(1), 1–20 (2008)
33. Prasad, T.S.K., Goel, R., Kandasamy, K., Keerthikumar, S.: Human protein reference database–2009 update. Nucleic Acids Res. **37**(1), D767–D772 (2009)
34. Ribeiro, P.: Towards a faster network-centric subgraph census. In: International Conference on Advances in Social Networks Analysis and Mining (2013)
35. Ribeiro, P., Paredes, P., Silva, M., Aparicio, D., Silva, F.: A survey on subgraph counting: concepts, algorithms, and applications to network motifs and graphlets. ACM Comput. Surv. **54**(2), 1–36 (2021)
36. Ribeiro, P., Silva, F.: G-tries: a data structure for storing and finding subgraphs. Data Min. Knowl. Disc. **28**, 337–377 (2014)
37. Schbath, S., Lacroix, V., Sagot, M.F.: Assessing the exceptionality of coloured motifs in networks. J. Bioinf. Syst. Biol. **2009**(1), 616234 (2009)
38. Salwinski, L., et al.: The database of interacting proteins: 2004 update. Nucleic Acids Res. **32**(Suppl. 1), D449–D451 (2004)
39. Schreiber, F., Schwöbbermeyer, H.: Towards motif detection in networks: frequency concepts and flexible search. In: Proceedings of the International Workshop on Network Tools and Applications in Biology, pp. 91–102 (2004)
40. Schreiber, F., Schwöbbermeyer, H.: Frequency concepts and pattern detection for the analysis of motifs in networks. Trans. Comput. Syst. Biol. **III**(3737), 89–104 (2005)
41. Schreiber, F., Schwöbbermeyer, H.: MAVisto: a tool for the exploration of network motifs. Bioinform. Appl. Note **21**(17), 3572–3574 (2005)
42. Squartini, T., Garlaschelli, D.: Analytical maximum-likelihood method to detect patterns in real networks. New J. Phys. **13**(8), 083001 (2011)
43. Tran, N., Mohan, S., Xu, Z., Huang, C.: Current innovations and future challenges of network motif detection. Brief. Bioinform. **16**(3), 497–525 (2015)
44. Varshney, L., Chen, B., Paniagua, E.: Structural properties of the Caenorhabditis elegans neuronal network. PLoS Comput. Biol. **7**(2), e1001066 (2011)

45. Wernicke, S.: A faster algorithm for detecting network motifs. In: Proceedings of the 5th International Conference on Algorithms in Bioinformatics (WABI 2005), pp. 165–177 (2005)
46. Wernicke, S.: Efficient detection of network motifs. IEEE/ACM Trans. Comput. Biol. Bioinform. **3**, 347–359 (2006)
47. Wernicke, S., Rasche, F.: FANMOD: a tool for fast network motif detection. Bioinform. Appl. Note **22**(9), 1152–1153 (2006)
48. Wong, E., Baur, B., Quader, S., Huang, C.: Biological network motif detection: principles and practice. Brief. Bioinform. **13**(2), 202–215 (2012)
49. Xia, F., Wei, H., Yu, S., Zhang, D., Xu, B.: A survey of measures for network motifs. IEEE Access **7**, 106576–106587 (2019)
50. Yu, S., Feng, Y., Zhang, D., Bedru, H., Xu, B., Xia, F.: Motif discovery in networks: a survey. Comput. Sci. Rev. **37**, 100267 (2020)
51. Yu, S., Xu, J., Zhang, C., Xia, F., Almakhadmeh, Z., Tolba, A.: Motifs in big networks: methods and applications. IEEE Access **7**, 183322–183338 (2019)

Author Index

D. Cantone and A. Pulvirenti (Eds.): *From Computational Logic to Computational Biology*,
LNCS 14070, p. 269, 2024.
https://doi.org/10.1007/978-3-031-55248-9

Printed in the United States
by Baker & Taylor Publisher Services